Agrarian Systems and Rural Development

The World Employment Programme (WEP) was launched by the International Labour Organisation in 1969, as the ILO's main contribution to the International Development Strategy for the Second United Nations Development Decade.

The means of action adopted by the WEP have included the following:

- short-term high-level advisory missions;
- longer-term national or regional employment teams; and
- a wide-ranging research programme.

A landmark in the development of the WEP was the World Employment Conference of 1976, which proclaimed inter alia that 'strategies and national development plans should include as a priority objective the promotion of employment and the satisfaction of the basic needs of each country's population'. The Declaration of Principles and Programme of Action adopted by the Conference have become the cornerstone of WEP technical assistance and research activities during the closing years of the Second Development Decade.

This publication is the outcome of a WEP project.

Agrarian Systems and Rural Development

Dharam Ghai
Azizur Rahman Khan
Eddy Lee
Samir Radwan

A study prepared for the International Labour Office within the framework of the World Employment Programme

M

First published 1979 by
THE MACMILLAN PRESS LTD
London and Basingstoke
Associated companies in Delhi
Dublin Hong Kong Johannesburg Lagos
Melbourne New York Singapore Tokyo

Filmset by Vantage Photosetting Co. Ltd,
Southampton and London

Reproduced from copy supplied
printed and bound in Great Britain
by Billing and Sons Limited
Guildford, London, Oxford, Worcester

British Library Cataloguing in Publication Data

Agrarian systems and rural development
1. Underdeveloped areas – Agriculture – Case studies
I. Ghai, Dharam Pal
II. International Labour Office
World Employment Programme
338.1'09172'4 HD1417

ISBN 0-333-27343-5

The responsibility for opinions expressed in signed articles, studies and other contributions rests solely with their authors, and publication does not constitute an endorsement by the International Labour Office of the opinions expressed in them. The designations employed and the presentation of material do not imply the expression of any opinion whatsoever on the part of the International Labour Office concerning the legal status of any country or territory or of its authorities, or concerning the delimitation of its frontiers

Contents

List of Tables

Preface

The present volume incorporates the major results of research carried out at the Rural Employment Policies Branch of the ILO on the alternative forms of agricultural organisation for rural development. Several points need to be noted about the purpose and contents of the volume. An attempt was made to include in the case studies examples of the major forms of institutional organisation that prevail in the contemporary developing world. However, the editors recognise that the collection of case studies does not fully succeed in achieving this objective in as balanced and comprehensive a way as they would like. For example, it would have been desirable to have a better, and geographically more balanced, representation of cases belonging to the category of private farming. All the three cases included in the present volume are from Asia, and it would clearly be desirable to include one or more studies on some Latin American experience. While this was the intention, the actual organisation of the studies was obviated by the lack of the necessary technical manpower and the availability of suitable collaborators at the time. In comparison, the representation of the communal forms of organisation would appear to be proportionately greater. One should, therefore, note that in comparison with those of private agriculture the case studies of communal agriculture have been able to capture more of the variations in performance. However, two out of the three cases of private agriculture studied in this volume are those which are claimed to be outstanding successes.

The purpose of the present volume is not to try to identify the best agrarian system. This is why comparison between systems has generally been avoided. The focus has rather been on the identification of the conditions of success under each system. To do so it is much more important to compare the performances of the cases within each broad system. This is what the editors have tried to do in Chapter 1 while allowing the individual studies to concentrate on the successes and failures of each case.

The justification of this approach is obvious. The choice of an agrarian system is the result of a multitude of complex factors – historical, political and social – and it would be highly presumptuous to try to identify (on the basis of a set of criteria which is inevitably narrower than would be dictated by objective reality) the best system that each developing country should adopt. Furthermore, as clearly emerges from the case studies and the discussion of Chapter 1, even on the basis of the limited criteria that have been employed in the present set of studies, one cannot unambiguously identify the system that performs best.

Six of the nine case studies were actually carried out by the members of the staff of the Rural Employment Policies Branch of the ILO. The remaining three – those on India, Guyana and Cuba – were carried out by collaborators who are recognised experts of outstanding ability in the respective fields. Every effort was made to ensure a basic uniformity of approach throughout the case studies but, understandably, this is a matter on which complete success is virtually impossible to ensure.

The book is being published on the eve of the World Conference on Agrarian Reform and Rural Development of the FAO (which will be held in July 1979). The FAO maintained close contact with the ILO in the planning and preparation of the Conference. It is hoped that this volume will represent a major contribution to the Conference.

In organising the present studies the editors received a great deal of help from their colleagues in the ILO. Mrs. Antoinette Béguin, the Chief of the Employment and Development Department, provided constant help and encouragement – moral, intellectual and material. A large number of colleagues, too numerous to list individually, contributed to the quality of the volume by helpful comments and suggestions.

The Swedish Agency for Research Co-operation with Developing Countries (SAREC) deserves special gratitude for its generous grant from which much of the external collaboration and travel cost was financed. Those outside the ILO who deserve particular mention for their help on one or more of the chapters are: Amiya K. Bagchi, Keith Griffin, L. R. Khan, Atiur Rahman, Justinian Rweyemamu and M. Solaiman.

The editors owe a special debt of gratitude to Sandra Berlinka, Anne-Marie Causanillas, Frances Kaufmann, Jean Millar, Mercedes Richards, Deborah Russell and Jane Stanleick who helped with the technical preparation and typing of the manuscript. Evelyn Schaad went beyond the ordinary call of duty in providing extensive help on documentation.

International Labour Office   The Editors
December 1978

1 Alternative Agrarian Systems and Rural Development in the Third World

The Editors

1.1 Objectives and contents of the studies

The idea of rural development as the cornerstone of the growth strategy for the Third World, as distinguished from growth through industrialisation, has become a part of the 'new-orthodoxy' in development literature as well as in official plans. It has also become increasingly recognised that rural development cannot be viewed as a narrow technical problem of increasing production by combining the skills of the plant geneticists, agricultural chemists and transport engineers. The problem is a much wider one, of developing appropriate organisations and institutions to mobilise and induce members of the rural society to greater productive effort, to help them to overcome the constraints in the way of utilising available resources and to enable them to distribute the results of their effort equitably among themselves. The attention of the development planner is naturally turning towards the question of the appropriate forms of organisation and institutions within which the objectives of rural development can be achieved most rapidly in a given situation.

The purpose of the present volume is to analyse the performance of some of the different forms of organisation and institutions – the alternative agrarian systems – in promoting rural development in given historical circumstances in recent times. With this objective in view a number of experiences varying from private peasant agriculture to communal land systems have been studied.

The nine case studies in the present volume can be divided into three broad categories of agrarian systems. First, is the system of individual farming carried out under private ownership of land and other means of production. The distinctive feature of this category is the predominance of peasant farming in the framework of a private agriculture. In our study South Korea, India and Bangladesh are included in this category. Secondly, and at the other end of the spectrum, is the system of communal farming carried out under collective and/or social ownership of land which is studied with reference to three major experiences: China, the Soviet Central Asian Republics and Cuba. Finally, there is an imprecise intermediate category in which one can place Egypt, Tanzania and Guyana. In the latter two countries collective and private ownership of land and means of production coexist. In Guyana the dominant plantation sector has been nationalised and is now organised as state farms while a system of production co-operatives is being implemented in the peasant sector. Similarly in Tanzania, the estate sector is now largely run as parastatal farms and in the peasant sector communal production under the Ujamaa system is gradually being extended. Egypt, however, presents a borderline case. While there is no communal ownership of land, there is, nonetheless, a comprehensive system of supervised co-operatives through which the state enforces land consolidation, regulates certain aspects of agricultural production such as crop rotation and implements a tight procurement system. Moreover, membership of these co-operatives is compulsory, distinguishing the Egyptian system from the usual pattern of voluntary co-operatives with partial coverage and limited functions which prevail in most developing market economies. In addition it could be claimed that particularly in the 1960s, the Egyptian agrarian reforms occurred in the context of a substantial extension of state ownership in the economy and that there was reason to believe that the system of co-operatives would develop into a fully-fledged system of producer co-operatives.

In spite of these imprecisions, it could be argued that the typology that is being suggested provides a convenient framework of analysis. It conforms to the broad categories in terms of which most people think and in which the countries concerned prefer to project their own experiences. Thus, individualistic private farming, as a polar case, is identified with either a pure or mixed form of agrarian capitalism. Communal agriculture, another polar case, is the form of rural organisation under revolutionary socialism. The intermediate category is frequently offered as a middle course which avoids the extremes of the two polar cases.

While most people think in terms of the above broad typology, in reality each type includes a wide variety of agrarian structures. This is exemplified even by the experiences studied in this volume. Of the three cases that have been included as examples of individualistic private farming, South Korea, India and the Comilla experiment in Bangladesh, the first is an example of private farming which followed a highly egalitarian land reform, while in the other two cases the distribution of landownership remains quite unequal. The examples in the intermediate category included in the study are, as has been discussed above, also considerably different from one another. Even the three cases of communal agriculture included in our study differ substantially in organisation. In Cuba about a third of arable land is cultivated by private peasant households while in the remainder state and collective farms operate. In Soviet Central Asia and China peasant agriculture as a separate sub-system has long ceased to exist. But these two latter cases differ significantly in terms of the degree of decentralisation, the importance of private plots and the systems of payments and incentives.

The examples chosen almost always refer to cases which might be construed as attempts to overcome the problems of underdevelopment and poverty by way of organisational and institutional innovation. Perhaps India constitutes a case that does not quite fit this description. But it is unthinkable to exclude completely from a study of this kind a country of India's size and importance. Also, India experimented with a major shift towards a growth-orientated strategy besides trying out some innovative projects (for example, the Comprehensive Area Development Programme in West Bengal) on a smaller scale.

The purpose of these studies is to evaluate the performance of the alternative agrarian systems in order to identify the major factors conducive to agrarian change, the main problems faced by each system and, by so doing, perhaps suggest policies that might improve the performance of each. In planning the present studies it was intended to ensure a certain uniformity of approach by evaluating the performance of each case on the basis of a number of common criteria. The first criterion is productive efficiency: the extent to which each system succeeded in promoting growth on the basis of an efficient use of resources. Secondly, we asked to what extent each system succeeded in providing employment and promoting an egalitarian distribution of income. Thirdly, we looked into the ability of each system to generate surplus to sustain its growth on the basis of self-reliance. Besides these criteria an attempt has occasionally been made to determine the performance of the system in promoting popular participation and in satisfying basic human needs.

While an attempt was made in most of the case studies to adhere to the above criteria, it will be obvious to the reader that in some cases the emphasis differs quite markedly. Thus the performance of the strategy in India, itself a stylised summary of many different sub-strategies in a vast country, is evaluated very differently from the other standard cases. In evaluating the Cuban experience the lack of information on certain aspects has resulted in a different balance between various criteria. In analysing the Chinese performance a much greater emphasis has been placed on the system of incentives which constitutes an overwhelmingly important aspect of managing a collective agriculture. We believe that in spite of these differences the case studies incorporate a basic unity in the evaluation of performance and that the differences in emphasis and methodology from one case to another more often constitute interesting variety than irritating inconsistency.

Having outlined the rationale for the choice of case studies and the criteria that have been used to evaluate the performance in these cases, it still remains for us to explain the term 'agrarian system' which we have used as an organising concept throughout most of this study. In using the term 'agrarian system' we wished to draw attention to four key elements which recur in the analysis in several of the case studies. These are the pattern of landownership, the organisation of agricultural production, the structure of incentives and the overall economic context in which agriculture is situated. Although the relative importance of each element varies from case to case, together they constitute a common analytical core which runs through the case studies. Thus our scheme of analysis focuses on these key elements and their interrelationships. In the following we shall consider each of these elements in turn.

The pattern of landownership determines the basic character of an agrarian system but is by no means a sufficient basis for analysis. For instance, within a given system of landownership, whether private or collective, there can be major differences in the organisation of agricultural production, in the structure of incentives and in the relationship with the non-agricultural sectors. Such differences can therefore influence crucially the performance of a given system of ownership. Within a system of private ownership, the degree of concentration in landownership is a crucial variable. A highly concentrated pattern of ownership implies that the organisation of production will be based on either a system of small tenancies or on large estates using landless agricultural labourers. In either case, a large pool of landless labour will be present. It is only with a relatively egalitarian distribution of landownership are we likely to find owner-operated family farms as the dominant form of productive organisations. Similarly, under collective ownership of land, the organisation of production units can vary considerably. Production units can vary in scale, in the degree of decentralisation, in decision-making and in the payments system that is used.

It is also clear from the foregoing that the pattern of ownership is linked to the form of productive organisation that is in force and that this in turn is linked to the structure of incentives. In a system of private ownership of land, it is the market which determines incentives or the lack of them. An unequal pattern of landownership often implies unequal market power and consequently a heavy burden of rent on tenants, wages at or below subsistence levels and the existence of a large pool of landless labour. The 'incentive' structure does therefore depend as much on the sanction of a loss of subsistence as on the prospect of private gain. In contrast, under a system of collective ownership, employment and subsistence are guaranteed (even though at times it may mean sharing of under-employment) and the problem of incentives reduces itself to one of finding an appropriate balance between private and collective incentives and the necessary degree of dependence on incentives which would satisfy the objectives of overall growth, accumulation and income distribution. The efficacy of a given structure of incentives is thus a central element in analysing the performance of alternative agrarian systems since it impinges upon the issues of poverty and income distribution, accumulation and growth of production.

A major constraint on the structure of incentives is the nature of the relationship between agriculture and the rest of the economy. In all systems, the rate of surplus extraction from agriculture has an important effect on the total income that is available for distribution under a given payments system, and hence on production incentives and income levels. Furthermore, the rate of labour-absorption into non-agricultural sectors and the flow of investment funds between sectors can either aggravate or alleviate the development problems of the rural economy. The impact of these 'external' relationships of the agricultural sector must therefore feature in any overall evaluation of the performance of a given agrarian system.

It should also be clear from the foregoing that these elements can be viewed as interlocking parts of a system. The interrelationships between these parts can be seen most clearly in cases of transition from one agrarian system to another. For instance a change from private to collective ownership will involve the introduction of new forms of organisation of production and a new structure of incentives to replace the old. The rate of surplus extraction and the mechanisms by which it is achieved will also have to change. Similarly, any major change within a given system such as the introduction of a new technology will require simultaneous attention to all these elements. The diffusion of new technology under a system of private ownership will be thwarted by extreme inequality in landownership and can lead to increasing inequality under existing payments systems. Thus a thorough shift to a new technology may require a change in the pattern of landownership, the organisation of production and in the structure of incentives if growth and equity are to be preserved.

Thus the term 'agrarian system' is, we believe, more than merely a portmanteau. It is a convenient organising concept which has enabled us to ensure at least some measure of analytical consistency throughout the case studies in this volume.

1.2 Alternative agrarian systems: performance, problems and policies

In this section, we shall present, separately, overall evaluations of the performance of each of our three types of agrarian systems. The intention is to synthesise the observations about performance as well as problems and policies that are contained in the individual studies.

(a) The system of individual farming

Under this category we have encompassed the polar cases in the conventional scale of economic achievement; the South Asian and the East Asian 'models' of economic development. On the one extreme, we have Bangladesh with declining levels of per capita income and, on the other, South Korea, a superstar in the GDP growth league. In what sense do they belong to a common category?

The answer is that they both have agrarian systems based on the private ownership of land and agricultural production is based on individual farming. The economic distance between them does not destroy the validity of this classificatory principle; rather, it demonstrates, as we shall argue, the widely divergent outcomes that are possible within a given system of landownership.

In Bangladesh, the rate of growth in agricultural output per capita during 1950–73 has been negative (−1.1 per cent per annum). Over the same period, India has had only a moderate rate of growth (0.8 per cent) while, in contrast, the rate of growth has been three times higher in South Korea (2.3 per cent).[1] At the same time, there is little doubt that the numbers in absolute poverty (defined conservatively as inadequate food intake) in the rural areas of Bangladesh and India have increased whereas it is highly likely that absolute rural poverty no longer exists in South Korea.

The dangers of drawing a naive morality tale from these stylised facts are considerable since they subsume within them complicated economic and political factors. Without pretending to provide more than a partial explanation, however, we would single out the difference in the degree of land concentration as a central differentiating characteristic.

In the two countries on the Indian subcontinent, landownership is highly concentrated and powerful landowning interests exist. The agricultural surplus is also highly concentrated and a substantial proportion of the rural population live in poverty as small tenant farmers and landless labourers. As is well known, under such an unequal structure, institutional monopolies exist in the land and labour markets and the rural rich can influence the political and administrative structure in their favour. There is thus a strong tendency for inequality in the distribution of wealth and income to increase, and for poverty to be perpetuated.

Within such a structure and in the absence of a major shift in political power in the countryside, there are limited options in rural development strategy. Maximising the rate of growth and hoping for 'trickle down' to occur is often seen as the only way out of the 'poverty trap' within the given economic and political constraints. Such a strategy, however, as the Indian case study demonstrates, rarely succeeds in bringing about the requisite broad diffusion of a new technology and in increasing the rate of labour-absorption in agriculture. The new technology is beyond the means of the poor majority and its introduction exacerbates inequalities and often leads to an absolute deterioration in their welfare.

The need for full diffusion of the new technology and to counteract its disequalising and deleterious effects is usually recognised, but effective instruments for attaining these objectives do not exist. The usual instrument adopted is some form of administrative intervention designed to counteract the exclusion of the poor due to highly imperfect markets. Thus co-operatives are instituted as distribution channels for new inputs and for credit, and extension services are created as the sales network for the disembodied part of the new technology. 'Comprehensive', 'integrated' approaches are launched in the hope that a 'big push' will do the trick better.

But as the Bangladesh study shows clearly, such improvisations cannot easily overcome the combined forces of entrenched economic and political interests. 'Co-operative capitalism' as the means to broad-based rural development is difficult to achieve in a situation where assets are very unequally distributed and a substantial minority are dispossessed and destitute. It is true that the Comilla experiment does show that a massive concentration of resources in a small area and generous subsidies on new inputs can generate impressive increases in output. But it also shows that even in such an artificially favourable situation, an equitable distribution of income was not achieved, marginalisation and further misery were not arrested and the use of the co-operatives in the interests of the rich was not prevented. Worse, it is even doubtful if the co-operative infrastructure *per se*, as opposed to the infusion of heavily subsidised new inputs, made any independent contribution to the observed growth.

The studies on Bangladesh and India also show that, even the simple growth objective, when it is directed to the rural economy as a whole and not merely to an experimental enclave, is extremely difficult to attain. In India, the various mutations of a growth strategy over a quarter of a century have not achieved an acceleration in the rate of agricultural growth, while in Bangladesh the attempt to replicate Comilla has had no perceptible effect in stemming the fall in per capita agricultural output. Thus even the feasibility of promoting sustained growth on a national scale has yet to be demonstrated.

A central contradiction inherent in a growth strategy within a highly unequal agrarian structure is that such a strategy requires, and obtains, the lion's share of investment funds for agricultural development. The logic of a growth maximisation strategy dictates that resources should be concentrated on progressive farmers and regions. There is consequently relatively little in the way of development resources that will be left over to deal with the mass poverty that remains unalleviated by growth. Thus, the opportunity cost of growth that does not 'trickle down' is the benefit that could have been derived from using these investment funds in a direct attack on poverty. It is little wonder therefore that resources devoted to 'small farmer programmes' and to public works schemes have been puny in relation to estimated requirements. Thus, a Paretian justification for a growth strategy is patently unconvincing; those who lose absolutely are not compensated, both because benefits do not 'trickle down' and because the demands of growth leave little for the implementation of compensatory, 'poverty-orientated' schemes.

If the impression we are left with thus far is that an agrarian system based on the private ownership of land can be a developmental morass, then the case of South Korea constitutes a timely counter-example. It shows that a radical land reform can, at a stroke, remove the deadweight of landlordism and install a system of egalitarian peasant farming which ensures growth without mass destitution.

Prior to land reform, South Korea suffered from the subcontinental syndrome of today: endemic rural poverty, a tenancy system characterised by burdens of high rent and an extreme concentration of agricultural surplus. It also had one of the most unfavourable land to man ratios in the world. Against such a background, the land reform needed to be drastic, and drastic it was. Expropriation was almost total; the ownership ceiling was set at a level that ensured near-universal access to land; and firm provisions were instituted to prevent the re-emergence of disequalising tendencies. There was thus a drastic reshuffling of the claims to land and the rules of the game were then redefined to exclude extreme polarisation. Yet within these constraints, the system remained one based on individual farming.

The post-reform system based on very small family farms has obviously worked. The egalitarian base created by land reform has resulted in the easy diffusion of new inputs and there is consequently a broad uniformity in production conditions across farms of all sizes. Growth in ouput and productivity has been impressive while the broad access to asset ownership guaranteed by the reform has ensured that the benefits from growth have been widely shared. The South Korean case also shows, however, that disequalising forces will still continue to operate even after a drastic reform and restrictions on land tenancy are required.

The simple juxtapositions of the preceding paragraphs need, however, to be qualified. There were several factors specific to the Korean case which made the transition to a more egalitarian system feasible. The reforms took place in the immediate aftermath of a major social upheaval and the pre-reform system of small tenancies and the absence of widespread landlessness made the conversion to a system of family-operated farms easier. Furthermore, the level of agricultural productivity was already high at the time of the land reform and thus favourable preconditions for further growth already existed. Finally, the task of rural development in South Korea has been greatly facilitated by its extraordinary rate of export-led industrial growth. Labour-absorption in industry was extremely high and since 1969 the rural population has been falling.

In conclusion, we would say that our case studies show that gross inequality in landownership is the dominant obstacle to broad-based rural development. Thus, where feasible, a radical land reform offers a swift way out from mass rural poverty and the post-reform system, as the Korean case has shown, can still be based on individual farming. Where such a reform is not possible the task will be very much more difficult.

(b) The intermediate category

Within this category Guyana and Tanzania provide interesting points of comparison in terms of the problems of transforming underdeveloped agrarian structures, so in what follows we shall deal first with these two countries before discussing the case of Egypt.

The agrarian structures which were inherited by Guyana and Tanzania from their colonial past could broadly be described as dualistic and dependent, but with significant differences on several points. Although plantations existed in both countries, Guyana was much closer than Tanzania to the model of a pure plantation economy.[2] One crop (sugar) and one multinational corporation (Booker McConnell Ltd) attained a prominence in the Guyanese economy which had no parallel in Tanzania. In addition, the Guyanese economy was built on the basis of importing African slaves and Indian indentured labour and little survives of the indigenous Amerindian population. The peasant agriculture that has developed in Guyana originated from the attempts of the imported plantation labour to break away from a strong dependence on the plantation system. In Tanzania, by contrast, a plantation sector intruded into a large pre-existing indigenous system of peasant agriculture without destroying it. Thus, although by 1961 the plantation sector had come to account for 20 per cent of total cultivated land and the bulk of agricultural exports, the vast majority of the Tanzanian population still earned their livelihood in traditional agriculture.

These differences in initial conditions account for the difference in emphasis in their respective strategies for agricultural transformation, although both countries espoused a form of 'socialist' ideology. In Guyana the focus was on nationalising the plantation sector, whereas in Tanzania it was on transforming the peasant economy through villagisation and the fostering of communal production on the Ujamaa model. At the same time, however, Guyana did have plans for introducing 'co-operative socialism' in the peasant sector and Tanzania did nationalise her estate sector.

It is difficult to evaluate the achievements of these strategies for agricultural transformation in these two countries. In neither case has there been an established alternative agrarian system for long enough to permit a meaningful evaluation; the main changes have been too recent. All that can be done is to give a qualitative assessment of the direction of change that occurred under these strategies and to identify the key problems that remain to be overcome.

One problem that both countries face is that of growing differentiation among the peasantry. In Tanzania this process was still not very marked at the time of the Arusha Declaration in 1967, but the trend was clearly towards increased differentiation. This was one of the major factors which prompted the attempt to pre-empt this development through the institution of a system of Ujamaa villages. Private land on new settlements was equally allocated and communal production in both agricultural and non-agricultural activities was actively encouraged. Furthermore ceilings were imposed on private landownership and restrictions were placed on the hiring of permanent agricultural labourers. The principle of equal distribution of land on new settlements was also followed in non-Ujamaa 'development villages'. Although the land distribution remained unchanged in settled villages, the massive extent of villagisation that had occurred by 1976 indicates that this principle of distributing new land would have made a significant contribution to ensuring equal access to land among the Tanzanian peasantry. Moreover, communal production on the Ujamaa model, although still of a relatively limited scale, together with communal action in non-agricultural activities would have also made some contribution to greater equality in income and asset distribution. The quantitative impact of these changes cannot be estimated precisely. It remains true, however, that if these measures are applied consistently, and if steady progress were made towards increasing communal production, then the Tanzanian strategy would succeed in ensuring a far greater degree of equality than would prevail under a system of *laissez-faire*.

It is also important to note that the Ujamaa and villagisation programme in Tanzania does not appear to have caused any serious disruption in agricultural production in spite of the vast resettlement of the rural population and the other problems of transition that were involved. Villagisation often meant the shift from an extensive to an intensive mode of cultivation and the adoption of new agricultural technology. Furthermore, in Ujamaa villages communal production meant the introduction of an entirely new institution with many attendant organisational problems.

In Guyana, on the other hand, little real progress has been made so far towards building 'co-operative socialism' among the peasantry. According to Thomas, 'Co-operatives have resulted primarily in the creation of barely disguised agencies of state or estate control.'[3] Differentiation of the peasantry has proceeded to a much greater extent in Guyana and in spite of the fact that, like Tanzania, it is a land-abundant country (with a population density of 10 persons per square mile) landlessness and land-hunger among the peasantry were very real problems. This situation arose from the need of plantations to preserve an adequate labour supply by discouraging the emergence of an independent peasantry, the pre-empting of the best land and of investment funds by the plantations and the fact that the heavy cost of opening up new land ruled out spontaneous settlement by peasants. In addition, the nationalisation of 'the plantation sector' in Guyana has not been followed up by any moves for land reform in the peasant sector, or to reduce land-hunger through a vigorous resettlement programme. Moreover, the unfavourable relationships which bound the sugar-growing section of the peasantry to the plantations have as yet been unchanged. Similarly, production relations within the nationalised sector remain unchanged and a 'confrontationist' system of industrial relations, a high degree of absenteeism, low wages and poor working conditions still prevail. There has thus been as yet little tangible change in the condition of the Guyanese peasantry and plantation workers arising from nationalisation and the inauguration of co-operative socialism.

In considering this record of agricultural transition in these two countries, certain fundamental problems can be identified. One is that of devising appropriate incentives in the process of agricultural transformation. In Tanzania, this issue is crucial for inducing faster growth in communal production, which at present accounts for only a small fraction of total agricultural output. Since private ownership of land still predominates, it is important to find means other than mere moral suasion to induce communal action if the Ujamaa ideal is to be achieved within a reasonable period. Similarly, in Guyana, 'co-operative socialism' among the peasantry can only be attained if appropriate incentives are created to replace the present unfavourable system which binds peasant producers to plantations and the huge inequalities in landownership within the peasant sector are redressed.

A second problem is that of reorganising and transforming the relations of production within the nationalised sectors in both countries. If nationalisation is to have a far reaching impact on the welfare of the rural population as a whole, then more than a simple legal transfer of ownership is required. In particular, the structure of production may have to be reorganised to meet the requirements of an overall rural strategy. Continued dualism and the concentration of resources on the plantation sector as a convenient means of extracting a surplus for the non-agricultural sectors, may be incompatible with the overall goal of egalitarian rural development.

The final consideration relates to the policies that are followed in relation to settlement patterns and the allocation of new land in land-abundant countries such as Tanzania and Guyana. In the latter country, the historical obstacles to new peasant settlements have not been overcome and consequently land-hunger as well as what Thomas calls the 'disintegration of rural communities' remain as serious problems. Since plantations occupy the best land and the extension of the land margin, in spite of the huge potential, is difficult, attention would have to be focused on a comprehensive land reform (covering both plantation and peasant land) to ensure equitable access to land and a balanced pattern of rural settlement. In Tanzania, on the other hand, the extension of the land margin has occurred far more easily. Under the traditional pattern of extensive cultivation, expansion of cultivated area has been possible without infrastructural investment. The control of the state over access to land is thus much looser than in a land-scarce country or one where new settlement is not possible without heavy infrastructural investment. The justification for altering the pattern of settlement through villagisation thus lies less in the need to 'grant' land than in the imperatives of increasing agricultural productivity, reaping economies of scale in the provision of social services and bringing about the eventual socialist transformation of rural society. This implies that, unless the benefits of villagisation are immediately obvious to the rural population, either progress would have to be gradual or a certain degree of compulsion would have to be applied. The Tanzanian experience is indeed one where a reliance of gradualism gave way to a greater reliance on compulsion in the programme of villagisation. We have noted earlier that villagisation in Tanzania offered an opportunity for influencing the distribution of land towards greater equality and that the principle of equal distribution of new land ensured that this was realised. It remains true, however, in Tanzania as in Guyana, that the extension of the land margin cannot be a substitute for comprehensive land reform. In Tanzania, the inequalities in ownership in the regions with long established settlements (and which also happen to be the more prosperous regions) would have to be dealt with in the course of the socialist transformation.

Turning to the case of Egypt, we observe that successive agrarian reforms beginning in 1952 have virtually eliminated the very large landownerships and created an extensive network of co-operatives, but left ownership almost exclusively in private hands. However, the state control over the agricultural sector was extended and this enabled it to manipulate the inter-sectoral terms of trade in such a way as to ensure the transfer of agricultural surplus to the modern sector.

In evaluating the performance of the post-reform agrarian system we shall concentrate on growth, distribution and the disposal of the agricultural surplus.

The growth performance of Egyptian agriculture over the last twenty-five years has, on the whole, been slow. The average annual growth rate of agricultural value-added did not exceed that of population. Certainly, the 1950s and 1960s witnessed an acceleration in growth, mainly reflecting the impact of improved co-operative services and the effect of the new crop rotation system. But by the end of the 1960s there was a slow-down in growth due to the combined effect of a number of factors, most important of which were the lack of continued investment in agriculture infrastructure, the decline in co-operative services and a lack of resources to transform the technique of production (which in turn was due to the squeeze to which agriculture was subjected).

As regards land and income distribution, the initial period of agrarian reform witnessed a definite reduction in inequality and a drop in the magnitude of absolute poverty. But by the beginning of the 1970s the equalising effects of the reform were thinning out, and this, together with a rapid population growth, explains the increase in inequality and impoverishment. After 25 years of agrarian reform, we find that the initial inequalities in rural Egypt were not fundamentally diminished: 40 per cent of the households were landless and 35 per cent lived in poverty. Moreover, the slow growth of agriculture, on the one hand, and the inability of industry to absorb the new entrants to the labour force, on the other, aggravated the employment situation. This led to the rapid 'tertiarisation' of both the rural and urban sectors as the rural poor who face limited access to land and have no prospect of employment in agriculture, try to earn their living by joining the ranks of the so-called 'informal sector' in urban areas, and by engaging in low productivity activities such as handicrafts and petty trade.

Government policies towards the rural sector have compounded the problems of ensuring growth and equitable distribution. A slight improvement in real per capita rural income during the 1950s and 1960s was followed by stagnation, and the rural-urban differentials in real per capita incomes have been widening. A key factor explains these disparities. In Egypt, the government has, until recently, manipulated the inter-sectoral terms of trade in such a way as to be able to squeeze agricultural surplus. The effect of these policies was a net transfer of agricultural surplus to finance capital formation in the rest of the economy. After allowing for government investment in agriculture, the net outflow of surplus from agriculture amounted to 15 per cent of agricultural income in Egypt over the period 1960–75.

The attempt to achieve a rapid and egalitarian rural development through major institutional change in Egypt no doubt resulted in initial gains, but since these reforms did not continue, the forces of the system, mainly the inherent inequalities, acted in such a way as to erode these gains.

(c) The system of communal farming

The stereotype of a communal agriculture that most people have in their mind is based largely on a popular version of the Soviet collectivisation experience. It conjures up a picture of sluggish growth, egalitarian income distribution and a high rate of involuntary accumulation imposed by the state policy of transferring resources from agriculture to industry. The three experiences analysed in the present volume substantially contradict this stereotype. They also make neat generalisations impossible.

The growth performance of Cuban agriculture, except in very recent years, has not been encouraging. But in Soviet Central Asia collective agriculture served as a vehicle for rapid growth especially after the first half decade of collectivisation. In China, in spite of some vicissitudes, the communal agriculture achieved a reasonably, though not a spectacularly, high long-term rate of growth. Clearly, collectivisation does not necessarily condemn agriculture to stagnation. Nor is it a sufficient condition for rapid growth. One must go beyond the mere fact of collectivisation and look into the details of organisation and policies within the framework of which the system works. As we shall discuss below and as is evident from the case studies, the performance depends largely on the proper evolution of what may broadly be called a system of incentives.

In the provision of employment and egalitarian distribution of income, the performance of the system, as exemplified by the three cases, is uniformly good. Employment is guaranteed and sometimes imaginative provisions are made to overcome the problems of seasonality and to ensure greater participation of women. Basic human needs are by and large satisfied; in the areas of health, education and social security especially, the production units serve as the organisational framework for the provision of services. Poverty, in the sense it prevails in large parts of Asia, Africa and Latin America, is virtually unknown.

It is, however, not true that collectivisation automatically and easily overcomes all inequalities in the distribution of income. For Cuban agriculture the relevant information is not available, but in both Soviet Central Asia and China significant, even considerable, inequalities in rural income distribution persist. Such inequality derives from difference in earning capacities of individual members (aggravated by the difference in their dependency ratios) *within* a basic production unit, differences *between* basic production units and differences between regions. Once again, as emerges from the experience of China, the minimum feasible inequality in any given historical and cultural situation is determined by the need to provide material incentives to ensure a reasonable rate of expansion of production ('development of productive forces') without which the very basis of an egalitarian social system would flounder.

On accumulation the evidence is less certain. The limited information about the Soviet Central Asian collective farms shows that a farm can often save at a high rate. The case of the Chinese communes is even more convincing: the total contribution to capital formation is frequently of the order of 30 per cent of a commune's net output. It is, however, not true that collectivisation is an inevitable mechanism of transferring vast resources out of agriculture. Neither in Soviet Central Asia nor in China has such large transfer of resources been a feature of State policy towards agriculture. In both cases the ability of the system to provide reasonable incentives derived substantially from the policy of not extracting a large surplus out of collective agriculture.

The success of a collective agriculture depends crucially on the provision of a well-balanced system of incentives. As is evident from the experience of Cuba in the 1960s and of China during the Great Leap Forward, an attempt to curb material incentives, without a proper development of the political and cultural consciousness of the population (the possibility of which is itself unproven), can have disastrous consequences for productive efficiency. The provision of incentives in a collective economy is not a simple task that can be administered easily. As is discussed in the case study on China incentives at three different levels – individual, collective and sector – need to be provided on a consistent and comprehensive basis.

The proper configuration of the basic collective unit ('the basic accounting unit') is crucial to the development of a system of incentives for the individual workers. Such a unit should be small enough to make a proper assessment of individual performance possible. It also means that the hierarchy of organisation should be such that the activities involving indivisibilities and economies of scale are handled by a higher level of collective organisations. The organisation of the Chinese communes perhaps constitutes what may be called a successful example. In the three tiered organisation, involving the team, the brigade and the commune, the team is almost universally the basic accounting unit. Its small size – usually about 30 households – is highly conducive to a proper evaluation of individual performance so that rewards can be made proportional to such performance. The larger land improvement or irrigation works, as well as the industrial enterprises embodying economies of scale, are naturally beyond the ability of a team to handle efficiently. These are organised by the brigades and/or the commune. A proper configuration of the production units and an appropriate degree of decentralisation of decision-making among different levels of production units are essential pre-conditions for the evolution of the system of incentives.

Incentives must also be adequately provided for the various collective units – the teams, the brigades and the communes, to name the Chinese ones – to induce them to achieve greater productive efficiency. Here the systems of taxation, procurement and distribution of inputs play a crucial role. Both in the Soviet Central Asia and in China the state authorities have demonstrated considerable imagination in recent years in evolving an incentive-oriented system of taxation and procurement. The main ingredient of this policy is a two-tier procurement system with a lower price for a basic procurement quota and a much higher price offered to any voluntary sale over and above the basic quota. The system effectively renders the marginal rate of taxation (subsidies) lower (higher) than the average.

Finally, the question of incentives for the agricultural sector as a whole is linked up with the question of the extraction of resources out of agriculture through taxation or the imposition of unfavourable terms of trade. Unless such resource extraction is modest the agricultural sector would be subject to serious disincentive which would be transmitted downward to affect adversely the incentives of the collective units and individuals. In some early experiences of collectivisation, excessive extraction of resources through such means was undoubtedly the chief factor in explaining slow growth. It is, however, interesting to note that neither in Soviet Central Asia nor in China, has state policy favoured any large and continuous extraction of resources out of agriculture. This may have been one of the most important causes of the success collective agriculture has achieved in these two cases.

In all systems of communal agriculture some form of private activity continues to exist. In Cuba it takes the form of private peasant households who cultivate about a third of the arable land. In Soviet Central Asia and China it takes the form of private plots to which all members of the collectives have claim. The advantage of the continued existence of private activity is the flexibility that it provides the system: it is far simpler to allow individual households to grow their own vegetables and poultry than to set up a network of supply for the distribution of such perishable goods. But it also creates additional problems of operating the collective economy. Collective income is subject to deductions for taxation by the state, financing of welfare services and collective accumulation. It is virtually impossible to subject the earnings from private plots to such deductions, and it is at best difficult to subject the peasant producers, such as the ones in Cuba, to equivalent deductions. The result is a built-in bias in favour of labour allocation to private activities. Such bias is further aggravated by the fact that private production can frequently be sold freely while collective production is subject to procurement at fixed prices. Given the difficulty and high cost of taxing private activities, it would be hard to devise policies to correct the balance. Also, it would be undesirable to give up this source of flexibility by abolishing private activities altogether. All the state policy can do is to keep the distortion to a minimum by subjecting collective production to no more than modest taxation on surplus extraction and by raising productivity in collective activities by directing as much investment to them as is feasible.

1.3 Some concluding comments

In the above the achievements and problems of the three broad categories of agrarian systems, as well as the comparative analysis of the cases in each category, have been discussed. To conclude, let us briefly recapitulate the major conditions of success of each category of agrarian systems.

In a system of private farming the most important condition of egalitarian growth is an equitable land reform and its continued enforcement. In the present set of case studies this is the characteristic that distinguishes South Korea from the other two cases of private agriculture and largely explains its vastly different performance.

The question may be raised that the present case studies do not really include an example of successful private agriculture without an egalitarian land reform such as in the Punjab in India and Pakistan. Although no case study on these areas has been included in the present volume we have evidence from previous studies to serve as the basis for an answer to such a question. While the growth in output in the Punjab has been quite impressive, it has been reasonably well established that the performance in terms of income distribution has been unencouraging, even dismal.[4] Moreover, compared to many South and South-east Asian countries the Punjab has a far more favourable land to labour ratio.

South Korea, on the other hand, has about the lowest land to labour ratio in the whole world – even lower than that most archetypal of the land scarce countries, Bangladesh.[5] Nor is the South Korean experience an isolated example: similarly strong redistributive measures achieved similar results elsewhere – in Japan and in Taiwan to mention only the leading contemporary examples.

Perhaps the most important question is that of the feasibility of such a highly egalitarian land reform. In South Korea, Taiwan and Japan such reforms were initiated by occupying armies and directed against landed aristocracies, large sections of which were closely identified with the recently defeated forces. Also, in South Korea disruptions were avoided since the reforms left the actual operational units relatively unaffected – an outcome made possible by the fact that while the pre-reform distribution of ownership of land was very unequal, that of actual operational units was highly equal. This condition does not appear to pertain in areas like the Indian subcontinent.

Yet another condition of success is the continued enforcement of the ownership limit imposed by a land reform. An important question is whether such an enforced limit can be a viable framework for agricultural development in the long run, if the rest of the economy experiences a more or less unbridled capitalist expansion as in South Korea. One must continue to observe the Korean case for answer.

On a somewhat different plane one must worry about another kind of feasibility consideration. Can egalitarian peasant farming continue to serve as an organisation for growth in those densely populated Asian countries which cannot foresee anything but a continued absolute pressure of population on land? In many parts of Asia the prospect of an average farm size declining to the level of a fraction of an acre is real if egalitarian land distribution is to take place and continue to be in existence for a decade or two. Would this not render the peasant farms incapable of deriving any advantage of economies of scale and of coping with activities (e.g. the management of water) which involve a significant degree of indivisibility? Without doubt, agriculture in such societies would need to have a transition towards some form of joint farming, be it in the form of co-operative activities to cope with the indivisibilities or of full-scale communal farming.

While in a system of private farming the most important question is that of the distribution of ownership, in a communal system it is the question of incentives (broadly defined, as we have done, to encompass the issues of the configuration of efficient production units and decentralisation). Communal systems are built by destroying capitalism. Along with capitalism they destroy the capitalist system of incentives which is more advanced than all systems of incentives that historically preceded it. It is absurd to hope that collective agriculture can succeed without carefully building up an alternative system of incentives. One need not reject the vision of some socialist planners that mankind must ultimately be able to transform itself so that it no longer requires material incentives to fully utilise its productive potential. But no one can seriously claim that such a transformation immediately follows that of ownership. The most optimistic assumption is that such a transformation may be possible in the very long run. Until then, a solid and comprehensive system of incentives must continue to operate.

The most important fact about the cases in the intermediate category is that they all represent different stages of transition from private farming towards some form of joint farming. Essential features of such transition are the continued predominance of private ownership and the attempt on the part of the state to promote some combination of co-operative, collective and state farming by gradualist (that is to say non-revolutionary) means. These transitional characteristics define the basic conditions that need to be fulfilled if such a system is to succeed. Since private ownership predominates, an essential condition of egalitarian distribution of income is an egalitarian distribution of land and other means of production. In the three cases reported in this volume the main source of inequality derives from that of the distribution of land. In Egypt, the agrarian reform eliminated the very large landowners but paved the way for the consolidation of the position of the medium and rich farmers at the cost of a high and rising degree of landlessness. In Tanzania the distribution of landholdings in the newly settled villages is equitable but that in the much richer, older villages continues to be less equal. In Guyana the peasant sector is highly stratified.

The coexistence of private and joint farming underlines the importance of balancing incentives for the two forms of organisation. In Egypt collective ownership is virtually non-existent and the problem of incentive consists of the effects of the state action towards agriculture in the form of the imposition of a heavy (and perhaps asymmetrically distributed among income groups) concealed tax through the manipulation of prices. In Guyana and Tanzania the state–collective sector is extensive so that the problem in these countries is to ensure that relative incentives for the different forms of organisation do not create obstacles in the way of appropriate balance of effort. Thus the vision of making Ujamaa villages the vehicle of growth and collectivist transformation will not be realised unless the incentives for collective work are properly organised to match that for private farming. Nor will the objective of egalitarian growth be promoted in Guyana if the relatively poorer peasant producers of sugar-cane are discriminated against in order to augment the state sector's surplus which is not deployed to achieve particularly egalitarian purposes. In the final analysis one must continue to watch these to determine if the phenomena of persistent inequality of the peasant sector and the inconsistent balance of incentives between communal and private activities are mere mistakes to be corrected with the accumulation of experience or if they are the manifestations of the logical inconsistency of the gradualist path which ostensibly attempts to redistribute income and wealth away from the rural rich without fundamentally altering the balance of their power in the society.

In every system the question of surplus generation and extraction is of overwhelming importance. It determines the incentives for the sector and ultimately for the production units and individuals. As the dominant sector of a developing economy, agriculture must save at a reasonable rate if the hope of growth is to be realised. But there is no need to extract massive proportions of such surplus out of agriculture to finance industrialisation. There are clear feasibility and desirability limits on such a transfer. The question of how much resources can be transferred without any serious disincentive effect can only be answered with reference to each specific circumstance. In China, for a period immediately after land reforms, a significant resource transfer out of agriculture took place. But such resources were only a part of the rent and profit previously accruing to the landlords. Thus the peasants' incomes also increased at the same time as resources were being transferred and the problem of disincentive was not a serious one. The Soviet policy of resource extraction out of agriculture (to which the Central Asian experience reported in this volume was an exception) had, on the other hand, serious effects on the incentive of the sector.

It is not, however, the overall level of resource transfer out of agriculture that alone is important. Equally important is the distribution of the burden of surplus extraction among different groups of the rural population. Thus, there are cases in systems of individual farming where the overall terms of trade have improved for agriculture but, because of asymmetrical effects on different groups, this has led to a deterioration in rural income distribution. In general, therefore, a simple reduction in the level of surplus extraction out of agriculture is not by itself a sufficient condition for attaining balanced rural development.

The question of the desirable limit on resource transfer is perhaps even more important. Given the objectives of the society and the production (including export) possibilities it is a question of which sectors should receive investment in what proportion? In few of the contemporary cases of industrialisation through the extraction of resources from agriculture were decisions made on the basis of a proper investigation of this question.

Notes

1. K. Griffin, *Growth and Impoverishment in the Rural Areas of Asia* (mimeo, Oxford: Queen Elizabeth House, 1978) Table 3, p. 7.
2. See G. L. Beckford, *Persistent Poverty: Underdevelopment in Plantation Economies of the Third World* (New York: Oxford University Press, 1972).
3. See Chapter 6.
4. For example, see the evidence in Chapters 2 and 3 of ILO, *Poverty and Landlessness in Rural Asia* (Geneva: 1977).
5. The amount of land per member of an agricultural family is slightly lower in Bangladesh than in South Korea but per agricultural labourer land is lower in South Korea because of a higher labour participation rate. See FAO, *Production Yearbook* 1976, for the data on which these statements are based.

2 Egalitarian Peasant Farming and Rural Development: The Case of South Korea

Eddy Lee

2.1 Introduction

Economic growth has largely been accompanied by stagnation, growing inequality and even absolute impoverishment in the rural areas.[1] The exceptions to this trend are few and far between. They are believed to lie in a few socialist developing countries and in a few developing market economies which have combined very rapid growth rates with stability around a distribution of income that is more egalitarian than that of most other developing countries. The latter set of countries are exemplified by Japan, Taiwan and South Korea, sometimes referred to collectively as the 'East Asian' model of development.

The fascination that the 'East Asian' model exerts for liberal development studies is easy to understand. It is seen as the vindication of the viability of the neo-classical route to development, that the free market can generate rapid growth without floundering in the impasse of growing inequality and mass poverty.[2] Furthermore, given the natural propensity to believe that what is possible must be replicable, these models have been dissected to see 'what made them tick' in order that the secrets of success might be more widely diffused in the rest of the developing world.

The aim of this paper is to look at the South Korean experience from the perspective of the rural sector. Much of the previous work on South Korea has concentrated on the industrial sector and its extraordinary growth through a strategy of concentrating on export-oriented, labour-intensive industries. Yet the rural sector offers several interesting issues for analysis. The available data seem to indicate that the performance of the rural sector has also been exceptional in comparison to the rest of the developing world.

Output and productivity growth have been steady while there has been no trend of increasing inequality or landlessness, as has been the case elsewhere. This performance has occurred from a base of a thorough land reform between 1945 and 1957[3] and the resultant agrarian structure, being relatively egalitarian and homogeneous, has been a principal factor in explaining the widely diffused growth in the rural economy.

In the following sections we shall describe the main characteristics of the agricultural system of South Korea, evaluate its performance in terms of growth, distribution and accumulation and then go on to analyse the factors which explain the performance.

2.2 The agricultural system of South Korea

The dominant characteristic of South Korean agriculture is the extreme land scarcity. South Korea has one of the world's highest land to man ratios and the average holding size after the land reforms was only 0.9 hectares.[4]

This adverse resource endowment, combined with a harsh tenancy system, unequal distribution of land and a high rate of surplus extraction, resulted in conditions of widespread and extreme rural poverty during the Japanese colonial period.[5] From the end of Japanese rule in 1945, the alleviation of rural poverty through agrarian reforms was seen as an important element in the programme for restoring political stability in the southern half of the country. The first steps were undertaken in 1948 but it was not until the end of the Korean Year in 1953 that full implementation of land reform began.

The land reform was more an expropriatory tenancy reform than a redistribution of land from large units to small units or the landless. The pre-reform agrarian system was characterised by the absence of a dualistic structure of agricultural production in the form of large estates coexisting with small peasant farms and a pool of landless labour. Instead, the system was one of a high concentration of landownership but with most of the land rented out in small units. As a consequence, it was a very high degree of tenancy of small parcels of land rented out at extortionate rents that was the mechanism of exploitation. The main thrust of the land reform was thus the replacement of tenancy by owner cultivation and not a radical shift in the size distribution of production units. 60 per cent of all cultivated land was tenant farmed in 1945 and at least 50 per cent of farm households were pure tenants while 4 per cent of farm households owned 50 per cent of all cultivated land.[6] A further 20 per cent of the arable land was owned by the Japanese. The land reform changed this situation radically; post-reform, only 7 per cent of farm households were tenants, 69 per cent were full owner-operators and the rest were part owners; the figures in Table 2.1 illustrate this point clearly. At the same time, however, the size-distribution of operating units changed only slightly. The proportion of farms over three hectares in size dropped from 1.4 per cent to 0.1 per cent. The proportion of farms between one and three hectares dropped slightly while the number of farms of less than 1 hectare increased from 74.5 per cent to 79.1 per cent.

The impact of the land reform was thus primarily redistributive

Table 2.1 Tenancy Conditions and Size-distribution of Landholdings, 1947 and 1965

A

Type of Tenure	*1947 (%)*	*1965 (%)*
Full-owner	16.5	69.5
Part-owner	38.3	23.5
Tenant	42.1	7.0
Slash and Burn Farmers	3.1	—
Total number of farm Households	2172000	2507000

B

Size of Landholdings (hectares)	*1947*		*1953*	
	('000)	*(%)*	*('000)*	*(%)*
less than 0.5	895	41.2	1011	44.9
0.5–1.0	724	33.3	769	34.2
1.0–2.0	409	18.8	371	16.5
2.0–3.0	113	5.3	96	4.3
more than 3.0	31	1.4	3	0.1
	2172		2249	

Source Robert B. Morrow and Kenneth H. Sherper, 'Land Reform in South Korea', *Agency for International Development Spring Review* (Washington: June 1970) Tables 5 and 7.

with respect to ownership and without any substantial change in the configuration of production units. The effect of the redistributive change, however, also affected production incentives positively and widened the 'absorptive capacity' of the agricultural system for investment and technical change. *A priori*, a shift to owner-operated units with a relatively narrow dispersion in farm sizes would be expected to enhance the capacity of the system to respond to market incentives more uniformly and hence permit growth without a sharp polarisation in income distribution.

In relation to other developing market economies it is undeniable that the South Korean agricultural system is relatively egalitarian; the Gini coefficient of the size-distribution of landholdings (which also, as pointed out above, largely corresponds to ownership distribution) was only 0.38 in 1965 as compared to figures of 0.6 to 0.7 for many other developing countries.[7] This narrow dispersion in farm sizes, with an upper limit of only three hectares, results in a striking homogeneity in production conditions. For example, there is a basic scale neutrality in factor proportions and productivity. As Table 2.2 shows, the breakdown of the cost of production is broadly similar for farms in the different size groups and the same is true of yields per unit of land. The use of 'modern' inputs such as chemical fertilisers and pesticides is fairly uniform across farm sizes and this uniformity applies also to traditional inputs such as organic fertilisers and draught animals.

The only striking variation in the pattern of input use relates to labour. Larger farms have a progressively higher input of hired labour but the reverse is true when total labour inputs (hired and family labour) are considered. The variation in total labour inputs was not very pronounced according to the 1969 figures but had widened considerably according to the 1974 figures. This change might perhaps raise some questions about the degree of accuracy of the observed relation between the pattern of input use and yields across farm sizes for the latter year. It is possible, however, that part of the discrepancy could be accounted for by the fact that the cost item 'implements' does not fully reflect the influence of increased mechanisation. Data on the number of power tillers and threshers in use in 1974 show a significant increase from the very low level prevailing in 1969[8] and this increase could account for the observed relative reduction in total labour input on the larger farms.

Another important characteristic of South Korean agriculture is the relatively low level of technology. It is 'dependent almost

Table 2.2 Cost of Production of Paddy by Farm Size (*won* per *tanbo*)

Size of Farm (chungbo)	*Seed*	*Fertiliser* *Chemical*	*Fertiliser* *Organic*	*Pesticides*	*Irrigation*	*Taxes*	*Implements*	*Draught Animals*	*Hired Labour*	*Total Labour Input*	*Yield* (*kg. per* tanbo)
					1969						
less than 0.5	225	913	273	208	367	264	291	492	1065	4985	402.0
0.5–1.0	244	821	370	210	362	248	283	421	1160	4947	387.0
1.0–1.5	236	892	293	226	473	363	328	448	1253	4738	406.5
1.5–2.0	224	874	268	237	550	396	338	506	1931	4821	410.9
more than 2.0	229	897	262	231	553	495	349	491	2571	4670	399.2
					1974						
less than 0.5	665	1524	901	876	965	187	1681	1185	3690	15227	468.1
0.5–1.0	616	1388	854	759	1008	308	1304	1074	3611	13373	460.9
1.0–1.5	612	1298	576	686	945	413	1230	1047	3918	12462	448.7
1.5–2.0	585	1287	548	664	833	521	1032	1098	4018	11251	447.6
more than 2.0	569	1390	472	706	1142	709	1075	962	5384	9882	465.9

Source Korea, Ministry of Agriculture and Fisheries, *Yearbook of Agricultural Statistics* 1970, Table 131, pp. 310–11, and 1976, Table 88, p. 202.

Notes 1. 1 *chungbo*=0.9917 hectares; 1 *tanbo*=0.1 *chungbo*; the *won* is the South Korean unit of currency (in 1974 US$1= 406 *won*).

2. 'Irrigation' costs refer to the recurrent costs of obtaining irrigation services. The item 'implements' is not clearly defined in the source; most of it is accounted for by 'purchase and disbursement' and 'depreciation' of 'large tools'.

entirely upon human and animal draught power', and until recently 'lagged behind Taiwan and Japan in biological technology'.[9] However, there is substantial use of fertiliser and plant protection materials and irrigation facilities had been relatively well developed during the Japanese colonial period. The small farms are intensively cultivated and yields of rice in the 1960s were higher than those in developing countries but lower than those in Japan.[10] During the period 1952–6, the yield per hectare of sown acreage in South Korea was 3340 kilograms of cleaned rice which was considerably higher than the corresponding figures for India (1280 kilograms) and Thailand (1350 kilograms) but lower than the Japanese figure of 4340 kilograms.

There is also a very high degree of uniformity in cropping patterns across farm sizes and regions. Rice is the dominant crop, accounting for 60 per cent of the cultivated area and together with barley, the second most important crop, accounts for 68 per cent of total cultivated area. Thus, in addition to the level of technology and the farm size distribution, this is a further factor making for homogeneity in the rural economy. This predominance of grain production contrasts with the cropping pattern in economies which have undergone Western colonial rule, where export crops predominate and where also a 'dualism' between a plantation and a peasant sector is often reinforced by differences in cropping patterns.[11]

Lest the popular image of South Korea as the embodiment of 'free market' virtues misleads one into thinking that Korean agriculture operates in a *laissez-faire* environment, it is necessary to point out the significant extent of state intervention. Compulsory procurement of rice and barley was introduced in 1948 and the Grain Management Law of 1950 confers, until today, the authority for complete control over grain procurement and marketing. Although the degree of state intervention has diminished from the complete control of the post-war years, in 1975 the government still purchased 50 per cent of marketed rice and 90 per cent of marketed barley. The supply of fertiliser and pesticides is likewise completely controlled by the government through the National Agricultural Co-operatives Federation. It is also important to note that the ceiling of three hectares on landownership dating from the land reform is still in force. There is thus also, legally, no free market in land transactions.

Finally, a feature of considerable importance distinguishes Korean agriculture from that in many other developing countries.

The rate of out-migration has been extremely high since the industrialisation drive which began in the early sixties, and the total farm population has in fact begun to decline since 1969. The share of farm population in total population has declined from 58 per cent in 1960 to 38 per cent in 1975 and, in spite of an overall population growth rate of 2.1 per cent per annum, the absolute numbers of the rural population fell from 14.5 million to 13.2 million over this period. There has thus been an easing, rather than an aggravation, of the demographic pressure that bedevils rural development in many developing countries.

2.3 Performance of the rural economy

(a) Growth

The growth rates in South Korean agriculture have been respectable by international standards. The growth rate in the value-added of crop and livestock production was 3.5 per cent per annum for the period 1952–71; the corresponding figure for Japan was 2.2 per cent per annum and for India, 2.9 per cent per annum. Taking sub-periods, we find that the growth rate was higher during 1961–71 (3.7 per cent per annum) than during 1956–61 (3.1 per cent per annum).

It might be argued that although these growth rates were higher than those in other developing countries, they were not substantially so. It has to be borne in mind, however, that the growth in the South Korean case occurred from a base of much higher productivity in relation to other developing countries and within an extremely land-scarce country. The growth achievement is thus qualitatively different from that achieved by introducing technical change (for example, irrigation and high-yielding varieties) into 'traditional' agriculture, or through the rapid expansion of cultivated area in land-abundant countries.

The growth performance can also be gauged by the trend in the yield per sown hectare of the predominant crop, paddy. Yield per hectare increased from the already high level of 3340 kilograms in 1952–6 to 4550 kilograms in 1970 and 4700 kilograms in 1974. The yield increase between 1952 and 1970 is remarkable in that high-yielding varieties were introduced in Korea only in 1968 and by 1971 only 100,000 hectares had adopted the new varieties. The increase in yields in this period was thus largely linked to the increase in the application of chemical fertilisers (fertiliser inputs increased at a rate of almost 9 per cent per annum between 1956 and 1970)[12] and to improvements in irrigation facilities. Since 1970 the rapid spread of high-yielding varieties has been an additional factor in pushing up rice yields. By 1976 600,000 hectares (47 per cent of the cultivated area under rice) were growing the new high-yielding Tongil variety and there were euphoric claims of world records being established in yields per hectare by some farms.[13]

Unlike physical output, however, it is not clear that the average real incomes of rural households increased smoothly over this entire period. (Table 2.3). While the average real income of rural households was 51 per cent higher in 1975 than in 1963, most of the increase was concentrated in the period after 1970. The data from the Farm Household Surveys show that real incomes remained about 10 per cent below their 1963 levels until 1968. This pattern of

Table 2.3 Average Real Income per Rural Household

Year	*Nominal Income* (won)	*Rural Price Index*	*Real Income* (won)	*Index of Real Income*
1963	93179	100	93179	100.0
1964	125692	129	97435	104.6
1965	112201	144	77917	83.6
1966	130176	161	80855	86.8
1967	149470	182	82126	88.1
1968	178959	209	85626	91.9
1969	217874	227	95980	103.0
1970	255804	259	98766	106.0
1971	356382	291	122468	131.4
1972	429394	325	132121	141.8
1973	480711	346	138934	149.1
1974	679451	492	137084	147.1
1975	872933	618	141251	151.6

Source Korean Ministry of Agriculture and Fisheries, *Farm Household Income Surveys* 1970, Table 187, p. 430; 1976, Table 78, p. 171.

Note Our real income series differs from that given in Korean Economic Planning Board: *Major Statistics of the Korean Economy 1977*, Tables 4–10, p. 62. In that series the Seoul Consumer Price Index was used instead of the Rural Price Index. The same mistake is repeated in Parvaz Hasan, op cit, p. 52.

change in rural incomes is consistent with movements in the terms of trade for agriculture, which we shall discuss in Section 2.4.

These rates of growth of output in agriculture were, however, substantially lower than those in the non-agricultural sectors and the share of agriculture in GDP dropped sharply over this period. This, of course, is not a significant yardstick for measuring agricultural performance and, as we have shown, by the more appropriate criterion of international comparisons, South Korean agriculture has performed well in terms of output growth. Moreover, the stagnation and even recent decline in the size of the rural population meant that these output gains could be expected to be transmitted into gains in per capita *and* per worker terms. Nonetheless, because of the wide intersectoral differences in the rate of growth; growth in value-added per worker in agriculture lagged behind that in the non-agricultural sectors.[14]

It is surprising, therefore, to find that claims have been made, on the basis of household survey data, that the gap in average income per household between rural and urban areas has been narrowing since the late sixties to such an extent that by 1974 average rural incomes are claimed to have surpassed those in urban areas.[15] Considerable doubt must, indeed, be attached to this latter claim. First, the data on urban household income are known to be understated because of deficiencies in the sampling procedure.[16] Secondly, real wages in the industrial sector have been rising faster than agricultural wages. A recent International for Bank for Reconstruction and Development study estimates that between 1963 and 1972 real wages in manufacturing increased by 7.2 per cent per annum while the corresponding figure for agricultural wages was 6.5 per cent.[17] The sources for the latter figure are not, however, specified and our own estimate (Table 2.4) is that the correct figure would be only about 4 per cent, considerably lower than the figure for manufacturing wages. Finally, as Table 2.5 shows, although the per household and per capita income figures may have narrowed, the differences between rural and urban areas in income per worker still remain very wide. It should also be noted that it is the level of the ratios between rural and urban incomes that we are questioning and not the direction of changes in these ratios. The improvement in the relative income of rural households is explainable by movements in the terms of trade for the agricultural sector (which we discuss in Section 2.4) but we doubt if average rural incomes were actually higher than average urban incomes in either 1963 or 1974. It

Table 2.4 Agricultural Wages

Year	Money Wage (won *per day*)	Rural Consumer Price Index	Real Wage (won *per day*)
1963	167	100	167
1964	199	129	154
1965	221	144	153
1966	256	161	159
1967	307	182	169
1968	381	209	182
1969	463	227	204
1970	579	259	224
1971	695	291	240
1972	803	325	247
1973	886	346	256
1974	1141	492	232
1975	1467	618	237

Source Korean Ministry of Agriculture and Fisheries, *Farm Household Income Surveys* 1970, Table 193, pp. 440–1, Table 187, p. 430; 1976, Table 144, pp. 380–1, Table 138, p. 368.

Note Wages refer to those for adult males.

remains true, nevertheless, that agricultural output and average rural incomes have increased significantly, although these have lagged behind the achievements in the non-agricultural sectors.

(b) The distribution of rural incomes

The only source of data on the rural income distribution in South Korea is the Survey of the Farm Household Economy, conducted annually by the Ministry of Agriculture and Fisheries since 1963. Although these data are comprehensive and are based on closely supervised daily entries by the sampled farm households, there are deficiencies in the coverage and sampling procedure which reduce their reliability. Nonetheless, these data have been widely used and we shall therefore survey this 'received doctrine' before going on to discuss the statistical limitations and offer our assessment of what the true picture looks like.

The usual claims are twofold: that the size distribution of incomes in South Korea (for the country as a whole as well as for rural and urban areas taken separately) is lower than in most other developing countries and that this relatively favourable distribution has

Table 2.5 Comparison of Incomes of Rural and Urban Households (*won*)

Category	*1963*	*1967*	*1968*	*1969*	*1970*	*1971*	*1972*	*1973*	*1974*
Rural									
A. Household	93179	149470	178959	217874	255804	356382	429394	480711	674500
B. Per capita	14582	24423	29727	36373	43210	61129	75200	84000	118000
C. Per worker	29209	47907	59653	73606	87905	122049	144092	164000	218000
Urban									
D. Household	80160	248640	285960	333600	381240	451920	517440	550200	644520
E. Per capita	14417	45538	52566	61550	71393	85591	98186	105000	123000
F. Per worker	67361	192744	219969	254656	286647	339789	394992	405000	476000
Ratio									
A/D	116.3	60.1	62.6	65.3	67.1	78.9	83.0	87.4	104.7
B/E	101.1	53.6	56.6	59.1	60.5	71.4	76.6	80.0	96.0
C/F	43.4	24.9	27.1	28.9	30.7	35.9	36.5	41.0	46.0

Source Parvaz Hasan, op cit, Table 16, p. 52.

remained essentially unchanged during a period of rapid structural change and economic growth. The explanation that is usually offered for this remarkable conjunction of events remains essentially that which has been advanced by Irma Adelman. It attributes the low initial level of inequality to the 'static asset redistribution (mainly the land reforms)' that occurred after the Korean War and the subsequent stability in the distribution in the face of rapid growth to the effects of this initial equality and of 'dynamic redistribution' in the form of a vast expansion of educational opportunities. In addition, the high labour absorption in the rapidly growing labour-intensive export industries is held to explain the weak disequalising effect during the high-growth process.[18]

The statistical base, however, does not permit such firm theorising. The data on urban incomes are probably more unreliable than that for rural incomes and there are problems about merging the two distributions to arrive at a reliable overall size-distribution of incomes. These points, however, are outside the scope of this paper and we shall concentrate solely on the problems with the rural data.

The basic problem with the rural data is that of coverage. Its sampling frame consists only of households who farm land and whose holdings are larger than 0.1 hectares in size. Single-member households are also excluded. The effect of this is that the survey data give an incomplete picture of the distribution of incomes and one which is biased towards showing greater equality. From the 1970 Census of Agriculture, it emerges that the sampling frame excludes about five per cent of the total agricultural population; the near-landless with holdings of less than 0.1 hectare, the landless and the agricultural labourers. These are precisely the groups which one would expect to be at the bottom tail of a rural income distribution; consequently their exclusion leads to a probable understatement of inequality. The exclusion of single-member households is also likely to add to this understatement as is the exclusion of non-agricultural households, since the latter would include groups like shopkeepers, government servants, artisans, etc., who would tend to have relatively high incomes.

The other major flaw of these survey data is that they are based on a fixed sample of households that is changed only every five years. Since the sample is stratified by the size of landholdings, this procedure introduces a built-in bias towards stability in the distribution during each five-year period. This fact is not mentioned in published sources and only emerged during discussion with statisticians responsible for the survey. This procedure implicitly assumes that there are no significant movements in the size distribution of holdings and hence begs the central question in the analysis of rural income distributions. It could be argued, however, that while these arguments are valid for each five-year period, they do not necessarily invalidate comparisons made for each year that the sample changes. Nonetheless, the problem remains because the sample design has been changed since 1973 and this raises problems of comparability.[19]

Bearing these limitations in mind, we can proceed to evaluate the changes that are described in the distribution of rural incomes. As shown in Table 2.6, the distribution appears scarcely to have changed between 1963 and 1975 while the average farm income increased by 51 per cent in real terms. Deflating the incomes in each income group by the consumer price index for farm households shows that the incomes of all groups increased significantly over the period, with the two lowest income groups showing the greatest increase. There was consequently a slight decrease in inequality, with the Gini coefficient falling from 0.212 to 0.188. Applying an income-group specific cost-of-living index (to take into account the differential impact of inflation on income groups with different consumption patterns), however, removes this apparent reduction in inequality. The Gini coefficient for 1975 now becomes 0.206, almost the same as for 1963. The indices used and the resulting distribution in real incomes are shown in Table 2.7.

It must be pointed out, however, that the above discussion is based on an unsatisfactory classification of data. The income figures are classified by the size of landholding and not directly by income class. Furthermore, only five class intervals are available and this increases the standard error around the estimates of the Gini coefficient. Unfortunately, this is the only form in which the data are available for the entire period of our study.

Data classified directly by income class are available for some intervening years and the measures of inequality based on them are presented in panel B of Table 2.6. These show that even without allowing for the differential impact of inflation, there is a slight increase in inequality between 1965 and 1971.

It should be emphasised that even if relative inequality did increase, it was not by a large margin. Given the 50 per cent increase in the average real income over this period, the real incomes of the bottom two income classes would have still increased significantly in

Table 2.6 Distribution of Incomes in Rural Korea

A. Distribution of Rural Incomes by Size of Landholding, 1963 and 1975

Size of Landholding (chungbo)	*1963* *Number of Households*	*1963* *Average Income per Household ('000 won)*	*1975* *Number of Households*	*1975* *Average Income per Household at 1975 Prices*	*1975* *Average Income at 1963 Prices*	*1975/1963*
less than 0.5	316	57.5	634	532.5	86.2	1.48
0.5–1.0	455	77.4	944	776.2	125.6	1.63
1.0–1.5	228	118.0	517	975.0	157.7	1.34
1.5–2.0	94	146.0	240	1253.9	202.9	1.39
more than 2.0	79	195.3	182	1734.8	280.7	1.44
Total	1173	93.2	2517	872.9	141.2	1.51

Source Korean Ministry of Agriculture and Fisheries, *Farm Household Income Surveys*: 1976, Table I.1, p. 45; 1964, Table I.2, p. 42.

B. Estimates of Gini Coefficient Based on Data Classified by Income Class

Year	*Gini Coefficient*
1965	0.2985
1967	0.3186
1968	0.3176
1969	0.3163
1970	0.2995
1971	0.3247

Source T. Mizoguchi, Do Hyung Kim and Young Il Chung, 'Over-time [sic] Changes of the Size-Distribution of Household Income in Korea (1963–1971)', Table 4, p. 64, in *Papers and Proceedings on the Income Distribution in South Korea* (Tokyo: Hitotsubashi University, 1976).

Table 2.7 Distribution of Real Income by Size of Landholding, 1963 and 1975

	1963		*1975*			
Size of Farm (chungbo)	(1) *No. of Households*	(2) *Average Income* (*'000* won)	(3) *No. of Households*	(4) *Average Income at Current Price*	(5) *(4) Deflated by Income-group Specific Index*	(6) *(5)÷(2)*
less than 0.5	316	57.5	634	532.5	85.0	1.48
0.5–1.0	455	77.4	944	776.2	124.4	1.61
1.0–1.5	228	118.0	517	974.9	159.0	1.35
1.5–2.0	94	146.0	240	1253.9	205.7	1.41
more than 2.0	79	195.3	182	1734.9	288.0	1.47
Total	1173	93.2	2517	872.9	141.2	1.51

Note The income-group specific index was calculated from data on expenditure weights given below. The procedure was to apply the base-year consumption weights of each income-class to the separate commodity cost indices in order to obtain the increase in the cost of living for each income-class. The indices, in ascending order of income-group were 626.7, 623.8, 613.0, 609.5, 601.9. An alternative calculation using terminal year consumption weights was also carried out but this did not yield significantly different results in terms of the change in income distribution.

Table 2.7—continued

	Expenditure weights for households by size of landholding, 1963				
	Food	*Housing*	*Fuel and Light*	*Clothing*	*Miscellaneous*
less than 0.5	0.6572	0.0342	0.1052	0.0565	0.1407
0.5–1.0	0.6306	0.0333	0.0979	0.0589	0.1792
1.0–1.5	0.5731	0.0369	0.0855	0.0695	0.2350
1.5–2.0	0.5592	0.0357	0.0800	0.0726	0.2525
more than 2.0	0.5369	0.0328	0.0749	0.0766	0.2787
Average (%)	60.3000	3.4000	9.2000	6.5000	20.6000

Source 1963 weights from 1964 ***Farm Household Survey***, 1964, Tables 1–54, p. 70.

Note The disaggregated rural cost of living index by commodity group (1963=100) was as follows: Food and Beverages 685; Housing 599; Fuel and Light 533; Clothing 518; Miscellaneous items 472. The source for this was Jae-Joon Park, 'Inflation and Income-distribution in Korea' in ***Papers and Proceedings for Seminar on the Income Distribution in South Korea*** (Hitotsubashi University, 1976) Table 4, p. 16, and the ***Farm Household Income Survey 1976***, pp. 368–9.

spite of an increase in relative inequality. This is in sharp contrast with other developing countries where absolute rural poverty has shown a tendency to increase. This observation probably holds true even after we allow for the exclusion of the landless and near-landless to which we referred earlier; the data on rural real wages, as mentioned earlier, showed a steady increase over this period and employment opportunities in the rural areas are unlikely to have decreased since there was heavy out-migration.

These trends are to some extent confirmed by the data on household consumption expenditure data from the same source. Table 2.8 shows that total consumption in real terms increased for all income classes. A puzzling feature of the consumption data, however, is that consumption expenditure on food increased only very slightly for the bottom two income classes and not at all for the other income classes. Data on the physical quantities of grain consumption are consistent with this trend; Table 2.9 shows that total grain consumption fell for all income classes but with bigger proportionate falls by the top three income classes. While the fall in grain consumption could be explained by a switch to superior foods, such as meat and poultry in response to income increases, the fall in total food consumption remains problematic and must be counted as another indication of the unreliability of data.

Table 2.9 also reveals another feature which might, at a stretch, be interpreted as an indication that relative inequality increased over this period. While the total amount of grain consumed decreased, the amount of grain marketed increased by more than 20 per cent for every income class. The magnitude of increase was greater for the bottom two income classes than for the topmost but, due to the higher proportion of the marketable surplus for the higher income classes, total grain availability (i.e. consumption plus sales) showed a different pattern of change. It declined for the bottom three income classes and increased for the top two. Apart from being consistent with a widening of relative inequality among farm households, it also raises some questions about the nature of the income increase. The relative price of grain (both rice and barley) increased sharply between the two years as a result of a sharp increase in procurement prices and this could account for the fact that consumption was cut back in favour of grain sales. This would imply that most of the income increase could be explained by this relative price effect and not by the increase in physical output. This would then run counter to the general trend in productivity

Table 2.8 Change in the Distribution of Food and Total Consumption Expenditure by Size of Landholding, 1964, 1975 ('000 *won*)

	Expenditure on Food			*Total Expenditure*			
Size of landholding (chungbo)	*(1)* *1964*	*(2)* *1975*	*(3)* *1975 at 1964 Prices*	*(4)* *1964*	*(5)* *1975*	*(6)* *1975 at 1964 Prices*	*(6)÷(4)*
less than 0.5	35.9	227.4	36.6	54.6	420.9	67.8	1.24
0.5–1.0	42.9	277.6	44.7	68.2	567.3	91.4	1.34
1.0–1.5	54.9	320.8	51.7	95.9	683.3	110.0	1.15
1.5–2.0	61.2	358.2	57.7	109.4	822.1	132.4	1.21
more than 2.0	71.1	410.3	66.1	132.4	1071.3	172.5	1.30

Source Korean Ministry of Agriculture and Fisheries, *Farm Household Income Survey*, 1964 and 1976.

Table 2.9 Average Annual Grain Consumption per Household by size of Landholding (in litres of hulled grain)

Size of Landholding (chungbo)	*Rice Consumption*		*1975/1964*	*Total Grain Consumption*		*1975/1964*	*Total Grain Sold*		*1975/1964*
	1964	*1975*		*1964*	*1975*		*1964*	*1975*	
less than 0.5	680.5	650.1	0.955	1317.0	1149.8	0.873	267.6	350.8	1.31
0.5–1.0	911.7	816.5	0.896	1658.1	1405.8	0.847	624.2	869.5	1.39
1.0–1.5	1245.6	988.9	0.794	2057.8	1631.8	0.793	1206.7	1476.4	1.22
1.5–2.0	1472.9	1142.7	0.776	2366.3	1815.0	0.767	1550.6	2252.4	1.45
more than 2.0	1849.1	1361.2	0.736	2621.9	1994.4	0.761	3307.1	4019.7	1.22

	Grain Sold and Consumed		*1975/1964*	*Percentage Marketed*	
	1964	*1975*		*1964*	*1975*
less than 0.5	1584.6	1500.6	0.950	16.9	23.4
0.5–1.0	2282.3	2275.3	0.997	27.3	38.2
1.0–1.5	3264.5	3108.2	0.952	37.0	47.5
1.5–2.0	3916.9	4067.4	1.038	39.6	55.3
more than 2.0	5929.0	6014.1	1.014	55.8	66.8

Source Korean Ministry of Agriculture and Fisheries, *Farm Household Income Survey*, 1964 and 1976.

increase that we have documented in the previous section and counter also to the data in Table 2.2 which shows that yields per unit of land increased relatively uniformly for farms across all holding sizes. These observations would remain true even after we check for the possible effects of changes in the composition of farming output and in the composition of total farming income. As shown in Table 2.10, there were no marked changes either in the composition of farm output or income. The proportion of receipts from rice in net farm income remained virtually unchanged for farms of all size groups and there was no change in the proportion of total income coming from non-farming sources except for farms in the smallest size group.[20] There has, moreover, been no significant change in cropping patterns over this period 'because of the physical constraints on expanding rice output'.[21] It could thus be argued that the increase in the relative price of grain probably contributed to a widening of relative inequality among farming households. The richer farmers gained more because of their higher initial marketable surplus, higher share of income from grain in total income and because they could afford to divert more from consumption to sales in response to the sharp increase in the relative price of grain. Indeed it has been claimed that 'within the farm sector high support prices tended to have a regressive influence on income distribution'.[22]

In view of the above, it would not be unreasonable to assert that the quality of the data leaves much to be desired. It certainly cannot be used to establish definitively that even relative inequality did not increase in the South Korean economy. Indeed, it has been argued that the reverse has been probably true. Paul Kuznets has argued that, for the economy as a whole, relative inequality did increase due to a widening of inter-sectoral differences in value-added per worker.[23] It has also been argued that an increase in relative inequality has been true of the rural sector, taken separately.[24]

On balance, we are inclined to agree with the judgement that relative inequality has increased. The data base – bearing in mind the various limitations we have indicated – is not firm enough to lead to the conclusion that the degree of inequality has remained unchanged. Instead we would argue that there are several indirect indications that relative inequality has probably increased in the rural economy. First, there is the internal inconsistency in the data from the Farm Household Surveys which show that total grain

Table 2.10 Source of Farm Income by Size of Landholding, 1964 and 1975 average per household ('000 *won*)

Size of Landholding (chungbo)	*Income from Rice*		*Income from Barley*		*Net Agricultural Income*		*Total Net Income*	
	1964	*1975*	*1964*	*1975*	*1964*	*1975*	*1964*	*1975*
less than 0.5	16.8	146.0	10.1	n.a.	39.3	310.4	57.5	532.5
0.5–1.0	35.1	336.9	14.0	n.a.	62.6	611.3	77.4	776.2
1.0–1.5	60.6	497.5	17.3	n.a.	100.4	846.9	118.0	975.0
1.5–2.0	81.1	723.9	18.2	n.a.	128.7	1138.2	146.1	1253.9
more than 2.0	132.5	1163.2	16.6	n.a.	178.3	1601.6	195.3	1734.9

	Rice/Net Agricultural Income		*Net FarmIncome/ Total Net Income*	
	1964	*1975*	*1964*	*1975*
less than 0.5	0.427	0.470	0.683	0.583
0.5–1.0	0.561	0.551	0.809	0.788
1.0–1.5	0.604	0.587	0.851	0.869
1.5–2.0	0.630	0.636	0.881	0.908
more than 2.0	0.743	0.726	0.913	0.923

Source Korean Ministry of Agriculture and Fisheries, *Farm Household Income Survey*, 1964 and 1976.

availability has moved in the opposite direction to that suggested by the income change. Secondly, the sharp upturn in the terms of trade for the agricultural sector could have had a regressive impact due to the unequal distribution of marketable surplus amongst farm households. Thirdly, the data on savings, which we discuss in the next sub-section, show that the average savings ratios have been higher for larger farmers and have increased since 1963. It is likely, therefore, that these substantially higher savings margins over this period have been translated into widening inequality in income levels. Finally, the data on the re-emergence of tenancy and other disequalising tendencies that we discuss in the next section are consistent with our contention that relative inequality has increased.

While relative inequality might have increased, it cannot be seriously argued that the same has been true of absolute poverty. It seems indisputable that real incomes have increased significantly for all income groups in the Korean rural economy. It still remains true, therefore, that the South Korean rural economy performed reasonably well by the criterion of the alleviation of absolute poverty. This latter point can be verified in an approximate way by the figures of the distribution of caloric intake of farm households that are available for 1973 (Table 2.11). This shows that although the bottom 38 per cent of households had an average per capita daily caloric intake that was less than the estimated requirement, the shortfall was no greater than 1.5 per cent. Although we must allow for the fact that the variation around this average level could mean that the shortfall would be greater than 1.5 per cent for many households, it would still appear that there is no pervasive absolute poverty in rural South Korea as measured by a nutritionally based poverty line. Since the data on real food consumption presented earlier show that there was no increase between 1964 and 1975, it could be claimed that this was broadly true even in the earlier period. The average income level around which we have been discussing the changes in rural income distribution is thus significantly higher than in other Asian countries such as India and Indonesia. It is important to bear this in mind in our evaluation of the performance of the Korean agricultural system.

(c) Savings and accumulation

We have seen from the previous two sections that production has increased steadily and that there was no increase in absolute

Table 2.11 Average per Capita Daily Caloric Intake of Farm Households by Income Class, 1973 (in calories)

Income Class / *Food Item* ('000 won)	*Less than 100*	*100–220*	*220–340*	*340–460*	*460–580*	*580–700*	*700–830*	*830–940*	*940–1060*	*More than 1060*	*All Groups*
Rice	1135.6	1021.2	1144.9	1284.9	1314.0	1446.1	1494.0	1484.8	1796.7	1858.6	1329.0
Barley	684.4	689.2	579.2	515.2	573.2	480.0	460.4	667.5	436.0	562.8	557.4
Wheat	71.5	92.8	89.3	109.2	111.7	117.1	104.7	120.0	118.4	108.3	104.9
Miscellaneous grains	55.9	47.6	46.1	39.3	44.7	67.3	37.4	54.7	29.5	28.1	45.6
Pulses	20.0	23.7	19.4	19.5	26.8	35.5	35.1	28.4	32.0	42.6	26.5
Potatoes	26.2	67.0	65.8	71.2	84.8	72.4	62.0	69.2	65.5	76.9	70.6
Fruit	2.0	2.1	1.8	1.9	2.0	2.7	2.3	3.7	3.3	4.1	2.3
Vegetables	17.1	20.8	21.4	21.7	26.5	27.2	26.3	31.3	33.0	30.7	24.7
Beef	1.1	0.7	0.9	0.9	1.2	1.5	1.6	1.9	2.7	2.9	1.3
Milk	—	—	—	—	—	—	—	—	—	—	—
Pork	4.8	5.8	5.9	6.7	7.3	8.6	9.1	10.6	13.5	12.3	7.7
Chicken	0.5	0.3	0.4	0.4	0.5	0.7	0.6	0.8	1.1	1.1	0.5
Egg	0.4	0.4	0.4	0.5	0.3	1.0	1.0	0.7	1.5	1.1	0.6
Fish and seaweed	31.8	30.3	28.3	27.3	30.1	32.5	34.5	39.5	46.6	43.0	27.8
Total calories	2051.3	2001.9	2003.8	2098.7	2223.1	2296.0	2269.0	2513.1	2579.8	2772.5	2198.9
% of households in each income class	2.24	14.31	22.45	16.84	14.31	10.76	7.02	4.21	3.10	4.77	100.0

Notes 1. Calculated from data in Alan R. Thodey, *Food and Nutrition in Korea 1965–74* (Korean Agricultural Sector Study, Special Report No. 11, Michigan State University, East Lansing, Michigan 1975), Annex Tables 11.3 and 18.1.
2. The average daily per capita requirement given in the same source was 2030 calories.

poverty. In this section we shall briefly show that savings ratios rose and that there was a significant increase in the capital stock in the rural economy.

Table 2.12 shows that average savings ratios rose substantially for farm households between 1963 and 1973, and that this increase occurred for all income groups. In 1963 the bottom income classes had an almost negligible level of saving, but by 1973 the average saving ratio had increased to a respectable 15 per cent. The ratios for the top income classes in 1973 were extraordinarily high. It should be noted that the very high marginal propensity to save shown by these average ratios was implicit in our earlier discussion of the low increase in food consumption and, to some extent, in total consumption expenditures.

In table 2.13 we see that data on the changes in the value of assets of farm households are consistent with the change in savings ratios described above. Between 1963 and 1973, the average real value of net assets per farm household almost doubled while the value of liabilities declined by 16 per cent. The value of draught animals and of large farm implements in constant prices increased by 270 per cent and 78 per cent respectively, indicating a significant amount of capital formation over this period. There was also a significant increase of 70 per cent in the real value of liquid assets.

It will also be seen that the increase in the real value of net assets was diffused across all income classes, although the margin of increase was progressively greater for higher income classes.

We could thus conclude our evaluation of the performance of the South Korean agricultural system by saying that it adequately met

Table 2.12 Savings Ratios by Size of Landholding, 1963 and 1973 (*won*)

	1963			*1973*		
Size of Landholding	*Average Income per Household*	*Savings*	*S/Y*	*Average Income per Household*	*Savings*	*S/Y*
less than 0.5	56048	735	0.013	283989	42061	0.148
0.5–1.0	74757	4842	0.065	400390	74881	0.187
1.0–1.5	114284	15703	0.137	542591	141501	0.261
1.5–2.0	141512	29417	0.208	685943	228757	0.333
more than 2.0	189215	54902	0.290	971573	372760	0.384

Source Korean Ministry of Agriculture and Fisheries, *Farm Household Income Survey*, 1964 and 1974.

Table 2.13 Distribution of Fixed Assets (value terms) (average per household, '000 *won*)

	Fixed Assets						
Year	*Total*	*Land*	*Animals*	*Large Implements*	*Liquid Assets*	*Liabilities*	*Net Assets*
1963	327.2	277.1	6.9	2.3	45.6	6.7	371.1
1964	323.5	277.6	6.9	1.5	58.2	7.6	380.9
1965	392.2	340.6	9.8	1.7	56.8	10.6	452.5
1966	441.0	374.6	11.1	1.8	67.5	10.0	510.3
1967	496.3	421.6	14.8	1.9	77.7	11.4	578.9
1968	547.4	455.7	25.2	2.7	99.6	14.0	658.0
1969	616.6	500.4	26.5	3.6	116.1	12.5	753.9
1970	743.0	601.8	34.6	3.7	131.2	15.9	899.3
1971	858.3	671.4	49.9	3.8	195.8	10.3	1090.1
1972	1204.8	929.2	86.5	10.1	260.8	13.9	1507.5
1973	1783.4	1483.7	87.2	12.7	303.0	13.8	2135.4
1974	2737.9	2247.9	98.5	19.2	386.5	26.1	3200.5
1975	3963.2	3374.2	114.6	25.7	469.3	33.4	4555.0
1975 values at 1963 prices	(638)	(543)	(18.5)	(4.1)	(76)	(5.9)	(733)
1975 constant value÷1963 value	1.95	1.96	2.68	1.78	1.70	0.84	1.97

Table 2.13—continued

By Size of Holding										
	1969					*1975*				
Size of Landholding	*Fixed Assets*	*Animals*	*Large Implements*	*Liquid Assets*	*Net Assets*	*Fixed Assets*	*Animals*	*Large Implements*	*Liquid Assets*	*Net Assets*
less than 0.5	339.3	9.1	0.7	49.2	413.2	1738.6	64.5	11.8	209.2	2019.3
0.5–1.0	500.6	23.3	2.4	89.1	604.2	3106.3	110.5	21.5	400.9	3591.5
1.0–1.5	645.8	32.6	4.3	133.6	798.9	4479.3	130.8	25.3	566.7	5167.6
1.5–2.0	1067.4	45.8	7.7	208.3	1304.7	6492.1	156.8	41.4	734.4	7438.4
more than 2.0	1355.5	54.9	10.2	283.9	1673.7	10792.7	199.0	73.6	1128.5	12294.3

Source Korean Ministry of Agriculture and Fisheries, *Yearbook of Agricultural Statistics*, 1970 and 1976.

Note The figures at 'constant prices' were obtained by deflating the current price figures by the index of 'prices paid for inputs' by farmers from the *Yearbook of Agricultural Statistics.* Admittedly this is a crude deflator, but it does suffice to indicate the orders of magnitude of change in the fixed asset position.

the tests in terms of production, distribution and savings and accumulation. Although relative inequality in income probably increased and the pattern of change in savings ratios and in the value of assets would tend to reinforce inequalities, there was no increase in absolute poverty and the gains from growth were adequately diffused. In the next section we shall proceed to examine the factors which accounted for this performance which, by comparison with a majority of developing countries, was exceptionally good.

2.4 Factors explaining the performance of the agricultural system

To explain the functioning and the resulting performance of the South Korean agricultural system in rigorous quantitative terms is virtually impossible. The explanation would need to separate out a complex set of interlocking factors and the statistical base is simply not sufficient to allow such an exercise to be performed. Thus what we shall attempt to do in this section is to offer a set of qualitative arguments which are consistent with the available data on the central aspects of the Korean agricultural system and which, taken as a whole, constitute a plausible general explanation of the functioning of the system. The structure of the argument will be as follows. First, we consider the effect of the 'egalitarian base' created by the land reform, then we analyse how the system coped with disequalising tendencies, and finally we move on to evaluate the impact of government policies and the growth of the non-agricultural sectors of the economy.

In Section 2.2 we had referred briefly to the fact that the land reform was indeed thorough in its effects. Here we shall offer a few more details and evaluate its impact on productivity and the distribution of income. The thoroughness of the land reform is attested to by the fact that in the pre-reform period almost 60 per cent of the total cultivated area[25] was under tenancy and 50 per cent of farm households were pure tenants, whereas after the reform these percentages fell to 14 and 5 per cent respectively.[26] Thus almost half the total farmland changed hands, either by direct distribution by the government or through private sales induced by the land reform legislation, while two-thirds of all farm families received land under the land reform provisions.[27] Under some circumstances such a drastic change in the land tenancy system could be expected to lead to serious dislocations in production but this did not happen in

South Korea. The main reason was the fact that the reform was essentially a 'land-to-the-tiller' programme in a system where poverty was widely shared through a system of small tenancies; the number of landless agricultural families without any access to land was estimated at fewer than 3 per cent of the total in 1945. This meant a minimal disruption in the configuration of production units and there were few of the problems arising out of the creation of new producers that would have been involved had Korea been a predominantly plantation economy. It is not very surprising, therefore, that the land reform did not lead to any immediate fall in productivity; indeed as we have shown, productivity increased steadily in the post-reform period.

Nonetheless, it has been claimed, erroneously in our view, that

> ... the implementation of the land reform not only changed drastically the type of ownership, but also changed the size distribution of farmland. The number of farm households having less than 1 *chungbo* increased from 72.1 per cent to 74.1 per cent between 1945 and 1955, while the amount of land owned by this group increased from 10.4 per cent to 47.2 per cent of the total, resulting in fragmentation of farm holdings.[28]

This claim is, in fact, based on a confusion between the distribution of operating units and that of landownership. The figures for the percentage of farm households are for operating units but those for total area refer to landownership. The figure most often quoted for the existence of large operating units refers to 1945 yet it is admitted that

> ... because the status of the legal ownership of land just after World War II was in a state of considerable confusion there is no data on distribution of farms by size and ownership from which degrees of ownership concentration may be calculated.[29]

Moreover, contradictory assertions are made on the basis of this uncertain data. Morrow and Sherper interpret the figures as referring to 'management scale' but then go on to say that 'the actual farming operations must have been carried out by tenants or hired labour because the existing technology required a family farm for every couple of hectares'. Relying on the same data source, Perkins, in a recent study, states that 'although only 4 per cent or so of farm families ran farms whose size was greater than the limit [of three hectares], these farms cultivated 30 per cent or more of the total acreage'.[30] He then goes on to assess the productivity effects of the land reform by considering whether or not there were pronounced economies or diseconomies of scale in Korean agriculture. The rest of the text, however, suggests that Perkins may have repeated the past errors. Since he cites Pak's figures, the percentage of total acreage should be 50 per cent and not 30 per cent; there is also a large discrepancy in his figures for the distribution of cultivated area by size of farm for 1945 in his Table 9.8. Moreover, he contradicts himself by saying that:

> Land reform was also not the cause of the decline in small farms nor had it created them in the first place. Nearly 50 per cent of all farms in Korea in the Japanese period were under 0.5 hectares in size, a figure even higher than that in 1953.[31]

The confusion is in our opinion adequately resolved if we accept that the 1945 figures represent approximate ownership distribution figures while the pre-reform distribution of operational units are shown by the figures for 1947 which are usually quoted together with the 1945 figures. The 1947 figures together with the post-reform figures for 1953 are as follows:

Size in hectares	*Percent of farms 1947*	*Percent of farms 1953*
Under 0.5	41.2	44.9
0.5–1.0	33.3	34.2
1.0–2.0	18.8	16.5
2.0–3.0	5.3	4.3
More than 3.0	1.4	0.1

Source Dwight Perkins, *op. cit.*, Table 9.7.

It will be seen that the change in the size distribution of operational units is slight although the transfer in ownership was substantial. The impact on productivity from this source would thus have been very slight. Moreover, although Perkins, unlike us, implies that there was a large shift in operating units, he concludes that the elimination of farms of more than three hectares 'harmed farm output little if at all', and that there is no evidence that the continuation of the three-hectare ceiling after the reform has been an important negative influence on productivity.[32] In fact, agricultural output in the post-reform period grew at 3.6 per cent per annum between 1953 and 1961, a rate of growth which would clearly be inconsistent with the existence of a large negative productivity impact from the land reform. It is curious that Pak, who argues that there was a large negative impact, does not try to explain away these figures. This rate of growth of output, moreover, occurred in spite of the fact that there was inadequate provision of supporting services, to replace those previously provided by landlords, in the post-reform period.[33]

One other feature of the land reform should also be noted. It resulted in a substantial redistribution of assets and income in the rural economy. This arose out of the fact that the repayment burden by beneficiaries of the land reform worked out to be only 1.25 times the annual yield of the land and landlords were expropriated to the extent of between three-quarters and seven-eighths of the value of their former assets. Perkins estimates that this sharp redistribution of assets resulted in an '80 per cent or so decline in the income of the top 4 per cent that was matched by a 20 to 30 per cent increase in income of the bottom 80 per cent who had been tenants or owner tenants before 1948'.[34] Thus the South Korean land reform was drastic not only in terms of the proportion of the total cultivated land and of the farming population that was affected, but also in terms of the real redistribution of assets and income that was involved.

It should be clear from the foregoing that the South Korean land reform succeeded in creating a genuinely egalitarian base in the agricultural system. However, such a base is only a necessary and not a sufficient condition for explaining the favourable performance of the system that we have documented. The dynamic properties of the system are in fact crucial in any analysis which seeks to explain the performance. In particular, it could be argued that within a free-market system disequalising tendencies would begin to operate and hence nullify the egalitarian initial conditions. The explanation of why this did not occur to any significant extent in the Korean case has to do with the fact, mentioned in Section 2.2, that the system is not a *laissez-faire* one and also that there were important favourable exogenous circumstances created by rapid industrial growth.

The key factor in dampening any disequalisation was the effective continued enforcement of the three-hectare limit on landownership and the prohibition of tenancy since the land reform. Needless to say, there has been, even according to official statistics, some backsliding since the land reform and that the true situation would probably be worse than that shown by official statistics. However, there seems little basis for believing that the achievements of the reform have been undone in any significant way.

According to official statistics the percentage of farms of more than three hectares in size increased from 0.1 per cent of all farms in 1953 to 1.2 per cent in 1965 and to 1.5 per cent in 1975. The percentage of total cultivated area taken by such farms increased from 1 per cent to 7 per cent between 1953 and 1975. It should be noted, however, that most of the increase in the proportion of farms of over three hectares and in the total area they occupy occurred largely between 1953 and 1965. Since 1965 the increase has been very slight.

In terms of a reappearance of tenancy, official figures again show that this has not occurred to any significant degree. Immediately after the reform about 8 per cent of the cultivated area was still under tenancy, most of this reflecting the exemptions granted under the land reform legislation. Since then, the percentage has grown to 13.5 in 1960, 16.4 in 1965 and 17.6 per cent in 1970.[35] By 1975, however, the percentage had dropped to 14.6 per cent. The net revival of tenancy, after subtracting the exempted land, amounts to about 8 to 9 per cent of the total land area. Parallel to this, the proportion of farm households that were pure tenants increased from 5 per cent immediately after the reform to 6.7 per cent in 1960, 7.0 per cent in 1965 and 9.7 per cent in 1970. Thereafter it declined to 7.9 per cent in 1975.

Non-official estimates place the figures for tenancy at higher than the above figures but not substantially so. One estimate puts the percentage of land under tenancy at 20 per cent,[36] while another claims that it is about 25 per cent.[37]

It is clear, therefore, that disequalising tendencies have been operating but that, at the same time, the impact has not been sufficiently great to restore the system to the *status quo ante*. Part of the reason for this is the existence of legal restrictions on land transactions. That these restrictions are real and not merely fictional is shown by the fact that there have been periodic calls for their removal. In particular, the Korean Agricultural Sector Study strongly urged the removal of both the ceiling on ownership and the prohibition of tenancy in the interests of greater efficiency.[38] While it is, of course, clear that the ceiling will eventually have to be removed as the labour force continues to decline and labour-saving mechanisation is introduced on a large scale, it remains true that until the terminal year of our study, these economic forces had not as yet come into play.[39] The ceiling thus achieved its purpose of containing inequality without imposing a high cost in terms of allocative inefficiency.

It is worthwhile to note that while these disequalising tendencies have been pushing against the ownership ceiling and the prohibition on renting, it is by no means clear that it necessarily represents a trend towards a concentration of ownership. It is claimed by Kim that the 'the average size of farm land owned by landlords was very small, ranging from 0.3 ha. to 0.5 ha.',[40] while another study, reporting on the results of a survey on land transactions, concluded that '. . . land transactions appear to be taking place mainly within the boundaries of individual villages' and that there was no evidence that absentee landlordism was on the increase.[41] The latter study also showed that the expected pattern of land sales being concentrated among small farmers, with purchases being made mainly by the rich, was contradicted by the evidence.

These results are based on small sample surveys and there must be doubts about their representivity. Nonetheless, they do not appear entirely implausible in relation to other independent data. In Table 2.14 for instance, we see that the number of farms of less than 0.5 *chungbo* fell both absolutely and relatively, while for those in the 0.5–1.0 and in the 1.0–1.5 *chungbo* categories, the exact opposite was true. This pattern of change would not be inconsistent with the observation that the smallest farms have been buying land from other small farmers, who migrated, as entire families, to the urban areas. The migrants and the enlarged farms of those who bought land would account for the fall in the number of small farms, while the buyers would have moved up to increase the number of farms in the 0.5–1.5 *chungbo* category. Two facts are worth noting in this connection. First, there is evidence that a significant part of rural out-migration has been *family* migration.[42] Second, the figures on land distribution show no significant increase in the average size of farms in the more than 3 *chungbo* category. This, again, would be consistent with the observation that there is as yet no trend of increasing penetration of large capitalist farms into the South Korean agricultural sector.

Table 2.14 Size-distribution of Landholdings, 1965 and 1975

Size of Farm (chungbo)	*1965* (1) No. of Farms	(2) %	(3) Total Area	(4) %	*1975* (5) No. of Farms	(6) %	(7) Total Area	(8) %
less than 0.1	69843	2.78	5445	0.24	96245	4.04	873	0.04
0.1–0.3	360982	14.40	80909	3.58	307850	12.94	61207	2.96
0.3–0.5	470015	18.75	194862	8.62	381210	16.02	149755	7.25
0.5–1.0	793864	31.67	603026	26.68	828157	34.81	601655	29.13
1.0–1.5	414723	16.54	517775	22.91	430666	18.10	522945	25.32
1.5–2.0	228582	9.12	397331	17.58	187231	7.87	321016	15.54
2.0–3.0	139599	5.57	346021	15.31	111717	4.69	264624	12.81
more than 3.0	29291	1.17	114867	5.08	35982	1.51	144527	6.99
	2506899	100.00	2260236	100.00	2379058	100.00	2066602	100.00

Source Korean Ministry of Agriculture and Fisheries, *Agricultural Statistics Yearbook*, 1970 and 1976.

Notes 1. The category '0.1' includes landless agricultural families for 1975, but it is not clear from the source whether this is also true for the 1965 figures.
2. The Gini coefficients for the degree of inequality in land distribution were 0.384 for 1965 and 0.389 for 1975.

The successful 'containment' of disequalising tendencies after the creation of the egalitarian base was facilitated to a very large extent by the pattern of growth of the Korean economy. The very rapid growth in industrial output and the very high rate of labour absorption has ensured an absence of the population pressure which usually heightens disequalising tendencies in the agricultural sectors of developing countries.

In South Korea, the total rural population increased at a relatively moderate rate of 1.65 per cent per annum between 1959 and 1967 and thereafter it declined steadily. By 1975 the rural population was 18 per cent lower than what it had been in 1967. These figures imply a substantial rate of rural-urban migration and one which, moreover, has been accelerating. Taking sub-periods, the growth rate of rural population was 2.11 per cent between 1959 and 1964, 1.0 per cent between 1964 and 1967, and thereafter the growth rate was negative. The share of the rural population in total population decreased from 58 per cent in 1959 to 38 per cent in 1975. Perkins[43] has estimated the size of the population transfer out of the rural areas and states that

> The rural population in 1975 was virtually the same as what it was in 1953–55. If, instead of holding steady, the rural population had maintained its share in the total population at 1953–55 levels, there would have been nearly 22 million rural people instead of 13 million. The share of the 9 million difference who entered the non-agricultural labour force . . . represents the farm sectors' contribution to Korea's industrialisation.[44]

The increase in labour absorption in the non-agricultural sectors began around 1963 when the new military government's drive for industrial growth began to take effect and intensified after 1966 when industrial growth shifted to an export-intensive bias and to higher growth rates. Between 1963 and 1973 output in the industrial sector grew at 17.1 per cent per annum and employment grew at 9.3 per cent per annum. It is interesting to note that the high rate of growth in industrial employment was due mainly to the very high rate of growth in output; the labour absorption rate of income growth[45] in Korean industry was not significantly higher than other Asian countries like Thailand, Malaysia, Taiwan and the Philippines.[46]

Apart from providing relief from population pressure, it has also been argued that the rapid growth of the non-agricultural sectors has been favourable to agriculture in several ways. Perkins puts this case most strongly when he argues that 'Korea is . . . a model of how an industrial revolution can precede and help bring about an agricultural revolution rather than the reverse'.[47] He claims that the industrial growth was financed without squeezing agriculture and that, once successfully under way, industrial growth provided the enabling conditions for the growth in agricultural productivity and income. He illustrates the latter point by pointing to the growth in demand for food generated by industrial growth and the easing of the foreign exchange constraint which allowed the import of a key agricultural input (fertilisers) as well as food. Food imports were seen as contributing to a loosening of the imperatives to extract a surplus from agriculture and 'helped Korea in a fundamental sense to avoid policies that could have seriously cut into farmers' incomes and incentives'.[48] It has also been argued that the interest-rate reform of 1965 and the growth of investment opportunities in the industrial sector drastically reduced the attractiveness of farmland as an investment proposition and a hedge against inflation. 'Floating funds concentrated on the purchase of farmland were diverted to banks and investment in non-agricultural fields'.[49] A survey conducted in 1976 showed that 'the rate of return for investment in farmland was only 4.8 per cent, which is lower than the bank interest rate of 16.2 per cent'.[50] According to this argument, therefore, the incentive for capitalist penetration into agriculture was weak and hence contributed to the containment of disequalising tendencies in the form of land concentration in the agricultural sector.

The Perkins thesis contains a large element of truth but overlooks some important questions. The crux of his argument about the absence of a squeeze on agriculture rests on an analysis of rural household savings and direct financial transfers out of the agricultural sector. After analysing trends in rural household savings, capital formation and tax burdens, he concludes that 'while farmers saved a respectable proportion of their income in the 1960s (but not in the late 1950s), relatively little of this income found its way into the modern sector',[51] and that 'throughout most of the post-1960 period, particularly after 1965, rural savings financed only about 10 per cent of total capital formation, a share about equal to the percentage contribution of agriculture to the increase in national product'.[52]

What the above argument overlooks is that there are a variety of ways of extracting a surplus from agriculture and that, ideally, the problem should be analysed within a general equilibrium framework, encompassing the effects of inter-sectoral terms of trade, exchange rate policies etc. If one takes this wider view then it can be shown that agriculture was by no means the 'untouched' sector that Perkins portrayed it to be.

Ever since the end of the Korean War, agriculture has been contributing to economic stability and growth through bearing the burden of low procurement prices for food. In the 1950s, the motivation was primarily to contain inflationary pressures during the reconstruction of the war-torn economy. In the 1960s and 1970s prices remained low in order to keep urban industrial wage costs down as well as to maintain price stability. The latter motivation can be seen in the fact that rice and barley were not only acquired at prices below the free market price but were also released at prices lower than the cost of procurement by the government.[53] Moreover, the level of industrial wages in Korea was low in relation to the average productivity of labour when compared to other developing economies.[54]

The burden that this 'cheap food' policy imposed on the agricultural sector can be seen in Tables 2.15 and 2.16. The effect of government procurement policies is equivalent to a tax on agricultural incomes and the burden of this tax depends on the extent to which procurement prices are below market prices and the proportion of disposable output that is subject to procurement. In South Korea the share of compulsory government purchase of the marketed surplus has been increasing steadily since the 1950s and in 1975 it stood at 50 per cent for rice and 90 per cent for barley. This procurement price has consistently been below the free market price and, although there has been a tendency towards a narrowing of the difference between the two prices since 1964, the procurement price was still some 12 per cent lower than the free market price in 1974. Even more significant is the fact that the procurement price was below the average cost of production up to 1961. 'The Government grain was in effect requisitioned from farmers through local administration channels. As a result, farmers suffered much loss because of forced sales to the Government'.[55] Since then the ratio of procurement prices to costs has increased steadily, especially since 1971.

Another indicator that agriculture has borne part of the burden

Table 2.15 Government Purchase Prices. Market Prices and Cost of Production for Rice, 1948–75 (*won* per 80kg)

Year	*Purchase Price (A)*	*Cost of Production (B)*	*Market[a] Price (C)*	*A/B (%)*	*A/C (%)*
1948	2.47	3.72	7.10	66.3	34.8
1949	2.67	6.71	13.21	39.9	20.2
1950	16.40	15.88	52.30	103.6	31.4
1951	65.37	n.a.	157.50	—	41.5
1952	200.62	329.09	447.50	61.0	44.8
1953	200.62	330.94	350.00	60.6	57.3
1954	308.33	330.94	581.00	93.2	53.1
1955	390.56	838.44	962.00	46.6	40.1
1956	1059.00	1134.00	1591.00	93.4	66.6
1957	1059.00	1384.00	1311.00	76.5	80.8
1958	1059.00	1297.00	1157.00	81.6	91.5
1959	1059.00	1300.00	1368.00	81.4	77.4
1960	1059.00	1313.00	1687.00	80.7	62.8
1961	1550.00	1377.00	1768.00	112.6	87.7
1962	1650.00	1422.00	2801.00	116.3	58.9
1963	2060.00	1373.00	3470.00	150.0	59.4
1964	2967.00	1936.00	3324.00	153.3	89.3
1965	3150.00	2672.00	3419.00	117.9	92.1
1966	3306.00	2495.00	3750.00	132.5	88.2
1967	3590.00	2735.00	4289.00	131.2	83.7
1968	4200.00	3403.00	5140.00	123.4	81.7
1969	5150.00	3565.00	5784.00	144.5	89.0
1970	7000.00	4642.00	7153.00	150.8	97.9
1971	8750.00	4682.00	9844.00	186.9	88.9
1972	9888.00	6115.00	9728.00	161.7	101.6
1973	11377.00	6578.00	12175.00	173.0	93.4
1974	15760.00	7959.00	17821.00	198.0	88.4
1975	19500.00	n.a.	n.a.	—	—

[a] November–January average prices

Sources Korean Ministry of Agriculture and Fisheries, *Grain Statistics*, 1967–75.
Korean Ministry of Agriculture and Fisheries, *Cost of Production Survey*, 1967–75.

Note Cost of production includes the imputed cost of family labour.

for industrial growth is the movement in the terms of trade for agricultural products.

As shown in Table 2.16, the terms of trade improved by 63 per cent between 1959 and 1963, but this improvement was from a base

Table 2.16 Terms of Trade for Agricultural Products, 1959–75

Year	*(A) Index of Prices Received*	*(B) Index of Prices Paid*	*(C) (A)/(B)*	*(D) Ratio of Market Price to Costs*	*Three-Year Averages for (D)*
1959	100	100	100	1.05	
1960	120	107	112	1.29	
1961	141	116	122	1.28	1959–61 = 1.28
1962	156	128	122	1.97	
1963	231	142	163	2.52	
1964	288	181	159	1.72	1962–4 = 2.07
1965	300	209	144	1.28	
1966	318	243	131	1.50	
1967	365	265	138	1.57	1965–7 = 1.45
1968	427	318	134	1.51	
1969	487	350	139	1.62	
1970	575	403	143	1.54	1968–70 = 1.56
1971	698	461	151	2.10	
1972	850	526	162	1.59	
1973	944	577	164	1.85	
1974	1234	776	159	2.24	1971–4 = 1.95
1975	1538	959	160	n.a.	

Source As for Table 2.15.

that was very unfavourable to the agricultural sector. (In 1959 the market price of rice was only 5 per cent higher than its average cost of production.) From 1963 onwards the terms of trade moved against agriculture and this trend was not reversed until 1969. Nonetheless, even after 1971, when procurement prices of grain were raised sharply, the improvement in the terms remained moderate, being only 12.5 percentage points higher in 1975 than in 1970. Moreover, the gains by 1975 did no more than restore the position lost since 1963.

Taking all these factors into account, it seems clear that it would be inaccurate to claim that there was no transfer of resources out of agriculture to finance industrial growth. Indeed, it is possible to advance the opposite argument that the rate of growth of agricultural incomes and productivity has been held down because of the disincentive effect of low prices. It has been argued that the availability of PL 480 (Public Law 480 of the USA concerning the disposal of surplus farm production as food aid) food imports up until 1969 explains the relative neglect of agriculture until the inauguration of the third five-year plan in 1972.[56]

On balance, however, it would appear that the generous availability of foreign aid did act as a restraining influence on the extent to which agriculture was squeezed in the initial stages of the drive to industrialise in the 1960s. The high rate of labour absorption also facilitated the growth of rural incomes and helped to check disequalising tendencies. Nonetheless, it is by no means true that agriculture escaped completely unscathed.

It should be noted, however, that the degree to which agriculture was or was not squeezed is not a vital part of the explanation of the favourable performance of the system in respect of the distribution of income and the degree of poverty. It is relevant primarily to the rate of increase in the level of rural incomes, and it is unlikely to have had very pronounced assymetric effects on the incomes of farmers with different farm sizes because of the relative homogeneity of cropping patterns and farm technology that characterises the Korean agricultural system. It bears reiteration that income and productivity growth in South Korean agriculture was respectable by international standards in spite of the 'cheap food' policy and that the system was highly responsive to changes in the price – cost differentials. According to Kuznets, agricultural output growth accelerated since the early 1970s, due mainly to the favourable change in this price–cost differential.[57]

In summary, we would attribute the relatively favourable performance of the South Korean agricultural system to the egalitarian conditions created by the land reform, the efficacy with which disequalising tendencies were controlled and the favourable impact that growth of the non-agricultural sectors had in dampening the power of these disequalising tendencies. These factors are to some extent interdependent. The initial egalitarian distribution of land not only created a low degree of inequality in asset and income distribution, but also created the pre-conditions for a widely diffused adoption of new productive inputs and of productivity and income increases. At the same time the wide diffusion of growth would have contributed to the dampening of disequalising tendencies and the alleviation of poverty, as did the rapid growth in the industrial sector.

2.5 On specificity and replicability

In the previous sections we have surveyed the performance of the South Korean agricultural system and identified some of the main factors which accounted for its relatively favourable performance. We now have to address ourselves to the question of the 'lessons of the Korean experience' and its relevance for the developing world. It is not a question which one would voluntarily pose, given the rather obvious value-bias and the connotation of a naive belief in the possibility of global social engineering. However, the question lurks willy-nilly and some answer is called for.

There are two senses in which the questions of the 'relevance of experience' are usually posed. First, there is a concern with understanding the 'mechanics' of the system; how it functioned and why it led to particular outcomes. Secondly, there is a concern with knowing how a system came to be installed and what surrounding circumstances enabled this to be done. The first sense is the domain of the economic exhorter offering nostrums such as 'let the market work'. The second is that of the 'systems transplanter' eager to retrace the steps of recent economic history, to identify the sequential ordering of major steps and then to use this as the underpinnings of recommendations on development strategy. It should be noted that the two concerns are often connected in that adopting the second position often assumes prior acceptance of the first.

A basic difficulty in answering the question in either or both of the above senses is that of the degree of specification. An over-specified answer would not be very instructive. If one piled up the number of conditions and 'ifs', one could come up with a truistic answer that, subject to all these qualifications, the system can be made to work or is transplantable. At the other extreme, an under-specified answer would be hardly any more interesting. For instance, to say that 'rapid industrial growth was essential to the success of the Korean system' begs too many questions to be considered a useful statement. It is therefore an inherently difficult type of question to answer.

The first sense in which the question has been posed has already been implicitly answered in the previous section. The factors which account for the relatively successful performance of the system were also the key elements in understanding the functioning of the system.

The real question, therefore, is that relating to the replicability of the experience, in the second sense. A useful way of tackling this question is to start by an enumeration of the main elements of specificity in the Korean experience.

The first element of specificity revolves around the configuration of domestic and international events which made the thorough land reform both necessary and feasible. The nationalist uprising against a harsh Japanese colonial rule towards the end of World War II contained a strong strand of agrarian revolt against Japanese landlords, as well as against their Korean counterparts who were regarded as collaborators. At the same time, the partition of the country along the 38th parallel in 1945 installed an American military administration in the South which was impelled to initiate a process of land reform in order to crush communist ascendency in the Korean countryside. A powerful alien military administration between 1945 and 1948 in a war-torn country was faced with little political resistance in initiating a land reform. It moreover had a convenient 'degree of freedom' in the form of former Japanese-owned land, forming about 20 per cent of the total cultivated area, which was available for distribution. The Korean administration under Syngman Rhee, which took office in 1948, was totally dependent on American support and hence pliant to American wishes on land reform[58] as on other issues. Moreover, Syngman Rhee himself had a very limited political base and used the land reform to break the power of his strongest political threat, the landlord-dominated Korean Democratic Party.[59]

The Korean War also facilitated the process of land reform. The upheaval of war weakened the possibility of resistance by landlords, and it has been claimed, more generally, that the effects of war resulted in a levelling in the asset distribution in Korea through the destruction of many large fortunes. The continuing geo-political significance of the Republic of Korea after the end of the Korean War also ensured favoured access to American foreign aid (including PL 480 food imports) which, as mentioned in the previous section, eased the pressure for domestic resource mobilisation to finance the industrialisation programme.

A second element of specificity derives from the extraordinary growth in industrial output after 1965. Part of the explanation of why such rapid growth was achieved lies in the geo-political significance which South Korea enjoyed. This created not only favoured treatment in terms of American aid disbursement but also in terms of the inflow of American and Japanese direct investment and of favoured access to the large American and Japanese markets.[60] The Korean export boom was largely absorbed by these markets.

It has also been claimed that sound economic management played an important part in bringing about this rapid growth. It is beyond the scope of this paper to evaluate fully these explanations but it is worth mentioning in passing that there is an 'efficient authoritarianism' about the Park government. Rapid growth has been posed as an overriding national goal and its attainment has been pursued with the imagery and style of a military offensive. National greatness is equated with rapid economic growth.[61] Economic growth is also seen as the key element in the competition with North Korea and vital to the nation's survival in the face of the constantly invoked bogey of communism. There is thus evidence of efficiency, as reflected in the relatively short time-lags in project implementation and the frenetic export drive, but also ruthlessness. The political repression and tight control over the labour force are the reverse side of the coin of this efficiency.[62] Yet rapid growth, however forced, was achieved and had the favourable impact on the agricultural system that was discussed in the previous section and hence constitutes an important distinguishing feature of the Korean experience.

The third element of specificity was the agrarian system that prevailed prior to the land reform. All the available evidence indicates that the system prevailing in Korea was similar to that in Japan. Both systems were subject to the same constraints of high land to man ratios and the labour-intensive nature of paddy cultivation, the predominant crop. It has been said of Japanese agriculture that

> change occurred within relatively narrow technical limits; and of the relatively unchanging elements of farming, the most important were the small size of units, family organisation of production, and the unsparing use of hand labour [and that] family farming remained the almost invariable rule, preserving the pattern of compact settlements and blocking the growth of large capitalist farms. Land ownership did tend to concentrate, but large owners turned their land over to families of tenants or part-tenants rather than work it themselves.[63]

We believe that the similarity in technology, cropping pattern and factor-proportions and the effects of Japanese colonial rule would have ensured that a highly similar system prevailed in Korea. Moreover, this is the only system that would be consistent with the data on the pre-reform size distribution of holdings and the absence of landless agricultural labourers that we discussed in an earlier section.

The genesis of such a system and the reasons for its survival is extremely interesting to enquire into but there appears to have been little work done on this question. Johnson and Kilby speculate that it is the peculiarities of technology that were the main determinants:

> . . . with the development of more productive but labour-using techniques, small units employing family labour were more efficient. Large landowners, being reluctant to sell their holdings, responded to the situation by renting their land to tenants who were prepared to apply the more intensive techniques that had been evolved.[64]

Whatever the reasons for its existence, such a system certainly makes it easier to institute a system of small family farms after a land reform than one where land concentration gave rise to large production units and proletarianisation of erstwhile peasants.

These three elements certainly limit the extent to which one can speak of the replicability of the Korean experience. It should be noted, however, that these elements have varying significance for the 'replicability' proposition. The political vacuum created by war and the particular type of pre-reform agrarian system are things which cannot be created by conscious policy and hence constitute unique circumstances. The facts of rapid growth and favourable access to external resources are, however, of different significance since it is by no means obvious that high growth is attainable only through the particular configuration of circumstances which prevailed in South Korea.

If we abstract from these elements of specificity it is still possible to draw some general conclusions, albeit weak ones, from the Korean experience. First, it could be said that a thorough-going land reform yields positive benefits in terms of both distribution and growth. The land reform brought about a swift alleviation of absolute rural poverty and, far from resulting in a loss of output, created the pre-conditions for widely diffused growth in the rural economy. It also showed that egalitarian small-scale peasant production is one viable alternative to the growth with increasing poverty that goes in train with growing concentration in land ownership within a market economy. More generally, it might be taken as an illustration of the proposition that a major redistribution of assets does create the pre-conditions for rapid growth within a market system without leading to a sharp polarisation in income distribution. It should be remembered, however, that the Korean agricultural system also showed that within such a system disequalising tendencies were strong, even given the presence of unusual favourable circumstances and that strong countervailing measures were necessary on a continuous basis.

Put differently, it might be said that if agricultural production has to be organised on the basis of private ownership of land, then a relatively egalitarian system of peasant farming is clearly superior to one with a high degree of concentration of land ownership. The broad dissemination of new inputs, new technologies and high-yielding material is unhindered by the 'landlord-bias' that has thwarted attempts to modernise non-egalitarian agrarian systems. At the same time the Korean system also demonstrated the possibility of maintaining sufficient 'elbow room' for private material incentives to animate the system[65] without leading to growing inequalities and absolute impoverishment, provided legal limits to land ownership and tenancy are adequately enforced. In addition, the extremely high land to man ratio that prevailed in pre-reform Korean agriculture does disprove one conventional wisdom, i.e. that land reform would be self-defeating under conditions of extreme land scarcity.

Given the general difficulty in arriving at firm conclusions that has been evidenced throughout this paper, it seems only fair that the author be allowed to conclude by reiterating a number of caveats. At best, the 'lessons of the South Korean experience' can be interpreted only as necessary and not sufficient conditions. Moreover, nothing that has been said should be interpreted in a naive prescriptive sense. Apart from confirming the validity of some general propositions about the relationship between land reform and economic growth, we do not believe that the South Korean model does in any sense provide a pre-fabricated blueprint for rural development. Finally, it should be remembered that this paper by no means constitutes an overall evaluation of the Korean path to economic growth. The evaluation was within the narrow confines of quantitative criteria relating to growth of output, income distribution and savings and we have eschewed the larger question of comparison across economic systems. There ought to be nothing in what we have said which precludes a judgement that the costs in terms of political repression have been too high, that the provision of socially provided services, like health and education, has been inadequate[66] or that alternative institutional arrangements could not have generated superior outcomes.

Notes

1. ILO, *Poverty and Landlessness in Rural Asia* (Geneva: ILO, 1977).
2. Irma Adelman, 'Case Study on South Korea' in H. Chennery et al, *Redistribution with Growth* (London: Oxford University Press, 1974).
3. The land reform began in 1945 with the distribution of former Japanese land vested with the American Military Administration. The Land Reform legislation of the new Korean Administration was passed in 1948 but implementation was delayed until the end of the Korean War.
4. The amount of arable land per rural resident (i.e. total arable land divided by rural population was 0.134 hectares). Comparative figures for China and India are 0.164 and 0.442 respectively.
5. See Takeshi Hatada, *A History of Korea* (Olio Press, 1969) and Yunshik Chang et al, 'The Impact of Population Growth on Land, Labour and Productivity in Rural Korea' (Geneva: ILO, WEP Working Paper, February 1977).
6. Robert B. Morrow and Kenneth H. Sherper, 'Land Reform in South Korea' in Agency for International Development, *Spring Review of Land Reform*, June 1970, Second Edition, Vol. III (Washington: 1970) p. 6.
7. See ILO, op cit, pp. 47, 109; S. Radwan, *Agrarian Reform and Rural Poverty: Egypt 1955–75* (Geneva: ILO, 1977) p. 25.

8. The number of power tillers increased from 8,000 to 60,000 units and of power threshers from 34,000 to 108,000 between 1969 and 1974 (Korean Ministry of Agriculture and Fisheries, *Agriculture Statistics Yearbook 1976*, Table 23, pp. 68–70). In spite of this increase the level of mechanisation in South Korea is still very low compared to Taiwan and Japan. In 1974 there was only 1 power tiller for every 34 farm households in South Korea.
9. *Korean Agriculture Sector Study*, Summary Report (University of Michigan, East Lansing 1977) p. 5. See also Sungho Kim, 'A Socio-economic Analysis of Farm Mechanisation in Asiatic Paddy-Farming Societies with Special Reference to Korean and Japanese Cases' (mimeo) (Seoul: National Agricultural Economics Research Institute, 1971).
10. Dwight Perkins, *Rural Development in Korea*, Harvard Institute for International Development (forthcoming) Table 2.4.
11. For example, Sri Lanka and Malaysia.
12. P. W. Kuznets, *Economic Growth and Structure in the Republic of Korea* (New Haven and London: Yale University Press, 1977) p. 132.
13. Korean Office of Rural Development, *Rural Development Program in Korea* (Suwan: Korea 1977).
14. P. W. Kuznets, op cit, Table 2.4.
15. Korea, Office of Rural Development, op cit.
16. Bertrand Renand estimates that the understatement is in the order of 20–30 per cent. See his 'Economic Growth and Income Inequality in Korea' (Washington: World Bank Staff Working Paper No. 240, February 1976) pp. 17–22.
17. P. Hasan, *Korea: Problems and Issues in a Rapidly Growing Economy* (Baltimore: Johns Hopkins Press, 1976).
18. Irma Adelman, op cit.
19. Korean Ministry of Agriculture and Fisheries, *Farm Household Income Survey*, 1976, pp. 24–5.
20. We know only that the price of rice and barley rose sharply in relation to other crops but we do not have information on changes in the relationship between these crop prices and the returns from other farm and non-farm activities.
21. Parvaz Hasan, op cit, p. 53.
22. Ibid., p. 53.
23. P. W. Kuznets, op cit.
24. D. R. Snodgrass, 'Education and Economic Inequality in Korea' (*Discussion Paper No. 23*, Harvard Institute for International Development, February 1977, p. 4).
25. Robert B. Morrow and Kenneth H. Sherper, op cit, p. 41.
26. Dwight Perkins, op cit, p. 9.5.
27. Robert B. Morrow and Kenneth H. Sherper, op cit, p. 41.
28. Ki Hyuk Pak, 'Income Distribution in the Agricultural Sector in Reference to the Land Reform in Korea', *Yonsei Business Review*, (Korea) [1972] No. 9, p. 38.
29. Morrow and Sherper, op cit, p. 6.
30. Dwight Perkins, op cit, p. 9.15.

31. Ibid., p. 9.16.
32. Ibid., pp. 9.15–9.20.
33. Ibid., p. 9.14.
34. Ibid., p. 9.13.
35. Sungho Kim, 'The Changing Patterns of Farm Land Problems after Land Reform in Korea'. Paper presented at Seminar on Agrarian Reform and Development of Agriculture, University of Wisconsin, July 1977, Table 1, p. 6.
36. Sungho Kim, 'A Socio-economic Analysis of Farm Mechanisation in Asiatic Paddy-Farming Societies' (mimeo., Seoul: National Agricultural Economics Research Institute, 1971) p. 27.
37. Yunshik Chang, et al, op cit, p. 35.
38. Korean Agricultural Sector Study, op cit.
39. Sungho Kim, op cit.
40. Sungho Kim, 'The Changing Patterns of Farm Land Problems after Land Reform in Korea', p. 13.
41. Yunshik Chang, et al, op cit, p. 68.
42. Ibid., p. 56.
43. Dwight Perkins, op cit.
44. Ibid., p. 2.16.
45. Defined as the rate of growth of employment divided by the rate of growth of output.
46. P. W. Kuznets, op cit, p. 119.
47. Dwight Perkins, op cit, p. 2.2.
48. Ibid., p. 2.26.
49. Sungho Kim, 'The Changing Patterns of Farm Land Problems after Land Reform in Korea', p. 8.
50. Ibid., p. 13.
51. Dwight Perkins, op cit, p. 2.14.
52. Ibid., p. 2.15.
53. Pal Yong Moon and Byung Seo Ruy, *Korea's Agricultural Policies in Historical Perspective* (Seoul: Korea Development Institute, Working Paper 7704, April 1977).
54. David C. Cole and Princeton N. Lyman, Korean Development (Harvard University Press, 1971).
55. Pal Yong Moon and Byung Seo Ruy, op cit, p. 60.
56. For example, P. W. Kuznets, op cit, Pal Yong Moon and Byung Seo Ruy, op cit.
57. P. W. Kuznets, op cit, Chapter 5.
58. Gary L. Olsen, *US Foreign Policy and the Third World Peasant* (New York: Praeger 1974) Chapter III.
59. Joungwon A. Kim, *Divided Korea: The Politics of Development 1945–72* (Cambridge, Mass.: Harvard University Press 1975) pp. 115 and 125.
60. Frank Baldwin (ed.), *Without Parallel: The American–Korean Relation – since 1945* (New York: Random House 1973) pp. 179–219.
61. Park Chung Hee, 'Collected Speeches' (Seoul: 1977).
62. Frank Baldwin (ed.) op cit.

63. Thomas C. Smith, *The Agrarian Origins of Modern Japan* (California: Stanford University Press 1959) pp. 208–9.
64. Bruce F. Johnston and Peter Kilby, *Agriculture and Structural Transformation* (New York: Oxford University Press 1975) p. 165.
65. M. Lipton, 'A Theory of Land Reform' in D. Lehman, *Agrarian Reform and Agrarian Reformism* (London: Faber and Faber, 1974).
66. Frank Baldwin (ed.) op cit, pp. 233–303.

3 Organisation of Agriculture for Rural Development: The Indian Case

Ashok Rudra

3.1 Introduction

In the vast Indian sub-continent a static view would reveal a high diversity in the different kinds of agriculture – diversity not only in the cropping patterns and cultivation methods determined largely by geographical conditions, but also in the forms of agricultural organisation and the structure of the rural economy and rural society, which are determined by very diverse historical conditions. Subsistence agriculture is carried out by animal power and human labour aided by the most primitive of tools. Although production is wholly for the market, it is still enmeshed in extremely backward technological as well as institutional conditions, as in the case of crops like jute and sugar cane. Production of food crops as a matter of profitable business is with the help of all the paraphernalia of modern technological agriculture but on a scale which, by standards of most developed countries, is extremely meagre. Plantations are run on purely capitalistic lines and on scales large enough to be comparable to the estate farms in developed countries of the West; cash crops which are suitable for plantations (like coffee, rubber) are grown, cultivated on a garden scale along with such crops as coconut, areca nut and various spices. Land is cultivated by land-owning family's members themselves; land is cultivated by tenants with the owner taking no interest in production whatsoever, but land is also cultivated by the landowner himself but leaving the manual operations to servants, casually employed labourers or tenants. These are but some of the aspects of the kaleidoscopic diversity which the scene of Indian agriculture presents. There is no discernible common pattern except perhaps in the practically total absence of any collectivistic production; a role of minor importance is played by co-operatives of sorts in some fields like credit and marketing, but that too within an extremely uneven spread.

However, behind this apparent lack of any common pattern in a static view, it is possible to discover, if one takes a dynamic view, the clear outlines of a unity. Unity in diversity is a cliché much used in connection with Indian culture and the cliché is no less true of Indian agriculture when one looks at the directions of its movements. These movements are no spontaneous movements. They are the results of a strategy of rural development that is implemented in a consistent, concerted manner which is remarkable for its absence in all other spheres of developmental policy in the country.

As we shall see in Section 3.2, the foreign advisers to the Indian government have played a central role in the shaping of the strategy of development that is giving a unified pattern to the dynamics of Indian agriculture in all its aspects. They have exerted a considerable influence in shaping India's agricultural policy ever since the first days of economic planning in India. They did not, of course, start with a clean slate: the slate was drawn over by some well-defined thoughts on development policy and strategy which was a product of the Indian nationalist genius of the pre-independence period. And the advisers themselves did not arrive at the strategy in one day. Their strategic recommendations passed through various phases, diverging further and further away from the Indian nationalist framework, until they arrived at the final phase which turned out to be vastly different from the one that was the product of the early nationalist consciousness.

Of course no strategy could have been accepted nationally if it had not been supported by powerful domestic interests. The strategy has received increasing support from the rich farmers as well as the industrial entrepreneurs, the interests of both of whom are served by the strategy. But the rich farmers and the industrial capitalists of the country were less clairvoyant of their own interests than were these advisers, whose task was to open the eyes of these groups to their own interests in this matter.

3.2 The strategy

The following is an account of the strategy by one of the strategists, Cummings:

1. The new strategy focused on combining high yielding varieties of seeds with a 'package' of complementary inputs concentrating on selected water-assured acreages.
2. The development of high yielding varieties of seeds was the critical new element.
3. The fundamental policy change was to concentrate inputs on the highest productivity areas.
4. In addition, there were commitments to:
   a. make available required inputs, particularly chemical fertilisers, assigning priority to scarce foreign exchange when necessary;
   b. actively encourage domestic and foreign investment in fertilisers and other input areas;
   c. implement a price policy which was 'producer orientated';
   d. reorganise agricultural research into a co-ordinated programme of all India schemes;
   e. refocus extension activities;
   f. support a system of state agricultural universities which would make scientific activity directed towards practical application a permanent part of the India scene;
   g. extend the distribution of rural credit.[1]

This statement of the strategy is quite complete; but its flat tone manages to hide what constitutes its sharp cutting edges. After all, the use of fertilisers or the best seeds does not constitute any strategy: there cannot be any way of developing agriculture when land is in limited supply without involving the use of yield increasing inputs. The first cutting edge of the strategy was the total absence in it of any consideration of institutional reforms. The strategy is a strategy of increased production by the most effective use of the technological factors alone, which are assumed to be capable of being manipulated without changes in the institutional framework. It is a strategy that aims at changing the forces of production without aiming at any concomitant changes in the relations of production.

The second cutting edge of the strategy is the 'package principle' or the 'intensive approach'. Once again, the idea that fertilisers should not be wasted on dry land and that high yielding variety seeds should not be sown without there being adequate supply of water and fertilisers hardly constitutes a breakthrough in wisdom. Hence the package principle, in that it means complementarities in inputs are to be taken care of in their application, is a commonplace idea. What is not a commonplace idea is that as resources are scarce, it is wasteful to spread them thinly on all farms; optimum use of resources requires that they be applied in optimum doses, and in the right combination, only to those farms who can make the best use of them.

It is, however, not only a question of intensive application of resources to geographically limited areas. The selective and concentrative approach has also to be applied to selected crops. As one of the pioneers of the new technology, Borlaug, writes:

> Yield increases of 15 to 20 per cent will convince no one. Demonstrations showing increases from 100 to 600 per cent, such as those which have been achieved over widespread areas in India and Pakistan during recent years, have caught the peasants' imagination and have triggered off agricultural revolution. Such demonstrations destroy the peasants' attachment to archaic practices, and with them their 'conservatism'. They generate enthusiasm in farmers, scientists, and government officials and usher in an entirely new spirit of hope and agressiveness at all levels.
>
> In order to mount such an attack, strong food production campaigns should be initiated around one single crop or animal product for which new high yielding technology is available. It should be a product which is important to the economy of the country and for which the minimum production services, such as roads, market facilities, and sufficient water for substantially higher yields, are available.[2]

Yet another direction of concentration is in time. The strategy aims at obtaining spectacular results in a very short time horizon. The same author writes:

> Production levels that have taken centuries to reach in the developed countries can be achieved in decades in the developing countries . . . We reject the approach of a 'yield take-off' as a means of changing a traditional agriculture.

A corollary of the intensive approach is the selectivity with respect to farmers. If the principle is to concentrate resources in such regions and on such crops as promise the maximum return, it follows that they should also not be spread thinly over all farmers, but only on those farmers who can make the best use of the resources. The Indian strategy identified such farmers as 'progressive farmers'. However, it cannot but happen that the progressive farmers would be mostly big and rich farmers. That is because the technology of high yielding varieties with intensive use of irrigation water and fertilisers in optimum proportions call for a lot of working capital; and it is only the big and rich farmers who can command such working capital from either their own surplus or from sources of credit.

The strategy is thus one that aims at maximising short-run growth of commodity production at the cost of increasing the disparity between the richer and poorer farmers, and that between advanced regions and backward regions. It is one that explicitly rejects growth with social justice, which happens to be a declared objective of planning for economic development in India. It is not a matter of our making inferences or interpretations. It is a matter that is explicitly stated in so many words by the chief architects and protagonists of the strategy. Thus, Ladejinski, one of the experts closely associated over a long time with the formulation and implementation of the strategy, wrote:

> The aim of the Green Revolution as envisioned by the policy makers is productivity; social imperatives were not part of it. It is not the fault of the Green Revolution that the credit service does not serve those for whom it was originally intended, that the extension service is weak and ineffective, that the village panchayats, or councils, are essentially political rather than developmental bodies, that security of tenure is not given to the many, that rentals are exorbitant, that ceilings on land ownership are notional, that even rising wage scales are hardly sufficient to satisfy the basic essentials of the farm labourer, or that generally speaking in those conditions economic necessity and social justice for the village poor do not ride in tandem. Many of these are man-made issues of very long standing.[3]

Of course the well-known social evils cited are not the fault of the 'Green Revolution'. It is a strange way of speaking. Fault attaches to human actions, not to the products of human actions. Of course the social evils cited are age-old man-made issues. But the so-called 'Green Revolution' is also a man-made phenomenon and it is clearly admitted above that man, in setting about bringing in a

'Green Revolution', did not even aim at removing the age-old man-made evils!

This, of course, does not mean that the strategy ignores the objective of social justice. While the pattern of growth promotion is not meant by itself to lead to harmonious economic development of the rural sector or to improve the pattern of distribution (rather it is accepted as bound to worsen it), it takes up the problem of improving the lot of the poor farmers as an independent sub-strategy which is of the nature of providing certain reliefs and benefits to the poor farmers who are deliberately to be left out of the growth process. That this is indeed not an integral part of a unified strategy for the development of agriculture, but something merely appended to a strategy for growth, is indicated by the way the fifth plan used the expression 'second part of the strategy' to denote the special schemes meant for the relief of the poor farmers. We shall take a more detailed look at these schemes in Section 3.5.

An important aspect of the strategy is the stress it lays on making agriculture dependent on industries for its inputs. Traditionally, Indian agriculture was self-sufficient in the matter of its input requirements; for fertilisers it mainly used organic manures and green manures; the implements that were used could be, and were, manufactured by village carpenters and blacksmiths themselves. As we shall see in Section 3.2 below, the strategic framework of the pre-independence India attached positive value to this village self-sufficiency and this emphasis remained for some time in the strategy for agriculture in the first period of planning. However, the strategy that has finally been adopted attaches a predominant importance to industrial products as intermediate goods to be used in agriculture.

Thus the role of organic manures has been minimised; that of green manures completely overlooked; and the entire reliance for plant nourishment placed on chemical fertilisers. It has been shown (see, for example, Minhas and Srinivasan[4]) that irrigation combined with moderate doses of fertilisers and ordinary seeds promises larger and more widely distributed benefits than intensive use of chemical fertilisers in combination with HYV seeds; but this approach has not been accepted. In irrigation the policy has either been to go for cement-steel intensive, large-scale works involving the use of earth-moving and various construction equipment or for much smaller-scale irrigation that calls for such industrially produced goods as deep tubewells. The planners have systematically left unexplored the scope for use of labour-intensive irrigation works. About mechanisation in general the policy is of selective encouragement, the argument being that not all machinery is labour displacing and some machinery can help to increase production. The strategy recommends not only domestic investments in the fertiliser and agricultural machinery industries but also imports of these products. As domestic investments almost always involve a great deal of foreign participation, the strategy effectively involves encouraging an increasing direct role of the multinationals in Indian agricultural development.

Yet another important feature of the strategy is the importance it attaches to research directed at keeping up a continuous flow of innovations in the matter of seeds and husbandry practices. As a logical corollary it lays stress on an extension service manned by personnel sufficiently trained in agronomy to carry information to the doors of the farmer. Both imply heavy investments in agricultural universities and research institutions and such institutions as have come up are heavily dependent on foreign experts. This is thus another aspect of agriculture being made dependent upon the industrial sector as supplier, a salaried bureaucracy as conveyor and foreign experts as advisers. This is in marked contrast to the philosophy of village self-sufficiency and of extension work being done by committed social workers, whether of the type of the village worker of the Gandhian scheme or of the type of the party cadres in the socialist countries.

Price support policy constitutes a crucial element of the strategy. In theory the price policy is supposed to achieve an equitable balance between what are considered 'remunerative prices' for the producer and 'fair prices' for the consumer. The presumption is that this can be achieved by cutting down on the private traders' margin. In actual practice the policy as far as food crops are concerned has served to prevent prices for the producers falling low, but has not prevented prices for consumers rising high. The terms of trade have moved consistently in favour of agriculture.[5] Agriculture getting a more favourable treatment than industries in an early stage of industrialisation is perhaps an exceptional phenomenon with great political significance.

The way government policy contributed to the above phenomenon is as follows. The government set itself the task of procuring food grains from producers and distributing them to the consumers, both at pre-determined prices. But in actual practice the volume of government procurement and distribution has quite often fallen much below the requirements of the non-agricultural consumers, so that most such consumers have been obliged to obtain part of their requirements from the free market at much higher prices than fixed 'fair prices'. On the side of the producers, they never suffered. For whenever market prices tended to fall, government procurement came in to maintain the prices offered by the government. On the other hand, when prices tended to rise, the government failed to procure sufficient amounts. This, despite the fact that the government has always talked about procuring food grains through levies, that is to say fixed quotas for each farmer. While the above account is true of food grains and other food crops and therefore broadly of agriculture as a whole, the story is different when it comes to crops like jute where a large mass of poor farmers face a monopsonistic buyers' market of jute mill owners.

Underlying the entire strategy is a philosophy of agriculture developing under the propelling force of the profit motive of the relatively bigger and richer farmers, with the state providing the infrastructure facilities. The basic tenet of belief of the architects of this strategy is that 'farmers in developing countries are acting in a rational economic way', by which is meant that these farmers maximise profit according to the rules of the neoclassical theory of the firm. In other words, faith is being laid for achieving the task of developing agriculture on the acquisitive spirit which motivates private entrepreneurs. In traditional Indian agriculture this spirit was located and condemned in the professional money-lender, the speculative trader, etc. An important discovery of the proponents of the strategy is that the same acquisitive spirit may be found latent in the cultivators too and all the components of the strategy are aimed at further encouraging this spirit. Thus, one of the principal architects of the strategy writes the following in clear, explicit words:

> We should recognise that unless we take appropriate steps to create the necessary social conditions and in particular foster amongst our farmers a desire for a standard of living which is appreciably higher than their present level of living, they will not have the necessary incentive and urge for working harder, for taking greater risks and for trying out new techniques with a view to producing more and earning more so as to get the wherewithal for purchasing the goods and services for which they have a new demand. It is not, however, suggested that we should make them crave for a standard of living which is completely unrelated to

> their present level of living. In fact if it is too unrelated and appears unattainable to them, it will not produce the stimulating effect that is expected. All that is needed is that like the 'Holy Grail', the standard of living should be always a little ahead of the level of living and yet not so far ahead that it appears to be beyond their reach. It is only then that they will not merely have a desire to improve their present position but will also have the feeling that with a little enterprise and a little more effort it will be possible for them to reach that higher level.[6]

This clearly stated aim seems to have been achieved. The 'Holy Grail' which the richer farmers are pursuing is the way of life of the urban middle class; while the latter in their turn are craving after the comforts of the consumption society of the West.

3.3 The evolution of policy

The strategy that has been described and discussed in the preceding section is the result of an evolution over time that started in the pre-independence period and took final shape towards the late sixties. In this section we shall briefly trace the path of this evolution. The two end-points of these present striking contrasts (in some respects diametric opposites), so that the end-point cannot be understood properly if taken by itself without reference to the evolutionary path.

(a) Pre-independence thought

An idea about the social framework of agriculture as conceived by the nationalist ideology of the pre-independence period may be obtained from the proposals of the Congress Agrarian Reform Committee which submitted its report in 1949. This conception was based on the philosophy that the main hurdles on the path of agricultural development lie in the institutional arrangements as opposed to the philosophy that the principal bottleneck lies among technological factors. This philosophy was later caricatured by the supporters of the technology-first-and-foremost strategy in such words as the following:

> The view that conspiracy is a source of agricultural problems has played a major role in Indian agricultural development program-

ming. An extensive literature protests the exploitation of Indian farmers by landlords, money-lenders and merchants, all charged with conspiring to the disadvantage of the poor cultivator.[7]

The aforecited report gave highest priority to the abolition of the revenue collecting intermediaries; next in importance in its scheme came the conferring of occupancy rights to tenants. It also recommended ceilings on holdings. And it emphasised greatly the role that co-operatives could play not only in the matter of credit and marketing but also in production; by way not only of co-operative aids to farming to enable small farms to overcome the disadvantage of small-scale farming but also of joint farming, that is farming on pooled land with the help of pooled resources of labour and capital. Such was the strength of the pro-co-operative ideology at that time that co-operativisation was recommended to be enforced! The Committee even recommended 'Collective type of farming on a portion of reclaimed lands where landless agricultural labourers would be settled'.

The first five-year plan gave the leaders of the nationalist movement the opportunity of putting into practice their programmes. Zamindari (that is, the institution of revenue collecting landlords) was abolished wherever it existed in the country during the first years of planning. This apart, none of the other land reform proposals (tenancy abolition, ownership ceilings and co-operativisation) ever got implemented. However, the proposals regarding tenancy and ceilings have remained on the agenda all along. Each successive plan has talked about difficulties in their implementation and admitted failures both in legislations and in implementation and reiterated resolutions for henceforth carrying out the tasks more satisfactorily.

The story is different for co-operatives. The programme was dropped for all practical purposes. Talks of co-operatives, joint farming and co-operative aids to individual farming were kept up during the first plan and the second plan (1951–55 and 1956–60), but there was no further talk of any 'enforcement'. Even talk of co-operation in the field of production has ceased since then. The talk of co-operatives in the field of credit and marketing still continues – it occupies a lot of space in the fifth plan volume. Such co-operatives are indeed quite widespread, but it is widely accepted that these organisations mostly perform a function very different from the one visualised for them in the pre-independence nationalist ideology. Instead of being instruments for enabling poor farmers to overcome the obstacles arising from their lack of resources, they have become instruments for enabling the rich farmers to receive state support. The prevalent practice is for rich farmers to form co-operatives and receive aid from the state, thereby undermining an institution, the co-operatives, which under different circumstances might have helped the poor.

In the matter of rural development the first five-year plan (1951-56) marked one important departure from the pre-independence nationalist thinking. The latter, under the influence of Gandhi's thoughts, always believed that village-level 'constructive work' should be carried out by political workers, one for each village, who would be motivated by the desire to serve the people and who would live in the village among the villagers as one of them. This idea was dropped. The concept of 'extension services' was imported from the experience of developed countries and the task of village development was entrusted to a cadre of government employees for whom this was not a political task nor a social service but merely a low-paid job. An entirely new wing of the administration was set up for this purpose under the community development programme. A network of community development blocks was set up in the early fifties and covered the entire country in the course of the next decade with a density of one block per 100 villages.

Each block covered 100 villages and some 60–70,000 people. The block was conceived as a co-ordinated administration which treated simultaneously the diverse but interrelated needs of village development. It recognised the interdependence of village services and hence the need for their co-ordination. At the head of the operation was the block development officer, who directed a hierarchy of technical specialists and village-level workers. Directly under the block development officer were the technical specialists. There were normally about eight of them – one for agriculture, one for animal husbandry, and one each for such areas as co-operatives, social education (largely literacy training), rural industries and engineering, women's and children's activities and local government. At the bottom of the structure was the village-level worker. Through him the specialisations represented by the technical specialists were supposed to be brought together.

The task envisaged for the Community Development Programme (CDP) – as the very title of the programme indicates – was not oriented towards agricultural growth taken by itself. Its aim was rural development in all its aspects and agricultural growth was considered merely as an important part of such comprehensive rural development. Thus it was also concerned with the development of agriculture, village industries, health, sanitation and education, mostly on a self-help basis and with resources available within the village. Government capital was to be used only to finance the new staff, provide share capital and loan funds to primary co-operative societies, establish government seed farms and finance minor irrigation.

It only follows logically from the comprehensive rural development philosophy of the programme that it should be extensive, and not intensive in its coverage. The programme did not discriminate between more advanced areas and less advanced areas, more enterprising farmers and less enterprising ones. All villages of all kinds were brought under the umbrella of the block network; and in theory at least the programme was supposed to pay all the more attention to the poor and the less resourceful farmer so as to enable him to overcome his handicaps.

In the matter of technology, in this first phase primacy was given to irrigation; and in the matter of irrigation emphasis was laid on gigantic multi-purpose river valley projects. On the question of plant nutrients, emphasis was laid on organic manures, green manures, etc. Pesticides were supposed to play a very minor part.

The CDP was destined to fail. The block organisation was hardly capable of discharging the services expected of it. The block development officer was a general administrator who was largely concerned with paper work and with co-ordinating the operations carried on in the block. The technical specialists were also heavily burdened with administration and form-filling. Even when extension personnel ministered to the farmer, the specialists' service was more one of facilitating loan and subsidy assistance from the government than of extending new technical knowledge.

It was the village-level worker who was supposed to do all practical work: set up field demonstrations, initiate talks and group discussions, carry out the programmes of all eight assistant development officers. In actual fact, he was far from such a paragon; he was a young man with, at best, a secondary education. Most were not known as successful farmers in their areas and lacked the experience or the skill to excel in traditional cultivation methods.

The block development officer and his technical extension officers lived in nearby towns, worked between 10 a.m. and 5 p.m.

like any other white-collar worker and travelled about in jeeps which were put at their disposal. This very style of action ruled out any possibility of their being anywhere near a village worker of the Gandhian scheme or a grass-root party cadre as in China or the Soviet Union.

(b) Abandonment of the comprehensive approach

The Community Development Programme (CDP) was largely indigenous in its inception, the product of Indian politicians imbibing the world view of the bureaucracy of the British period. It was however supported by American aid in the form of money and expertise from the beginning. The US Government and the Ford Foundation provided more than $100 million for this programme during the first and the second plans. American experts even took part in the training of the village-level workers. But they had reservations about the programme from the very early days. It is therefore not surprising that the subsequent changes that took place in the rural development philosophy should have been moulded by foreign advisers.

In 1952 a US Technical Co-operation Mission, in its report 'Observations of Indian Agriculture', advocated giving special attention to agriculture at the cost of other aspects of rural development. It focused on the need for training agricultural scientists and for agricultural research. This was the beginning of the view that prevailed in the end: the view that it is neither institutional obstacles nor the lack of means of the poor Indian peasant nor any 'irrationality' of the Indian farmer that is responsible for the stagnation of Indian agriculture, but lack of technological knowledge. In 1955 a Joint Indo-American Team recommended the establishing of agricultural universities similar to the state land-grant universities of the United States. The recommendation was accepted and such universities were set up in due course, heavily staffed by agronomists trained abroad as well as by visiting foreign scientists. These universities have indeed played a very important role in the agricultural development from the mid-sixties.

The first five-year plan ended with two successive good harvests due to unusually good weather conditions and this created some complacency among the country's planners. The complacency evaporated when the good harvests were followed by some bad ones. Desperately looking for ready-made solutions from other countries, the country's planners talked for some time about the

Japanese method of paddy cultivation which, it was said by Indian experts coming back from visits to Japan, could multiply paddy yields several fold. Visiting teams to Japan were followed by visiting teams to China and the talk of the Japanese method of cultivation was followed by the talk of Chinese methods of cultivation which, for a brief while, held the stage.

Finally, in 1958 the Government of India set up a Committee (the Nalagarh Committee) which recommended the integration into a single ministry, at the state level, of all ministries and departments dealing with programmes for the development of agriculture, irrigation, land reclamation, soil conservation, animal husbandry, veterinary science, agricultural markets, forestry, co-operation, agricultural research, education, extension, public works, etc. It also recommended that the extension staff of the integrated department concentrate on production; the village-level worker was to be relieved of his rural development duties in order to concentrate entirely on agriculture.

These recommendations were quickly superseded by those of the now famous Ford Foundation team that originated the 'package principle'. The 'Report on India's Food Crisis and Steps to Meet It' contained the seeds of most of the principal components of the strategy that was finally adopted. It also talked of co-operative institutions, land ceilings and consolidation of holdings. This lip service to institutional issues was eliminated in the final programme that emerged out of the report, namely the Intensive Area Development Programme (IADP).

c. The intensive area development approach

The IADP contained all the points of the strategy finally adopted as detailed in the first paragraph of Section 3.2 above. But it still placed itself in a framework of overall rural planning. It emphasised farm-level planning, village-level planning, as also rural work-programmes for utilising the vast surplus manpower.

The IADP was officially adopted in the third five-year plan and its implementation was taken up with a speed and efficiency that could be legitimately associated with the determination to lift India's agriculture out of its 'low level equilibrium trap'. The officers concerned quickly discovered that such parts of the programme which had anything to do with the philosophy of economic planning could be dispensed with. Thus farm-level planning and village-level planning never got beyond the stage of paper planning and that too only in some of the first few IADP districts. Also rural works programmes rapidly became enshrined as one of the many well-meaning ornamental edifices not meant for any functional use. And it is this kernel of the IADP programme that in the course of the sixties developed into the strategy that was finally adopted.

While the IADP gave importance to water management, it also emphasised a shift from large-scale to small-scale irrigation. It became fashionable to criticise the large-scale irrigation projects of the first and the second plan periods as being monumental white elephants: too capital intensive, involving too long a gestation lag and allowing only a low utilisation ratio of the water flowing through the canals. There is some truth in all three points of criticism. Yet one cannot but detect more than a trace of prejudice. The multi-purpose projects were meant not only to make water flow for irrigation purposes but also to control floods which used to cause havoc at periodic intervals in different parts of the country and it has not been argued by any of the critics that the taming of some of the notorious flood-prone rivers could have been managed on any small scale. Similarly, these projects also contained important components of hydro-power generation, and that too called for large-scale planning.

As to the low rate of utilisation of the canal waters, the problem lies, in there not having been adequate development of a network of feeder canals and drains to make the water reach the individual farmer's plots. Utilisation of canal water, whether through feeder canals or by the flooding method, calls for co-operation among the owners of land in the command area. As the drive for co-operatives has been dropped and replaced by encouragement for the spirit of individual self-seeking, it is only logical that in this matter of irrigation one should emphasise small-scale irrigation which does not call for much co-operation. Irrespective of the objective merits and demerits of canal irrigation and small-scale irrigation, the important fact remains that small-scale irrigation and sale of irrigation water in a private market for water is perfectly compatible with private property in land.

To sum up, the evolution of developmental policy with regard to agriculture underwent the following major changes: (i) a shift in emphasis from rural development as a whole to growth of agricultural production taken narrowly; (ii) a shift in emphasis from institutional reform to reliance on the potency of technology; and (iii) the dropping of the ideology of co-operatives and its replacement by the ideology of individual self-seeking.

3.4 Growth performance

We shall in this chapter attempt an assessment of this strategy in terms of its objective, namely growth. The task is more difficult than it may look at first sight. The question has obvious political, and other, undertones, and a great number of loose claims are made by politicians, journalists and vocal sections of the public as to how Indian agriculture is passing through a period of 'Green Revolution' and recording remarkable rates of growth or, alternatively, how the same agriculture is stagnating under the weight of conditions described as 'semi-feudal' etc. Most of the statements are made without any supporting empirical evidence or any careful statistical analysis of the same. What is regrettable is that even competent economists have indulged in strong assertions on statistical procedures which are highly questionable.

One of the elementary mistakes one frequently makes is to calculate the rate of growth from a comparison of two arbitrary end-points. That the choice of end-points would seriously affect the calculation of the rate of growth by this method is easily understood and accepted. But that the calculation of the rate from a fitted trend curve can be equally misguiding has not been as well understood. Thus, a usual procedure is to fit a trend curve of the form:

$$f(t) = a(1 + r)^t,$$

where r is a constant rate of growth. When one fits such a curve, one necessarily gets some estimates of a constant r and one puts this forward as the 'trend rate of growth' for a period covered by the time series. The procedure betrays a lack of awareness that such a curve fitting cannot ever tackle the question whether the rate of growth has changed over time or remained constant.

It seems to us that in order to examine if there has been an acceleration or deceleration or none of the two in the rate of growth of agriculture, one would have to fit a trend curve from which one can take a constant, increasing or decreasing rate of growth and also use to examine for possible breaks in the trend. A built-in increasing rate of growth would indicate a dynamic agriculture: a break in the trend would indicate what could justifiably be called a 'revolution'.

Guided by this idea we present the results of some careful statistical exercises with the help of two trend curves to data relating to the growth of agricultural output as a whole as well as to that of food grains[8]. One of the curves is the semi-logarithmic, that is the curve with a constant rate of growth given by the formula above; the other is the Gompertz, a curve that can accommodate a rate of growth that may increase, decrease or be constant. The results for all India are presented in Figures 3.1 and 3.2 and some explanations are provided in the Appendix at the end of this chapter.

The statistical exercises show that both the functional forms give very good fits to data. This has two important implications:

1. Remarkable as it may appear, there does seem to have been a common growth trend for agriculture as a whole as well as for food grains over the entire period 1951/52 to 1973/74. This rules out any break in the trend at any point; so that for agriculture as a whole, and for India as a whole, one cannot recognise at any point any such drastic change as may qualify for the adjective 'revolutionary'. As such, one can say, as far as can be judged by statistical evidence, that there has been no such thing as a 'Green Revolution' as a result of the introduction of the new strategy or otherwise.

2. The implications of the two curves fitted, each giving very good fits, are however extremely different. According to the semi-logarithmic fit, agriculture has been growing before and after the introduction of the strategy at a steady constant growth rate of three per cent per year. If however one takes the Gompertz fit one has to conclude that the smooth trend that has marked agricultural growth over the entire period 1951/52 to 1973/74 is one that involves a declining rate of growth. The rate of growth was four per cent in the early fifties and reduced to one per cent by 1973/74.

Two curves each giving more or less equally good fits and yet with such drastically different interpretations pose problems of inference which we do not want to go into now. However, we may all the same make one positive statement and that is that the trend certainly does not incorporate an increasing rate of growth. This means that the new strategy has at most succeeded to preserve the same trend as before the application of the strategy; if the rate of growth were constant before, it has continued to remain so after; if it were declining before, it has continued to decline afterwards in the same way.

However, even if the growth of agriculture for the country as a whole reveals a smooth trend so as to rule out any drastic change in

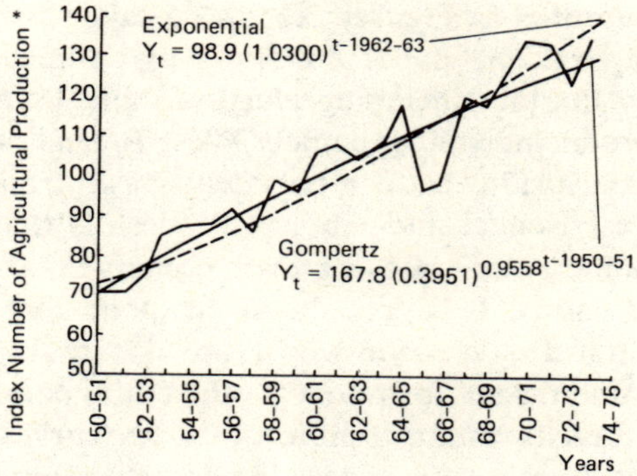

Figure 3.1 **Growth of Agriculture (all commodities)**

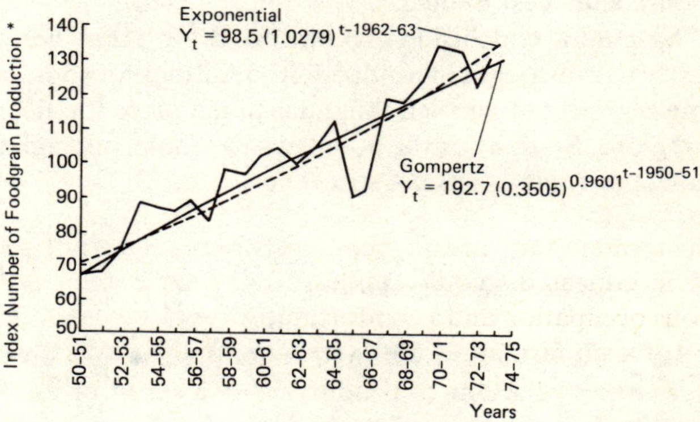

Figure 3.2 **Growth of Foodgrain**

the production conditions of agriculture in the country as a whole, a very different picture emerges if one looks at some of the different regions of the country. Trend analysis carried out on exactly the same lines as above yields interesting results as may be seen from the findings presented in Figures 3.3 to 3.6 taken from a highly illuminating paper.[9] The paper considers two wheat producing regions, namely Punjab (including Haryana) and Uttar Pradesh, and two rice producing states, namely West Bengal and Andhra Pradesh. The conclusion that can be safely drawn from the trend analysis for West Bengal and Andhra Pradesh is that there has neither been an increasing rate of growth nor any break in trend in these states for paddy. It is precisely on this point that results for Punjab and Uttar Pradesh for wheat are different. When one considers a common trend for wheat for the entire period 1950/51 to 1973/74 for these two states, neither of the two curves fit the data well over the entire time span, suggesting breaks in the trends. That it is indeed so is borne out by the fact that one gets very much better fits when one breaks the period into two at 1966/67 and fits semi-logarithmic curves separately to the two parts. It would, therefore, not be unjustified to conclude that there has been a revolutionary change in the production conditions of wheat in Uttar Pradesh and Punjab, but none whatsoever in the cases of paddy in Andhra and West Bengal.

The general conclusion that one can draw is that there has indeed been break-throughs in production conditions for some crops and in some regions but no such thing has taken place for all crops and in all regions. So that for the country as a whole, and agriculture as a whole, the strategy has not succeeded.

3.5 Impact on agrarian relations

In the present section we shall study the changes that are taking place in the relations of production as a result of the strategy. A large number of scholars have been arguing that this new strategy of agricultural development has been increasing the economic disparity among the agricultural population. They emphasise the increase in the inequality of income and wealth distribution – that is to say, the relative worsening of the economic conditions of the poorer farmers. Some go further and argue that there has been an increase in the poverty of the agricultural populations not only in

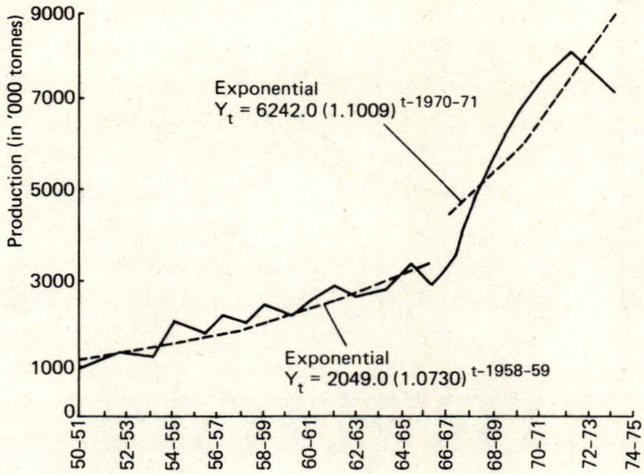

Figure 3.3 Growth of Wheat Production in Punjab and Haryana

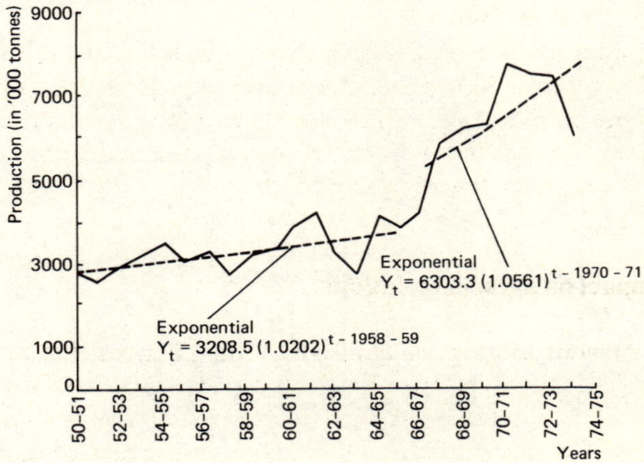

Figure 3.4 Growth of Wheat Production in UP

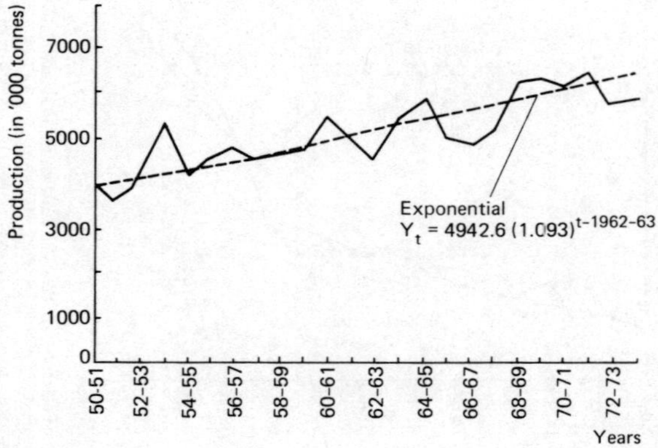

Figure 3.5 Growth of Rice Production in West Bengal

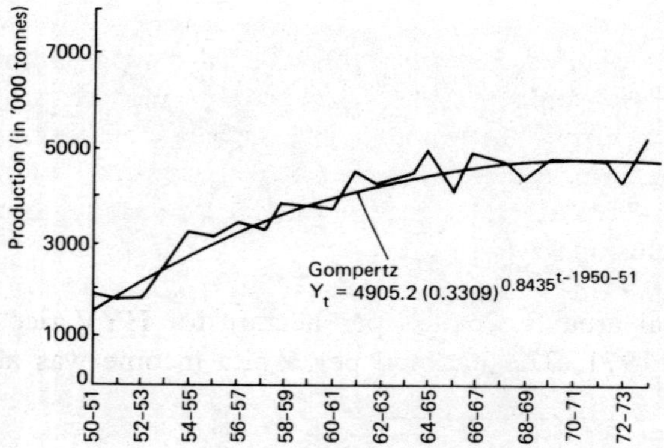

Figure 3.6 Growth of Production of Rice in Andhra Pradesh

relative, but also in absolute terms. Not many attempts have been made to establish the correctness or otherwise of this analysis. An exception is Pranab Bardhan who, in one paper has tried to demonstrate with the help of Lorenz curves that inequality of distribution has, indeed, increased in certain states.[10] He has also argued in a different paper that wage rates of agricultural workers have not kept pace with the rise in prices even in Punjab, so that there has been an absolute impoverishment of agricultural labourers. Bardhan's statistical exercises are not above criticism.[11] For example, it has been pointed out that the farm management data used by him for drawing the Lorenz curves are such as to invalidate the conclusions he draws. His contention on wage rates has also been challenged.[12]

This, however, is a matter on which one hardly needs sophisticated statistical exercises. That the rich farmer would be in a position to go in for the new agricultural technology and that the poor peasant would not be able to do so is a matter of simple commonsense and is accepted as a part of the strategy itself. It is naïve and ridiculous in this connection to talk of 'scale neutrality' of the new technology. What is meant by 'scale neutratilty' is that the responses to the divisible inputs like water, fertilisers, seeds, etc. are not found to be dependent on the size of plots. However, all these inputs call for economic resources and these resources are beyond the reach of the poorer farmers and, for reasons too well known to require to be emphasised, credit facilities are also not available to the poor farmers in the same degree as they are to the rich farmers.

That this resources constraint cannot but increase economic distance between the rich and the poor farmers has been illustrated by different students of the subject in terms of typical magnitudes. Thus, one conclusion shows

> For a typical area the inputs per hectare for HYV rice cost Rs.1125 in 1971. The national per capita income was about Rs.600 for the same year. In the area investigated, a small farmer who owned about two hectares could afford to spend only about Rs.350 per hectare on these inputs.[1]

Another calculation in a study carried out on a sample of traditional and modern farms (using HYV paddy and wheat), shows that the extent of acquisition of various items of fixed capital (excluding purchase of land) was on average Rs.407 per hectare for the traditional farms and Rs.1177 per hectare for the modern farms.[13] Ladejinski figured out that 'for one with seven acres must have a pair of bullocks worth Rs.2000, a tubewell demanding about Rs.6500 and inputs worth Rs.1500 or a total of Rs.10,000' and himself concluded: 'this is where neutrality to scale begins to break down'.[14] This apart, whether strictly called for or not by the new technology, mechanisation has started to play an increasingly important role in many regions; and machines being indivisible the advantages of large farm size and large plot size are obvious.

We may now turn to the question of the impact on employment. Much is made of the fact that the capital-intensive techniques of the strategy are supposed to be also at the same time labour intensive. The new technology, it is maintained, is characterised by frequent applications of water, fertiliser, insecticides and weeding, double cropping, bigger crops, large volumes of transportation, marketing and food processing. Hence, there is the temptation to think that if the impact of the new techniques on the owners of land is to increase the degree of inequality among them, this may be compensated by increased employment and increased income among the landless labourers and small farmers. Also one expects increased employment in industries associated with agriculture, in the secondary and tertiary occupations involving carpenters, masons, furniture makers, etc. However, the very same practices that are supposed to absorb more labour also involve rotary ploughs, barrows, and tillers to go with them, electrically-operated tubewells, sprayers and threshers, etc. which displace labour. The net effect on employment of the labour-intensive aspects of the new technology and the labour-displacing effects of some of the machinery (not all machinery, it is claimed, is labour displacing) is a matter of controversy. Some would discover a net labour-increasing effect and some would have it that only animal power is displaced, not human labour.

An interesting question in this connection is the change in forms of labour employment. In a study, based on data from a number of villages in Punjab, Kusum Chopra draws the conclusion that tractorisation leads to the increase in the employment of family labour as well as of casual labour, but decreases the employment of annual farm servants.[16] This, however, contradicts in some ways the findings of a survey carried out among the big farmers of Punjab in 1968.[17] Those findings were that tractorisation increases employment of farm servants at the cost of casual labour; and pumps and tubewells create demand for casual labour and replace permanent servants; and both permanent servants and casual labourers are substitutes for family labour.

We now turn to the impact of the strategy on the 'mode of production'. The new technology, by enchancing the profitability of cultivation, has made many a landowner who earlier leased out his land, or left it idle, into a commercial profit-maximiser. There has been large-scale eviction of tenants and resumption of cultivation by such owners. On the other hand, comparing the data of the Indian Censuses of 1961 and 1971, one finds a great increase in the proportion of agricultural labourers in the rural working force – from 15 per cent to 25 per cent.

A survey carried out in Punjab showed that between 1955/56 and 1967/68, farmers with between 100 and 150 acres increased their total land, mainly through purchase, by 38 per cent.[18] Ranadhawa presents a figure of a 23.8 per cent increase in size of farms above 24 hectares in the brief period of two years between 1967/68 and 1969/70.[19] Land acquisition through purchase is only one, somewhat restricted, means of increasing the size of holding. Alongside the eviction of poor tenants, there is the trend of large-sized farms leasing-in land from poor farmers. Ranadhawa provides some evidence about such leasing-in by bigger farmers during the years 1967–9. Kusum Chopra observed certain evidence of the big farmers indulging in consolidation of their fragmented holdings: the average size of fragments increased for all size groups of her sample of tractorised farmers.

Another matter of very great significance is the changes that are taking place in the tenancy arrangements. Widespread evidence shows that in the 'Green Revolution' areas in India, tenancy as such has not stood in the way of the adoption of the new technology. There has, of course, been in many places steep rises in rents or in owner shares. There have been in many areas changes in the mode of rent payment like share rent, fixed-cash rent, fixed-kind rent, etc.

A very important development is for the owners to share in the costs of cultivation so as to remove the disincentive of tenants to go in for expensive cultivation; also, to make advances, often interest-free, of seeds, fertilisers, etc. to enable the tenants to make use of these expensive inputs. The repayment of these advances in terms of crops, of course, means increases in the crop share of the owner. Along with this, there is a tendency on the part of the owners to become more and more the actual decision maker and the tenant to become more and more indistinguishable from a permanent farm servant. This provides an explanation for the observed similarity in the levels of inputs as well as yields between owners and tenant groups.[20]

An important associated question is what are the sections of the farmers who are at the forefront of implementing this strategy? A lot of evidence suggests that these dynamic farmers are owners of middle-sized farms and also are partly owners and partly tenants. In this connection, reference may be made to the study by Peter von Blanckenburg who makes the following important observations about the 'progressive' farmers in Punjab and Mysore who are found to be leading in the so-called 'new' agriculture:

> The real entrepreneur farmer, who is far ahead of the mass of other cultivators in rational resource allocation, in farm and labour organisation, in marketing practices, and in the adoption of agricultural innovations, is practically absent from our study.[21]

This somewhat agrees with the finding of a survey conducted by the author that there are farmers who are 'efficient' in that they excel in yield per acre of farm land and also make a good profit per acre of farm land.[22] But such 'efficient' farmers do not show, in comparison with the inefficient ones, any greater tendency to market their products, or employ hired labour against cash payment. That is, these efficient farmers do not reveal the features of full-fledged capitalist farmers. Also these farmers were only in a small proportion 'gentlemen farmers', i.e. drawn from educated urban, moneyed people.

Another question of interest is to what extent the adoption of the strategy has given rise to the emergence of a new class of 'gentlemen farmers'? If by 'gentlemen farmers' are meant people of urban background belonging to the educated middle classes taking up farming as a newly prestigious occupation and a comfortable way of living, then the importance of such farmers is much less than the impression one might get while touring the countryside. Given the strikingly different way of life of such people both from other cultivators and from other urban folk, they easily catch the eye. But in a statistical survey that the author conducted among the big farmers of Punjab the proportion of those with a non-farming background was found to be no more than eight per cent; and those with any education higher than the school level constituted only one per cent.[23] Hence the transformation in the Indian countryside cannot be ascribed in any significant measure to the role of 'gentlemen farmers'. The transformation is largely a phenomenon of erstwhile landlords or erstwhile rich peasants becoming enterprising farmers under the driving force of the profit motive encouraged by the strategy.

The fact that there has not been the emergence of any class of full-fledged capitalist farmers does not of course rule out the emergence of various capitalist features and trends in the farming sector. This leads us to the controversy that has been raging in the country on the 'mode of production in Indian agriculture' without there being any sign of any agreement being reached. On the one hand, there are those who point to the fact that the employment of wage labour is fast increasing and this, coupled with the accumulation of productive capital that is visibly taking place, is interpreted by them as unambiguously indicating capitalist development in agriculture.

There are serious opponents of this school who point out that labour in Indian agriculture is far from being a free seller of labour power; as such they deny that the existence or expansion of the wage labour force in Indian agriculture could be treated as an expression of capitalist relations. They further point to the continued prevalence of the institutions of sharecropping, usury, etc., and of the absence of any process of increasing concentration of landholdings in fewer and fewer hands in support of the view that the dominant mode of production is still pre-capitalist.

A third extreme view, which has unfortunately become fashionable in certain circles, has it that the prevailing mode is a 'semi-feudal' one characterised by big landowners being more interested in making investments in unproductive channels like usury and trade than in productive channels; not only that, they are opposed to investments being made by tenants and other poor farmers lest the rise in their income level makes them go out of their 'debt bondage'.

A fourth view has it that the mode of production in Indian agriculture is neither capitalist nor feudal; and among those who subscribe to this 'neither-capitalist-nor-feudal' view, some think it necessary to coin a new name like, say, the 'colonial mode of production'.

We on our part think that the controversy has been largely scholastic and futile. In our opinion the semi-feudal theory of landowners being opposed to productive investments does not deserve to be taken seriously as it flies in the face of all facts. The strategy under discussion is one primarily directed at promoting productive investments in agriculture and this could not have been adopted and implemented without the support of the landowning class. We further think that the capitalism versus feudalism issue is a non-issue.

We believe that criteria for feudalism and capitalism as used by Marx and Lenin have been applied in Indian conditions in a rather mechanical fashion. Thus, while we agree that labour in Indian agriculture is largely unfree, we are not prepared to go as far as those who use this fact to argue that on that account the production relations in agriculture are akin to feudal ones. The examples of restrictions on the labourers' freedom provided are mostly results of poverty leading them into relations of debt. But relations of debt and other relations of dependence of the poor on the rich are instances of economic coercion, whereas what characterises feudalism is *extra-economic coercion.*

Again we do not think that one should treat accumulation of land measured in surface areas as a necessary condition of capitalist relations. Capitalist relations get expressed in accumulation of capital and in the case of land it should be the value of land that should be expected to get concentrated under capitalist relations.

We believe that it is possible to discover different capitalist and feudal tendencies and traits in different combinations in the big landowners, but it is not possible to separate them into two classes – feudal and capitalist. Furthermore, among those landowners who reveal more capitalist tendencies and those who are more feudal in their comportment it is impossible to discover any conflicts of interests. As such it would be wrong in our opinion to think in terms of an emerging class of capitalist farmers in opposition to a declining class of feudal landlords.

The big landowners indulge in a variety of activities between which there are no conflicts of intersts. Thus, cultivation of land with the help of tenants, cultivation of land with the help of hired labourers, money-lending, trading – these activities are pursued in different combinations by different big landowners without thereby generating any conflicts among themselves. In our opinion these people constitute the dominating class in Indian agriculture who exploit the large masses of labourers, as well as poor owner and tenant farmers.

The effect of the strategy has been to make more and more members of this class resort more and more to productive investments. However, their unproductive investments continue to exist side by side. Also, given the vast supply of surplus labour the strategy has had no effect on making the unfree labourers any more free. Thus, the mode of production in Indian agriculture is characterised by the partially productive and partially unproductive reinvestment of surplus, generated by the exploitation of partially unfree sellers of labour power; and the effect of the strategy has been to aggravate further this very peculiarly mixed character of those who derive their economic and social power from the ownership of land; and to strengthen further this class in their exploitative relationship with the poor masses of labourers and peasants.

3.6 The sub-strategy for welfare

As has been mentioned before, the strategy for agricultural development which is being implemented by the government makes a conscious and explicit divorce between measures to promote agricultural production and measures to promote welfare of the broad masses dependent on agriculture. It is recognised that the growth-promoting measures necessarily increase the inequality of distribution; the sub-strategy is meant to compensate the majority of the agricultural population for the absolute or relative losses that they are to undergo because of the implementation of the growth strategy itself.

This recognition that welfare promotion and growth promotion cannot be integrated in to a common development strategy is based on the recognition of the following two hard facts:

1. The rate of development of the economy as a whole and the industries in particular are far from being adequate to absorb the growing surplus labour in agriculture; as a matter of fact it is inadequate even to absorb the growing labour force in the urban sector itself.

2. The land to man ratio in the country is such that there is no possibility whatsoever of distributing land among the landless labourers in any economically non-negligible amounts even if one were to impose and implement draconic ceilings.

This latter point is highlighted by Minhas who calculated that if all the land above a ceiling of 20 acres per holding were taken away and distributed among households operating fewer than five acres of land, it should be possible to have a uniform distribution of holdings of size 0.54 acres per person among this population of households.[24] To achieve this one would have to leave completely untouched the growing volume of the totally landless who accounted for about 26 per cent of the population in year 1969/70. In other words, even if the totally landless are not to be given any benefit at all, the land that could be obtained by imposing ceilings on the big-sized holdings is hardly adequate to convert the poorest landed peasants into cultivators of minimum-sized viable farms.

As the development of industries is not regarded as capable of absorbing the surplus among the landless peasants and there is no land to convert these landless peasants into landed ones, one is left with no other option, within the system of private property in land, than either to ignore the plight of the impoverished masses or to consider measures for them which are of the nature of relief and dole. It is understandable that no government can afford to declare officially the adoption of the first. Therefore, the government has chosen the second, namely relief and dole, which constitutes the welfare sub-strategy.

However, the dimensions of the relief and dole operations that have been incorporated in the fourth and fifth plans are so meagre in relation to the requirements that the chosen option is not really different from that of a relative neglect of the rural poor and unemployed. This should be clear from the following figures. Dandekar and Rath calculated in 1970 that if all the rural poor and unemployed (leaving out the poorest 10 per cent of the rural population) were to be brought up to a certain minimum level of per capita consumption, an additional income of Rs.822 crores per year at 1968/69 prices must be distrbuted to them.[25] These would call for works programmes of the dimension of about Rs.1000 crores per year in 1968/69 prices. This would mean a budget of Rs.5000 crores for rural works programmes alone.

The inability of the government to implement a programme of this order can be well appreciated with the help of the following two comparisons. The total outlay on agriculture in the fourth five-year plan was Rs.3466 crores. In other words, if the rural works programme to meet the minimum consumption needs of the rural poor was to be made a part of the fourth five-year plan (1969–74), it would have called for resources of 50 per cent more than the total amount that was allocated to agriculture and irrigation. The other comparison is with the amount that was actually allocated in the fourth plan to take care of these minimum needs, as also the similar allocations in the fifth plan. For example, the amount allocated to a scheme called Crash Scheme for Rural Employment during the fourth plan was Rs.50 crores. According to the fifth plan even this small amount (compared to the required Rs.1000 crores) could not be spent and even the bit of money that was spent was not fruitful in giving rise to the targeted amount of employment. Thus, the fifth plan writes 'the bulk of the expenditure, about 80 per cent, was incurred on communications'. Whatever might be meant by communications in this connection, it can hardly cover assets of a durable nature, in consonance with a local development plan which was the idea behind the scheme. This scheme has wisely been dropped in the fifth plan.

The official plans, the fourth and the fifth, put forward some other schemes described as constituting a 'special programme' for meeting minimum needs, i.e. the needs of the rural poor and unemployed. These schemes are grouped under the Small Farmers' Development Agency (SFDA), Marginal Farmers and Agricultural Labour (MFAL), Drought-Prone Areas Programme (DPAP), etc. Small farmers are identified as those with one and three hectares of land; marginal farmers are those whose holdings are even smaller. For the first two projects, namely SFDA and MFAL, the allocation in the fourth plan was only Rs.115 crores and the estimated expenditure was only Rs.49 crores. The fifth plan itself criticises these schemes for being based on 'inadequate project reports, unrealistic targets and above all lack of proper synchronisation of normal state plan activities and activities of these special projects'. As to the DPAP (for which the allocation in the fourth plan was Rs.111 crores and of which Rs.74 crores was spent), it was criticised in the fifth plan for tending to 'lack an appropriate order of priorities' and 'a sizeable deviation from the norms and yardsticks' of the schemes. According to the fifth plan, 'a large number of schemes are likely to be left incomplete by the end of the fourth plan'. Despite this poor record, the fifth plan allocates Rs.500 crores for this heterogenous bundle of schemes.

We may now attempt an assessment of what the substrategy for welfare amounts to. Even if the Rs.500 crores is going to be spent in such a fashion as to create productive assets and at the same time generate income to the tune of Rs.500 crores, it would meet less than 10 per cent of the minimum needs. We may remember that the minimum needs would call for Rs.5000 crores in the prices of

1968/69 and for the population of about the same date. The prices during the fifth plan would be substantially higher than those of 1968/69; so would be the population below the minimum-needs level. However, even this arithmetic of Rs.500 crores at certain prices against Rs.5000 crores at certain higher prices is unimportant to the real issue. As we have seen, the fifth plan itself assesses the schemes rather poorly. The political economic factors which make such schemes ineffective are not gone into by the plan and there is no likelihood of there being any change in them during the plan. While official sources do not show even rough estimates of how much money spent on these schemes went to meet the minimum needs of the rural poor, it is quite clear from the available accounts that much of the funds have gone to those who are much above the minimum needs level.

We have discussed in the above paragraphs schemes that have been proposed and somewhat tried out. However, there has been a view, correct in our opinion, held by many economists that such schemes can never possibly cope with the vast problem of rural poverty and unemployment. Thus, in the words of an economist who himself was for some time a member of the Planning Commission,

> setting up of separate agencies for tackling each of the many problems that small farmers face – the recent move towards a small farmer's agency may be taken as one example – is likely to make little dent on the total problem. We cannot pull vast millions out of abject poverty by passing out doles; there are far too many claimants and the cake is small. A make-work programme of rural works will have little lasting effects unless it is integrated into a comprehensive view of rural resource development.[26]

The same author correctly observes about the programme that 'generally, they end up doing nothing better than "digging holes and filling them up"'.

There have thus been thoughts amongst economists and policy makers about various ways in which the problems of the poor farmers and landless labourers can be approached in a 'comprehensive, integrated' fashion. Various proposals and schemes that have been put forward by various people have the following core of essentially sound ideas:

1. There are strict limits to the extent the productive qualities of land may be increased by actions taken on individual holdings, given the extremely small size of most such holdings and the even smaller sizes of most of the fragments in which the holdings are divided. There are many land improving actions which require to be carried out on the scale of a village taken as a whole. Thus, effective water management – whether with the help of a feeder network of canals, or by the sharing of small-scale irrigation water – flood control, contour-building, prevention of erosion, etc. can only be effectively carried out by planning at a village level for all the land taken together. In other words, development of land, taking into account only the constraints of geographical factors, calls for an action programme which involves a scale of operation much larger than that of individual holdings.
2. Such an action programme on a village scale can operate very largely with locally available resources and can absorb very large volumes of labour as well as result in high increases in production and thus high increases in income.

However, the rock on which all such schemes founder, either immediately after attempted implementation, or even before it, is the end left open in them, namely that of how to achieve co-operation among all the parties concerned, starting from the few big landholders to the many very small ones, not to speak of the numerous totally landless families. The problem arises from the fact that the different holders of land would benefit from village scale operations differently. Roughly speaking, the permanent benefits would be proportional to land owned; which means no permanent benefit to the landless, though the latter may temporarily benefit from increased employment. However, most of the schemes that have been put forward by different people take it as axiomatic that no kind of curtailment of private property rights in land are to be considered.

We present below one or two examples of such schemes by way of illustrations. Minhas put forward a scheme which he called 'an integrated programme for compulsory consolidation of landholdings and complementary public works'.[27] The principal idea was compulsory land consolidation and development. All the land holdings of a village are to be taken over by an agency, thoroughly surveyed, all possible improvements carried out on them and then divided into large homogenous fragments to be redistributed among the landholders so that no holding would consist of more than two fragments.

Minhas did not calculate the funds that would be required for the country as a whole nor the increased production and income that would follow from such a programme. However, he suggested that at least a part of the required funds could be collected locally in the form of consolidation fees with 'a considerable element of progression'. Minhas provided reasons for the landowners not agreeing to act co-operatively on their own in terms of such logic of contradiction between individual interest and collective welfare as illustrated by the 'prisoners dilemma' and argued in favour of coercive laws. However, he explicitly ruled out collective ownership of land. His preferred model is that of 'an *n*-person non-co-operative game-theoretic set up'.

While the above is an illustration of a scheme thought out by an individual economist, an example of a scheme which is backed by a state government is the Comprehensive Area Development Project of West Bengal. This scheme in its original configuration visualised an operational unit as being a compact block of 10,000 acres. Each such project would be under a Project Development Board which is supposed to receive co-operation from the different development agencies of the state government as well as the banks and is supposed to have statutory powers to plan and to implement all development measures such as installations of pumps and tubewells, digging of distribution channels, linking up of credit with marketing on the basis of hypothecation of crops, taking over of land of absentee landowners for settling them on the local poor peasants, etc.

The scheme incorporated a most original idea for making the project self-financing. It involves two measures: the first makes it obligatory that all sales should take place to the project board. The second measure is to make it obligatory for the board to supply all the input and credit needs of all farmers in the project area. These two measures logically rule out usury (because credit is supplied by the project) and speculative trade (because marketing is taken care of by the projects). Thus, at one stroke, the immemorial vices of usury, speculation and other forms of unproductive investment are to be liquidated. Not only that, the scheme proposes that a cordon should be imposed on the institutional savings of the area. Banks and other savings agencies would be under the obligation of investing in the project area the forced savings of the rich farmers (forced, because of the elimination of non-productive channels) which would be augmented because of the development carried out by the project.

The question that neither Minhas nor the people behind the West Bengal scheme just described face, is who is to bell the cat? Of course, all the needful things like imposing progressive levies, compulsory consolidations, compulsory confiscation of unused land for the settling of landless people, etc. can be done by an authority given statutory powers to do so. The statement cannot but be 100 per cent true, being nothing but a tautology. The underlying political philosophy is that the state and its various organs and other statutory bodies are above the influence of the rich and the powerful. This philosophy has inspired countless legislations and the vesting of various powers on various authorities for the implementation of various institutional reforms. However, the history of implementation and non-implementation of all these measures reveals a clear pattern when account is taken of the influence on the government exerted by different interest groups.

The government of India and the state governments have passed various legislations which, if implemented, would adversely affect the interests of the dominant groups in rural areas. It is easy to show that few measures of this kind have been effectively implemented to date. Examples are tenancy laws, ceilings on land, taxation on agricultural income – all of which have been legislated, but none, by the admission of the government itself, implemented. One merely has to be a hard-headed realist, and not a doctrinaire orthodox, to recognise facts for what they are and not to expect any miraculous exception in the case of the rural development schemes. Whatever be the statutory powers given to any local authority the latter would necessarily be influenced by the interests of the dominant local groups. As such, these authorities would find it difficult to take any steps which diverge from the interests of the rich landowners.

3.7 The emerging structure

The strategy that is being consistently and systematically applied by the government is resulting in deep changes in the agrarian structure of the country and leading it to a common pattern. This common pattern has not yet clearly emerged and the speed of the change is very uneven. All the same, its broad lines can be clearly delineated.

This emergent structure would be characterised by millions of farms as now, the overwhelming majority of the farms being of miniscule, small or medium size, an extremely small proportion of the farms being somewhat big – the biggest farms being nowhere near the big farms of the developed countries. The various land legislations that have taken place would prevent any large-scale transfer of land from the small farmers to the big farmers and prevent the emergence of really giant-size farms to any significant scale: they would remain exceptions. On the other hand, none of the half-hearted measures that have been taken to help the small farmers would lead to any reverse transfer of land from big farms to small farms or the formation of bigger-sized holdings from an amalgamation of the miniscule holdings. Hence, land distribution would remain highly concentrated but more or less as now; it would not become very much more concentrated.

There would, however, be a very largely increased concentration of assets and income among the bigger farmers, as a result of the heavy investments carried out by them.

Private property in land would remain universal. Land would not come under any kind of social ownership through any form of collective or co-operative ventures.

Tenancy would undergo serious modifications. Small tenant farmers would either be evicted or they would work under the owner's supervision and decision taking and be reduced to a status not very different from that of farm servants. Simultaneously, the other kind of tenants, the big-sized enterprising rich tenants, would gain in importance. Big and medium-sized farms would find it profitable to lease-in from poorer farmers on a fixed rent basis; leasing-in land by this class of cultivators from other medium- and big-sized farms and cultivation on crop-sharing cum cost-sharing basis also promises to become an important feature of this structure.

The bigger and medium-sized farmers would more and more generally and widely practise the intensive, modern technology-based cultivation as is being done by a small section of this class of farmers now. The geographical concentration, the fragmentary, scattered character which marks the phenomenon of the so-called new agriculture could be expected to be less marked. The rate of growth of output in those areas where the 'Green Revolution' has been most advanced could well decelerate but nonetheless the

'Green Revolution' could spread to hitherto less affected areas. It would only require some research breakthroughs in paddy for this to happen. The fact that the 'new agriculture' has as yet shown less important results until now for paddy than it has for wheat is a matter of technological accident, not explainable by any political economic analysis. The political and economic environment in many paddy-growing areas is no less favourable to technological advances in paddy seeds than have been some of the wheat-growing areas for the Mexican wheat.

The emergent agrarian structure would thus have a marked dualistic character. It would consist of a prospering, production-oriented, profit-oriented, technology-oriented thin crust of medium and big farms on a great mass of small and minuscule farms that would lag far behind in matters of production, profit, technology, etc. The thin crust of big- and medium-sized farms would operate increasingly on the lines of expanded production with the generation and reinvestment of surplus, but in respect of the employment of labour, they would retain feudal features; labour would remain a non-free commodity as a result of the vast surplus labour that would abound in the countryside. The vast mass of small farms would crawl along on pre-capitalist lines of simple reproduction.

The thin crust of big and medium farms would continue to swallow up the greater part of the resources that would be allocated by the state to the agricultural sector. However, to the extent the growth-oriented crust of medium and big farms would generate surplus within themselves, to that extent their dependence on surplus generated outside agriculture would diminish in importance. All the same, the structure, as it is emerging, would not play the historic role that agriculture has played in the process of industrialisation of the developed capitalist countries and the developed socialist countries; namely supplying industries with surplus generated in agriculture.

An important question is what kind of growth of agriculture as a whole can be expected to take place on the basis of such a structure? It would appear that the same mediocre, possibly declining, rate of growth that has characterised Indian agriculture as a whole during the entire post-independence period (which was examined in Section 3.4) could well continue into the future for some more years. It should be noted that even if the trend rate be a declining one, or a constant but a mediocre one, it is of great significance as long as it is neither negative nor zero. In other words, it would not be correct to say that agriculture in India has been marked by stagnation or decline. What the implications of such a non-zero but mediocre rate of growth for the prospects of industrial growth are is a matter that is outside the purview of the present study. However, the past trend of growth of agriculture was insufficient to pull the greater part of the population above the poverty line or to prevent unemployment taking vaster and vaster dimensions. It would be safe to say that the same would continue to hold for the future.

Appendix to Chapter 3

The Gompertz curve has the versatile property that depending on the range of values of the parameters it can describe both increasing and diminishing rates of growth.

When considering India as a whole, the Gompertz with a diminishing rate of growth gives reasonably good fits for agricultural production as well as for food-grains production, but the fits given by the semi-logarithmic function indicating constant rates of growth are also equally good.

Among the state data the Gompertz with a diminishing rate of growth gives a better fit than the semi-logarithmic for Andhra Pradesh, whereas for West Bengal the semi-logarithmic with a constant rate of growth of 1.9 per cent per year gives a better fit. (Figures 4.5 and 4.6)

For Punjab and Uttar Pradesh, however, careful trend analysis shows that one needs to fit two trend curves with a break at 1966/67. Prior to that year agricultural production was growing at the rate of 7.3 per cent per year in Punjab (and Haryana) and at the rate of 2 per cent per year in Uttar Pradesh. In the period following that year the growth rates were 10.1 per cent and 5.6 per cent per year respectively.

It may be mentioned that for Punjab (and Haryana) and Uttar Pradesh the fits involving two separate semi-logarithmic curves were tested against a single semi-logarithmic curve fit to the entire data using the formula for testing the homogeneity of two regression models. The test results were statistically significant.

Table 3.1 Annual Rates of Increase of Prices Received by Agriculture and Prices Paid by Agriculture (based on fitted trend curves)

	Per cent rate of growth per year (1951/52 to 1965/66)
Prices received by agriculture	
1. Prices of agricultural products purchased by non-agriculture for intermediate consumption	2.92
2. Prices of agricultural products purchased by non-agriculture for final consumption	3.23
3. Prices of agricultural products purchased by non-agriculture for all uses	3.14
Prices paid by agriculture	
4. Prices of non-agricultural products purchased by agriculture for intermediate consumption	3.01
5. Prices of non-agricultural products purchased by agriculture for final consumption	2.57
Terms of trade	
6. Prices of non-agricultural products purchased by agriculture for all uses	2.62
7. Net barter terms of trade: intermediate products	0.025
8. Net barter terms of trade: final products	0.65
9. Net barter terms of trade: all products	0.51
10. Income terms of trade: all products	3.40
Marketing and production	
11. Marketed surplus of agriculture (at 1960/61 prices)	2.90
12. Agricultural production	2.74

Source R. Thamarajakshi, 'Intersectoral Terms of Trade and Marketed Surplus of Agricultural Produce, 1951–52 to 1965–66', *Economic and Political Weekly*, June 1969.

Notes

1. Ralph W. Cummings and S. K. Ray, 'The New Agricultural Strategy', *Economic and Political Weekly*, 29 March 1969.
2. Norman E. Borlaug and Oddvar H. Aresvik, 'The Green Revolution – An Approach to Agricultural Development and Some of Its Economic Implications' in *International Journal of Agrarian Affairs*, Vol. V, No. 6, March 1973.

3. Wolf Ladejinski, 'How Green is the Indian Green Revolution', *Economic and Political Weekly*, 29 December 1973.
4. B. S. Minhas and T. N. Srinivasan, 'New Agricultural Production Strategy: Some Policy Issues' in A. M. Khusro, *Readings in Agricultural Development* (Bombay: Allied Publishers, 1968).
5. R. Thamarajakshi, 'Intersectoral Terms of Trade and Marketed Surplus of Agricultural Produce, 1951–52 to 1965–66', *Economic and Political Weekly*, June 1969. The table in the appendix has been taken from this carefully conducted study. The study is by now somewhat dated. But there is no doubt that during the period following the one covered by the study the same trend has continued, if not further aggravated.
6. S. R. Sen, *The Strategy for Agricultural Development, and Other Essays on Economic Policy and Planning* (2nd edn.), (London: Asia Publishing House, 1966).
7. John W. Mellor, 'The Evolution of Rural Development Policy' in J. W. Mellor, T. F. Weaver, U. J. Lele and S. R. Simon, *Developing Rural India – Plan and Practice* (Ithaca: Cornell University Press, 1968).
8. The results are taken from Amal Krishna Dey, 'Rates of Growth of Agriculture and Industry', *Economic and Political Weekly*, 28 June 1975, and 'Green Revolution Contrasts: Rice and Wheat', op cit, 4 June 1977. We thank the author for letting us have access to these results.
9. Amal Krishna Dey, 'Green Revolution Contrasts: Rice and Wheat', *Economic and Political Weekly*, 4 June 1977.
10. Pranab Bardhan, 'Inequality of Farm Incomes: A Study of Four Districts', *Economic and Political Weekly*, Annual Number, 1974.
11. Pranab Bardhan, 'Green Revolution and Agriculture Labourers', op cit. Special Number, July 1970.
12. Deepak Lal, 'Agricultural Growth, Real Wages, and the Rural Poor in India', op cit, 26 June 1976.
13. Bhagavan, Mr. et al, *The Death of the Green Revolution* (North London Haslemere Group, 1973).
14. Wolf Ladejinski, 'Green Revolution in Bihar', *Economic and Political Weekly* 27 September 1969.
15. Bhagavan, *The Death of the Green Revolution.*
16. Kusum Chopra, 'Tractorisation and Changes in Factor Inputs in Punjab', *Economic and Political Weekly*, 28 December 1974.
17. Ashok Rudra, 'Employment Patterns in Large Farms of India', *Economic and Political Weekly*, 26 June 1971.
18. Ashok Rudra, A. Majid, B. D. Talib, 'Big Farmers of Punjab', *Economic and Political Weekly*, 27 September 1969.
19. M. S. Ranadhawa, *Green Revolution* (New York: John Wiley, 1974).
20. See Aparajita Chakravarty and Ashok Rudra, 'Economic Effects of Tenancy – Some Negative Results', *Economic and Political Weekly*, 14 July 1973, and Harendranath Dwivedi and Ashok Rudra, 'Economic Effects of Tenancy – Some Further Negative Results', op cit, 21 July 1973.
21. Peter von Blanckenburg, 'Who Leads Agricultural Modernisation', *Economic and Political Weekly*, 30 September 1972.
22. Ashok Rudra, 'In Search of the Capitalist Farmer', *Economic and Political Weekly*, Review of Agriculture, June 1970.

23. Ibid.
24. B. S. Minhas, 'Rural Poverty, Land Redistribution and Development Strategy: Facts and Policy' in *Planning and the Poor* (New Delhi: S. Chand, 1974).
25. V. M. Dandekar and Nilkantha Rath, *Poverty in India* (Ford Foundation, 1970).
26. B. S. Minhas, op cit.
27. Ibid.

4 The Comilla Model and the Integrated Rural Development Programme of Bangladesh An Experiment in 'Co-operative Capitalism'[1]

Azizur Rahman Khan

4.1 Introduction

The inclusion of Bangladesh in a set of studies of agrarian organisations that constitute alternatives to the usual story of growth (or, worse, stagnation) with inequality needs some justification. In recent decades the rural economy of Bangladesh has not only failed to grow in terms of per capita output but also experienced remarkable increases in inequality and extreme poverty.[2] The agrarian system and the agricultural organisation that have produced such results are hardly the appropriate candidates for a study aimed at identifying the alternatives that could produce equitable and self-reliant growth.

While the global picture for the country has been dismal, it has been widely claimed that particular organisational experiments carried out within Bangladesh constitute solutions to the agrarian problems not only of the country itself but of similar countries in South and South-east Asia. A great deal has been claimed for the success of the Comilla type co-operatives in Bangladesh in promoting growth with equity and self-reliance. Some observers of the experiment have gone so far as to suggest that it represents a viable alternative to the two extreme solutions of the agrarian problem, namely socialist collectivisation and private capitalist farming. To quote from a recent publication:

> Currently in the developing nations, a number of carefully worked out approaches and experiments are under way, attempt-

ing to accelerate the economic and social transformation of rural society without the mistakes of attempting to directly transplant foreign models. The rural development programmes developed at the Pakistan (now Bangladesh) Academy for Rural Development are one set of these experiments.[3]

The Comilla co-operative system was started in Comilla Kotwali Thana in July 1961 as an experimental project of the Pakistan (since the independence of Bangladesh in December 1971 renamed Bangladesh) Academy for Rural Development (for which we shall use the acronym BARD even for the period before December 1971).[4] To this day the co-operatives in Comilla Thana remain a laboratory for the BARD which concentrates its action-research activities in this area.

This system was replicated elsewhere in Bangladesh in a number of steps. During 1963/64 it was adopted in three thanas far away from the Comilla District. Later in the 1960s, in two steps, it was extended to the remaining twenty thanas of the Comilla District. In 1971 the Comilla approach was adopted as the model for the entire country under the title of the Integrated Rural Development Programme (IRDP). Following the independence of Bangladesh the IRDP has been adopted rapidly over a wide area of the country. By 1976/77 the programme had spread to 200 of the 434 thanas of Bangladesh.

In this paper an attempt will be made to examine the contribution of this system of co-operatives in solving the problems of underdevelopment and poverty in rural Bangladesh. Since the system has been tried most intensively and over a long period in the Comilla Kotwali Thana, the primary emphasis of the paper will be on an evaluation of the performance of that programme. Later on some comments will be made on the IRDP programme.

4.2 The Comilla co-operative experiment

(a) The background

The basic organisation of the Comilla co-operatives consists of a two-tier structure: the village-based primary co-operatives called the KSS ('Krishi Samabaya Samiti' or Agricultural Co-operative Association) and their federation at the thana level called the ACF, Agricultural Co-operatives' Federation.[5] It was recognised that a village, consisting of about 200 households, should be the basic unit of co-operation if reasonable decentralisation and community interest are to be promoted. But the efficient operation of activities like credit, mechanisation, training and extension services, all of which were combined within the same co-operative structure, would require a much bigger scale of operation. These activities were concentrated at the thana level.

The basic philosophy behind the approach was a commitment to individual farming based on the private ownership and operation of land and other means of production. As Akhter Hamid Khan, the man who conceived and led the programme during its first decade of development, put it in a recent appraisal: 'There was no intention of putting an end to private possession. On the other hand we cherished the qualities of family farming. We admired the Japanese family more than the dispossessed commune workers.'[6]

The programme took the existing distribution of land ownership as given and made no attempt to bring about any change in it. Khan speaks of the recognition on the part of the organisers of the experiment that 'harmony had not been established among the three main classes left in the villages: the large proprietors, the peasant proprietors and the landless labourers'.[7] The basic objective of the co-operative programme, according to him, was

> to organise the peasant proprietors for production as well as protection. We had found that the peasants formed a 70 per cent majority. They also owned 70 per cent of the land and leased a good portion of the remaining 30 per cent from the large proprietors. The peasant producers, therefore, were the real agriculturalists.[8]

The large proprietors, about 10 per cent of the farming households according to the rough and illustrative estimate by Khan, were too powerful to be excluded. 'Initially we worked quietly around them, suggesting not that they should be excluded from the new co-operatives, but that they should not be allowed to dominate . . . .'[9] To the landless labourers, somewhat overestimated by him as being of the order of 20 per cent of the farm families, the co-operative programme had no clear role to offer. It had four main components:

1. The two-tier co-operative organisation served as a framework for the distribution of credit and other inputs and for the

mobilisation of savings. Beyond organising individual farmers for the absorption of institutional credit and modern practices the programme did not incorporate any pooling of individually owned inputs for co-operation in production activities.

2. A Thana Training and Development Centre (TTDC) was located at the higher tier to provide training to the leading officials of the KSS, the primary co-operatives, in advanced methods of farming.

3. A parallel irrigation programme, for which the capital cost was met by borrowing from the state, was also offered. Tubewells and low-lift pumps were made available to the individual KSS by the KTCCA through the ACF.

4. Another parallel activity was developed in the form of the Rural Works Programme which had the objective of utilising the idle manpower in the construction of infrastructural capital. Funds for this programme were provided by the Government.

To summarise, the vision of those who conceived and launched the Comilla experiment was one of protecting and developing the small and middle peasants who owned and rented land. The programme recognised that the bigger landowners could not be excluded from its benefits which consisted of credit and material inputs supplied at prices lower than would have to be paid otherwise. The originators of the programme recognised that they 'were the servants of a conservative government whose political, economic or administrative orientation we could scarcely change'.[10] There was nothing that the programme offered directly to those who owned no land without which the inputs that were made available were of little use. It was hoped that the increased demand for labour due to higher growth, the greater labour requirement of the high yielding varieties and the works programme would indirectly benefit the landless by raising wages. At the same time the perceptive leader of the programme recognised that

> better drainage, link roads and irrigation substantially enhanced the value of land and its rent. The unearned increment of the landowners was a hundred times more than the wages earned by the labourers. Even an elementary student of economics should

know that it would be so, as long as ownership is not transformed.[11]

(b) Increased production

Performance in terms of output can be judged reasonably accurately by examining the trends of production and yield per acre of rice which covers well over 90 per cent of cropped area.[12] Table 4.1 shows that output and yield of rice has increased rapidly in the decade since the Comilla programme got under way. The annual average rate of growth over the decade ending in 1973/74 was nearly 8 per cent, well over twice the rate of growth for the country as a whole. By 1973/74 the yield per acre was 1.17 metric tonnes, i.e. 144 per cent above the average of 0.48 tons for the country as a whole. Even in the early 1960s the yield of rice was higher in Comilla than the average for the country. But the difference was of the order of 20 per cent. In 1973/74 the yield per acre in Comilla Thana was nearly three quarters of that in Japan and of all the Asian

Table 4.1 Acreage, Yield and Production of Rice in Comilla Kotwali Thana in Selected Years

Year	*Cropped Area in Thousand Acres*	*Average Yield in Maunds per Acre*	*Total Production in Thousand Maunds*
1963/64	68.5	14.70	1007
1966/67	82.1	15.68	1287
1969/70	76.5	23.31	1783
1973/74	67.5	31.30	2113

Note For 1963/64 and 1966/67, acreage figures are from the *Annual Report* of the BARD, 1967 (p. 55), and average annual yield from the *Annual Report*, 1968/69 (p. 34). For 1969/70 and 1973/74 the estimates have been built up from the acreage and yield figures for the three seasonal rice crops, *Aus, Amon,* and *Boro,* shown in the *Annual Report*, 1973/74 (pp. 35–6). It is interesting to note that rice cropping intensity increased until 1966/67 and fell thereafter. This phenomenon has not been explained in the Reports cited above. One possible explanation is that the high yielding varieties expanded rapidly since the mid-1960s onwards. These varieties have a longer gestation lag. To adopt them it might have been necessary to reduce the area under the traditional variety of the following crop. It is also possible that the greater yield of rice permitted the diversification in favour of non-rice crops. The three seasonal rice crops are *Aus* (which is harvested in the mid-monsoon season), *Amon* (which is harvested in the late autumn) and *Boro* (which is harvested in the spring or early summer). A *maund* is 82.29 pounds or 37.33 kg.

countries below the national average of only Japan, Korea and Taiwan.[13]

It has frequently been claimed that the co-operatives led the way in bringing about this remarkable transformation. The evidence that is cited is the higher *level* of yield for the members of the co-operatives as compared to that of the non-members. It is usually conceded that the non-members also experienced rapid growth though the credit for their performance is attributed to the leading role of the members who are supposed to have shown the way.

The relevant data for the period up to 1973/74 are shown in Table 4.2. Several aspects of the evidence stand out clearly. First, the members had a higher *initial* yield per acre than non-members. Secondly, over the eleven years the geometric difference between the yield for the two groups, despite a good deal of fluctuation from one year to another, has been reduced to some extent. Thirdly, for the *Amon* crop (which, though declining in importance, is still the biggest of the three crops accounting for 48 per cent of the total in 1973/74) there is no significant trend in yield over time either for the members or for the non-members. Fourthly, for the *Aus* and *Boro* crops (both of which have been major users of irrigation and high yielding varieties of seed and technology) the trend rates of growth in yield are highly significantly positive for both the groups, and indeed, higher for the non-members than for the members of the co-operative societies.[14] It is also noteworthy that the variance of the estimated trend is lower (not greater) for the non-members than for the members in the *Aus* (*Boro*) season.

On the basis of this evidence it is difficult to justify the argument that the members of the co-operatives have been the leading agents of productivity growth.[15] If higher initial yield is the sole evidence for the claim that there is a positive association between productivity and membership, then it would appear more reasonable to reverse the direction of causation. As we shall show later, the co-operatives attracted the farmers with significantly higher than average holdings. It is likely that they also had the greater capacity to invest in productive inputs and thereby had already attained a higher yield. It is also probable that co-operative facilities (e.g. tubewells and pumps) were located near the areas of higher productivity (better land) which meant that the farmers with higher yield joined before the rest.

Since growth was common to both the members and the non-members of the co-operatives (and, perhaps, higher for the latter),

Table 4.2 Yield (in *maunds*) per Acre of Rice of Members and Non-members of Co-operatives in Comilla Thana

	Aus			*Amon*			*Boro*		
Year	*Member*	*Non-member*	*Ratio of Members to Non-members*	*Member*	*Non-member*	*Ratio of Members to Non-members*	*Member*	*Non-member*	*Ratio of Members to Non-members*
1963/64	17.60	13.70	1.28	27.70	22.20	1.25	—	—	—
1964/65	25.90	21.20	1.22	25.70	23.10	1.11	21.90	18.71	1.17
1965/66	25.30	20.00	1.27	22.10	19.80	1.12	18.10	11.80	1.53
1966/67	20.50	14.70	1.39	22.10	19.70	1.12	25.25	15.53	1.63
1967/68	16.16	14.62	1.11	29.60	25.30	1.17	40.00	29.35	1.36
1968/69	20.66	17.11	1.21	21.18	17.87	1.19	40.37	31.94	1.26
1969/70	20.60	17.10	1.20	22.20	18.90	1.17	47.50	44.00	1.08
1970/71	23.50	19.40	1.21	26.50	23.90	1.11	—	—	—
1971/72	25.16	25.07	1.00	30.70	23.90	1.28	40.00	36.30	1.10
1972/73	27.83	22.76	1.22	18.48	14.08	1.31	47.59	38.12	1.25
1973/74	33.43	32.20	1.04	28.87	27.24	1.06	41.30	35.94	1.15

Source Crop cutting Survey Reports of BARD quoted in the various *Annual Reports*.

one must look for an explanation in factors common to both groups (and, additionally, determine why the effect of such factors was somewhat lower in the case of the members). Such factors must be found in the extremely rapid expansion in the supply of inputs associated with the high yielding variety (HYV) of rice at a price that reflected very high subsidy by the government. With the setting up of the Pakistan Academy of Rural Development in Comilla and the emergence of a powerful lobby in support of channelling resources into the rural areas, Comilla Thana was able to attract a disproportionate amount of tubewells, pumps, tractors and other inputs to which both the members and the non-members had access. Indeed, the irrigation and related programmes elsewhere in Bangladesh had been going on without the framework of the Comilla type of co-operatives. The eagerness of the government and of the dynamic leadership of the movement to turn Comilla into a show-piece resulted in an abundant supply of the under priced inputs in the command area of the experiment. It is in this abundance of crucial inputs, rather than in any intrinsic ability of the co-operatives to mobilise the latent resources of the members, that one must seek an explanation for the remarkably rapid growth in production.

Table 4.3 shows the expansion of irrigation in Comilla Thana. In the early 1960s a negligible area was irrigated. For example, in 1962/63 only 36 acres (out of a total of 50,500) received irrigation water. This represented a level lower than the dismally low average prevailing in Bangladesh in those days. From the middle of the 1960s irrigation in Comilla Thana began to expand rapidly. By 1973/74 about a third of the cultivated area was under irrigation.

The availability of irrigation had a remarkable effect on the pattern and intensity of cropping. The availability of water during the dry winter made it possible to increase cropping intensity. But, more importantly, the ability to regulate water made it possible to adopt rapidly the high yielding varieties (HYV) of rice for which the water level needs to be controlled within a fairly narrow margin. Table 4.4 summarises some of these changes. The availability of irrigation made it possible to expand dramatically the *Boro* crop, the main vehicle for the adoption of HYV. By 1973/74 nearly all of this crop had adopted the HYV. The ability to control water also permitted the conversion of nearly half the *Aus* acreage to HYV. Admittedly, over 40 per cent of the acreage under the rain-fed *Amon* crop had also been converted to HYV, but the success of this

Table 4.3 The Expansion of Irrigation in Comilla Thana

	Low Lift Pumps		*Tubewells*		*Total*	
Year	*Number*	*Acres Irrigated*	*Number*	*Acres Irrigated*	*Acres Irrigated*	*Percentage of Total Acreage*
1964/65	3	129	34	1006	1135	2.2
1966/67	17	697	46	2384	3081	6.1
1969/70	110	3260	168	8001	11261	22.3
1973/74	76	5007	175	11194	16201	32.1

Note Except for tubewells in 1966/67 (which refers to the number available) the remaining figures for pumps and tubewells refer to the numbers actually in use. The figures have been compiled from the various *Annual Reports* of the BARD.

Table 4.4 Cropping Pattern and Cropping Intensity in Comilla Thana

Year	*Share of the Three Varieties in Total*			*Cropped Rice Acreage as Percentage of Total Cultivated Acreage*	*Total Cropped Acreage (all crops) as Percentage of Cultivated Acreage*
	Aus	*Amon*	*Boro*		
1960/61	46.9	51.6	1.6	127	128
1963/64	46.7	51.1	2.2	136	141
1966/67	40.2	57.4	2.4	163	172
1969/70	39.0	46.1	15.0	152	—
	(8.3)	(3.7)	(94.6)		
1973/74	25.0	53.0	22.0	134	—
	(48.5)	(42.9)	(97.3)		

Source *Annual Report* of the BARD, 1967, p. 55, and 1973/74, p. 35.

Note Figures in parentheses refer to the shares of high yielding variety in the corresponding crops (e.g. the figure in the first column for 1973/74 means that 48.5 per cent of the acreage under *Aus* crop in that year consisted of HYV).

particular variety in increasing yield has been far less pronounced.

On the question of overall cropping intensity the change seems less clear in view of our inability to locate data on the aggregate cropped acreage in more recent years. The intensity of rice cropping increased sharply through 1966/67 (a year in which irrigation had expanded only to a modest degree) and declined thereafter. It would seem paradoxical that the cropping intensity for rice should have declined during the period of most dramatic increase in irrigation. The explanation probably lies in the fact that the HYVs have a longer gestation lag than the traditional varieties. As a consequence an expansion of a given HYV crop meant a decline in the acreage of the following traditional variety of rice. This probably explains the sharp reduction in the area under *Aus* crop which follows the *Boro* season in which the HYV adoption has been most rapid. Over the years the share of rice has also declined a little in total acreage due to greater diversification of crops. Aggregate cropping intensity went up dramatically between 1960/61 and 1966/67. We do not have the estimates for more recent years, but it would appear to have declined, though perhaps to a lesser extent, for reasons similar to those outlined above.[16] Thus, the main contribution of the expanded irrigation facilities was not so much an increase in overall cropping intensity (which followed an inverted U-shaped path and by 1973/74 was only moderately higher than in 1960/61) as in permitting a radical change in cropping pattern in favour of the HYVs. While irrigation provided the main forward linkage, the supplies of fertiliser, new seed, credit and extension services, all provided by government agencies at high subsidy, were concomitantly expanded to make the 'package' programme a success.

The rest of Bangladesh lagged far behind Comilla Thana with respect to all these factors. While in the early 1960s the dismally low level of irrigation facilities was still a little higher than those in Comilla, by 1973/74 no more than 6.6 per cent of the cultivated acreage in the rest of Bangladesh received irrigation water from pumps, tubewells and larger projects. The share in Comilla was five times as high. Aggregate cropping intensity in Bangladesh followed the same general path as in Comilla though the amplitude of fluctuation was lower: from 127 per cent in 1960/61 it went up to 146 per cent in 1969/70 and then fell back to 131 per cent in 1973/74. The fall in the late 1960s and early 1970s was probably caused by the same factors that caused a similar, though more pronounced, shift in Comilla. By 1973/74 the *Boro* crop had expanded to just under 16 per cent of rice acreage (as compared to 22 per cent in Comilla) and only 56 per cent of that crop consisted of HYVs (as compared to over 97 per cent in Comilla). Only 4.3 per cent of *Aus* acreage was under HYV (as compared to over 48 per cent in Comilla). Thus, compared to Comilla Thana, the rest of Bangladesh had too small an expansion in irrigation to permit a radical restructuring of the cropping pattern in favour of HYVs.

Of the two sources of irrigation in Comilla Thana the tubewells have by far been the more important, accounting for 70 per cent or more of the irrigated acreage. It would, therefore, be useful to look at the method of financing this important source of irrigation.[17] By 1970 in Comilla Thana 179 tubewells had been sunk at an average capital cost of *taka* 23,900. The KTCCA, the central co-operative association, was in charge of these tubewells. In the year 1969/70 the KTCCA received a rental of Tk.300 per year per tubewell from the primary societies. Against this payment the KTCCA paid for the tubewell driver (Tk.500 per season) and undertook to repair and maintain the tubewells (for which no cost estimates are available). The primary society was to bear the remaining operating costs e.g. those for fuel. On the purely arbitrary assumption that annual repair and maintenance costs were approximately 2 per cent of the original capital cost, the KTCCA underwent a deficit of Tk.678 per tubewell on annual operation over and above earning nothing on capital investment. On the assumption of a 15 per cent rate of interest the total annual loss per tubewell would be Tk.4263.[18] Since a tubewell irrigated an average of 48 acres in that year, it amounted to a subsidy of Tk.89 per acre of irrigation. If anything, this is an underestimate of the subsidy.[19]

It is not clear what it cost the primary society to pay for the fuel and remaining cost of operation per acre. It seems that by 1970 the typical payment per acre made by the members amounted to Tk.40. The non-members paid more, often as much as Tk.100 per acre.[20] If one takes the payment made by a member as a reasonable approximation of the cost to the primary society, then such additional cost of operation to the latter would be about Tk.34 per acre (subtracting Tk.6 per acre, i.e. Tk.300 for an average of 48 acres per tubewell, that was paid directly to the KTCCA). Adding to this the cost to the KTCCA, we find that the per acre cost of tubewell water was about Tk.129 in those days.[21] Thus the subsidy amounted to nearly 70 per cent of the cost of tubewell water. For the low-lift pumps the subsidy was only a little lower.[22]

Thus it is quite clear that the spread of irrigation had little to do with the co-operative organisation mobilising its financial and organisational resources to invest in this crucial input. It was the heavily subsidised resources, made available by the government through the apex organisation, which was the main factor behind the expansion of irrigation facilities. The same is true of credit and other agricultural inputs such as fertiliser and pesticide. There is no doubt that the enthusiasm of the BARD and its remarkable and influential leader to see the Comilla experiment succeed had a great deal to do with attracting such massive government support for the programme. But it would be quite inadmissible to claim that the productivity gain in Comilla was due mainly to the organisation of the 'new type' of co-operatives. Elsewhere in Bangladesh the same types of services, at a much lower level of concentration, were being made available through other forms of organisation. There is no evidence to suggest that the productivity gains of these programmes, irrespective of the organisational forms, would have been less than those achieved in Comilla if the degrees of concentration of effort on all fronts were the same.

The inference seems justified by the evidence that irrigation and high yielding varieties were adopted by the non-members at a rapid rate. Unfortunately, we do not have enough information to make a careful comparison of the relative rates of adoption of irrigation and HYV by members and non-members. But the available evidence, though rudimentary in nature, is enough to disprove any claim that the lack of membership in the co-operatives was a significant obstacle to the adoption of these practices.[23] As early as 1966/67 nearly as many of the non-members as members had already adopted tubewell irrigation.[24] In the adoption of the HYVs the non-members had a slightly delayed start, but soon caught up with the members. The percentage of adopters in each group were as follows:[25]

Year	*1966*	*1967*	*1969*	*1970*
Members	7	31	87	98
Non-members	—	1	66	98

One must take into account the fact that non-members were not only disadvantaged in terms of the endowment of resources – a factor which will be discussed in greater detail later – but also suffered discrimination in the form of higher payment for water and other services. And yet they rapidly adopted the practices that were being popularised through the co-operatives at a lower price and attained a higher trend rate of growth in output per acre. One must, therefore, try to explain two apparent paradoxes: why did they not become members and obtain the same inputs at lower prices and why, apparently, did they succeed in achieving a higher rate of growth by using the same inputs? Our information base is not adequate to enable us to find reliable answers. All we can do is to speculate that there must have been difficulties in entry. Such difficulties could be due to the inability of the poor and the small farmers to satisfy the membership requirement (e.g. regular annual purchase of shares and regular weekly savings to be deposited with the society) or to the fear that an association with the bigger farmers would be detrimental to their interest. The higher trend rate of growth of output could be partly due to the lower initial level. But it is also possible that the non-members, who had much smaller holdings on the average, had a greater quantity of complementary input in the form of family labour.

Before ending the discussion on the achievements in production it is useful to look at the question of the efficiency of input use. We do not have much information about the degree of efficiency with which modern capital equipment was used in Comilla. In any given year a fair proportion of the machines and equipment remained out of commission, indicating an unsatisfactory state of maintenance and/or a low rate of use for other reasons. Thus, for example, in 1975/76, 43 of the 234 installed tubewells were out of commission and in 1973/74 of the 31 available tractors 10 were unused.

Table 4.5 shows the areas irrigated by an average pump and tubewell (each approximately two cusecs capacity) *in use* and the area ploughed per tractor *in use* per day. To compare how effective these rates of use have been one needs certain standards with which to compare them. For pumps one could use the average performance all over Bangladesh where a reasonably large programme has been in operation for some time. In the second half of the 1960s the average area irrigated by a two cusec tubewell in Bangladesh as a whole was nearly 40 acres. In 1973/74 it was 37 acres. It would, therefore, appear that the effectiveness of the utilisation of the low-lift pumps in Comilla was only marginally, if at all, better than in the rest of Bangladesh. On the other hand, if compared with the

Table 4.5 The Utilisation of Pumps, Tubewells and Tractors

Year	*Acres Irrigated Per Pump Used*	*Acres Irrigated Per Tubewell Used*	*Acres Ploughed Per Tractor in Use Per Day*
1963/64	n.a.	35	5
1964/65	43	30	n.a.
1965/66	45	45	n.a.
1966/67	43	51	n.a.
1967/68	35	43	n.a.
1968/69	35	49	6
1969/70	30	48	6
1970/71	44	57	n.a.
1971/72	42	45	n.a.
1972/73	40	57	7
1973/74	40	57	5
1974/75	41	49	n.a.

Source The calculations have been based on the data shown in various *Annual Reports* of BARD. In case of a contradiction between the data in two reports we have used the figure from the more recent Annual Report.

relatively modest target of 50 acres set in the First Five-Year Plan, the performance in Comilla would appear to have been less than satisfactory.

The average number of acres irrigated by a two cusec tubewell all over Bangladesh was low, only 34 acres in 1973/74, and the performance in Comilla was much better. But compared to that in Comilla the tubewell programme in the rest of Bangladesh is a much more recent phenomenon so that such a comparison tells us little about the relative efficiency of Comilla. The First Five-Year Plan estimated that a realistic and feasible target for Bangladesh as a whole, taking all the difficulties into account, would be 60 acres. In Comilla this modest target has never been reached.

One must, however, recognise some of the special virtues of the tubewell programme in Comilla Thana. First, Comilla pioneered the adoption of tubewells in the country. The KTCCA obtained the tubewells under a government grant and credit made available by the public sector financial institutions. Secondly, in Comilla the tubewells were sunk by adopting relatively cheap technology. The average cost of Tk.23.9 thousand per tubewell was a fraction of the cost incurred by the government agencies in the 1960s.

Tractors were used also for haulage, but, even after taking this into account, there were sharp reductions both in the proportion of the available tractors used and in the number of days worked per tractor in use. And yet new tractors were added to the fleet of the increasingly under-utilised ones. In the two years between July 1968 and June 1970 an available tractor in Comilla spent 128 days on the average in cultivation, haulage and other activities. By 1972/73 this figure fell to 40 and yet 14 new tractors were acquired in that year. As shown in Table 4.5 the number of acres ploughed per tractor in use per day of cultivation remained very low. One wonders if there could have been any reasonable economic calculations guiding such decisions.

(c) The distribution of income[26]

Table 4.6 summarises some important facts about the distribution of farm size in Comilla for members and non-members during 1964. These figures can be treated as reasonable approximations for the benchmark condition. The figures show that about 15 per cent of the population were landless. Perhaps a small minority of them had access to rented land on the basis of sharecropping[27] and this minority could still join co-operatives. But, for the vast majority of

Table 4.6 The Distribution of Farm Size for Members and Non-members of the Comilla Co-operatives in the late 1960s

Farm Size in Acres (above the lower limit and up to the upper limit)	*Percentage of Rural Population (1)*	*Percentage of Members (2)*	*Percentage of Non-members (3)*	*Percentage of Rural Population in Each Group who are: Members (4)*	*Non-Members (5)*
Nil	15	2	24	5	95
0–1	31	12	41	15	85
1–2	24	43	13	68	32
2–3	14	18	12	47	53
3–5	11	16	7	56	44
Above 5	5	8	3	60	40

Source F. Akhter, *Characteristics of the Members of Comilla Co-operatives* (Comilla, BARD, 1964), p. 4. It is not clear whether the size-groups refer to ownership units or operational units. We shall assume that they are ownership units. This is how one can explain that a small proportion of the landless became members. Presumably these are the sharecroppers. Although in Comilla there are special (non-agricultural) co-operatives, the above data presumably refers to the agricultural co-operative associations (the KSS) of which only farmers could become members. It is implicit in the above figures that at the time about 38 per cent of the rural households had become members of the co-operatives.

the landless who had no access to land the programme had nothing to offer directly.

The programme had little appeal to the very large number of very small peasants. Thirty-one per cent of the population belonged to the size-group above zero and up to one acre. Eighty-five per cent of them stayed out of the co-operatives. It is not possible to ascertain to what extent this was due to the difficulty of entry or to the fear that partnership with the bigger and more powerful farmers would prove to be disadvantageous.[28] Probably both factors exerted some influence. Whatever the reason, the co-operatives attracted very few of the small farmers during the first decade of their operation.[29]

Of the 46 per cent of the population who belonged to households which had either less than an acre or no land at all, the overwhelming majority – 88 per cent – remained out of the co-operatives. There was no direct benefit or protection that this vast majority of the rural poor could derive from the co-operative programme.

The majority of those who had more than an acre of land became members of the co-operative. Indeed, the appeal was greatest to the one to two acre group. Thus the leaders of the movement were successful in their objective of organising the relatively smaller peasant proprietors for their own protection. 'Evidently we were teaching them the principles of capitalism – thrift, saving and investment. We were not surprised when our penny capitalists responded to the call of co-operative capitalism.'[30] All that one needs to note is that in the conditions prevailing in Comilla this modal group of members was near or above the average [31] and that the proportion of the bigger farmers who joined the co-operatives also was very high.

Let us examine the scanty information on the change in the distribution of income in Comilla. First, let us try to determine which groups among the members of the co-operatives obtained the greatest benefit from the co-operative. The data in Table 4.7 present some useful facts although, unfortunately, we have only two sets of observations and the set relating to the more recent period is substantially out of date. The yield rates for the earlier year conform to the pattern that is by now quite familiar for South and South-East Asia: per acre yield is inversely related to size of farm. By the end of the period the difference was considerably reduced and virtually eliminated between the small and the middle groups. The rates of increase in yield per acre have been far greater for the medium and large farmers than for the small farmers. In terms of per capita

Table 4.7 Increases in Rice Yield and Per Capita Income of the Different Size Groups of the Members of the Comilla Co-operatives

	Rice Yield in maunds *per Acre*			*Per Capita Income from Farm and Non-farm Sources (current* taka *per year)*		
Size Group	*1963/64*	*1969/70*	*Percentage Increase*	*1963/64*	*1969/70*	*Percentage Increase*
Small farmers: Under 2 Acres	28.5	50.6	78	219	488	123
Medium farmers: 2 to 3.5 Acres	22.1	49.8	125	276	666	141
Large farmers: Over 3.5 Acres	17.3	38.8	124	242	573	137

Source S. A. Rahim, *Rural Co-operatives and Economic Development of Subsistence Agriculture* (mimeo), (Comilla: BARD, 1972)

income from all farm and non-farm sources the relative performance of the three groups are similar: the small farmers had the lowest rate of growth.[32] While the quality and quantity of information is far too inadequate to permit any firm conclusions there is some basis, however weak, to suggest that among those who became members of the co-operatives, the farms in the two larger groups made bigger proportionate gains in terms of yield and income than did the farms in the smallest group.

We do not have any information on the size distribution of the members in more recent years, and even if we had such information, we would not be able to judge the direction of change in the distribution of farm size among members by comparing it with the distribution in the earlier period. This is because the composition of membership has been changing over time and the incremental share of the smaller farmers has probably been greater than their average share. This would create a downwards bias in the estimated Gini coefficient of the size distribution of members over time.

We have some information, however, on the overall distribution of land ownership in some recent years. Tables 4.8 and 4.9 sum-

Table 4.8 Land Ownership Distribution in Comilla Thana *Amon* season 1973

Range in Acres (above lower limit and up to upper limit)	*Percentage of Families*	*Percentage of Land*	*Average Size of Holding for Each Group (acres)*
Nil	17.0	0.0	0.0
0.0–0.5	24.6	6.2	0.3
0.5–1.0	17.7	12.2	0.8
1.0–2.0	24.6	32.6	1.5
2.0–4.0	12.4	30.7	2.9
4.0–7.0	3.0	12.9	4.9
Above 7.0	0.7	5.4	8.8

Source M. Solaiman, *Landholding and Co-operatives in Five Comilla Villages* (Comilla: BARD, 1974).

Note Landholding includes land owned and cultivated plus land rented out (explained in a private communication from the author). The survey was based on the complete enumeration of landholdings in 5 villages, one from each of the 5 Unions of Comilla Kotwali Thana. Total number of households = 558, and land area = 650 acres. Average land per household = 1.16 acres.

marise some of this information. The first fact to note is the rather sharp difference between the distributions for 1973 and 1975. Landlessness amounted to 17 per cent according to the 1973 survey and 27 per cent according to the 1975 survey. Correspondingly there was a reduction in the number of those who had less than 2 acres of land from 67 per cent according to the 1973 survey to 54 per cent according to the 1975 survey. At the other end also the 1975 survey indicates a greater degree of polarisation. A strict

Table 4.9 The Size Distribution of Landholding and Landlessness in Comilla Thana

Amon Season 1975	
1. *Size distribution*	*Percentage of*
Range in acres	*Rural Families*
Landless	26.9
Up to 2.0	53.9
Above 2.0 and up to 3.5	14.2
Above 3.5	6.1
2. *Landlessness*	
Cultivating landless families	6.8
Landless labourers	20.1
3. *Membership*	
Families which are members of co-operatives	25.6
Landless families which are members of co-operatives	2.5 (i.e., only 9.1 per cent of the landless families)

Source A. Rahim, *Amon Crop Survey in Comilla Kotwali Thana* (Comilla: BARD, 1977).

Note From the way the landless have been classified into 'cultivating' and 'landless labourer' categories it appears that the holdings refer to ownership units although the publication itself does not explain. The Survey included 3702 households in 24 villages. Note that the Survey claims that only 26 per cent of families had joined co-operatives. In 1975 the population of Comilla Thana was about 320,000. Assuming that 75 per cent were farm families which, on the average, had 5.7 persons, this means 42,100 farm families. Thus, according to the Survey ratio there should have been only 10,780 member households. The 1975–76 Annual Report claims that membership was 14,722. The two figures can be reconciled if one assumes than an average member household had 1.37 members. Multiple member households are not uncommon.

comparison at the top end is obviated by dissimilar limits of the class intervals. But those with more than 2 acres increased from 16 per cent in 1973 to over 20 per cent in 1975. To summarise the changes between 1973 and 1975 distributions as suggested by these two surveys: there was an increase in landlessness, a reduction in the proportion of smaller peasants and an increase in the proportion of larger peasants.

One way to resolve the sharply different figures would be to claim that each survey was subject to sampling errors and errors of measurement. In that case it would be natural to attach greater confidence to the 1975 survey which had 6.5 times as many observations and nearly 5 times as many villages as the 1973 survey. But, on the face of it, it does not appear reasonable to try to explain away the difference in this way simply because the time interval between the two surveys was so short. This is particularly inappropriate in view of the fact that during the interval between the two surveys a major famine struck Bangladesh in 1974. It is probable that the famine rendered a substantial proportion of the very small peasant families completely landless through sale of land to the larger landowners.

Compared with Table 4.6 (which describes the distribution in the mid-1960s) the distributions for 1973 and 1975, especially the latter, indicate increased landlessness and an increase in the proportion of below subsistence (i.e. those with less than an acre) farmers. A more complete comparison of the distributions cannot be made due to their incompleteness (especially with respect to the percentage of land accruing to each size group). But it is clear that the programme failed to protect the small peasants from increased proletarianisation and near-proletarianisation. It would be no defence to claim that the members themselves have been protected from this process. It is not known if such a claim is made and, if so, whether it is valid. But, even if this were the case, the fact remains that a large number of small peasants were denied (or could not be persuaded to accept) such protection. It would indeed be interesting to find out the details of the process of polarisation. Were the small peasants bought up by the richer members of the co-operatives who accumulated many benefits as the major users of the subsidised inputs?[33] Unfortunately, the mushrooming action research programme of generating information at Comilla has not been directed to find answers to such urgent questions.

Finally, what happened to the wage rate for agricultural labourers? We do not have any time series of agricultural wage rates in the Comilla Thana, but the same for the Comilla District is available.

This may serve as a reasonable approximation in view of the fact that the Comilla type co-operatives had expanded to the entire district by 1968. Another difficulty is that the time series is a relatively short one. The daily money wage and real wage rates for Comilla District and for Bangladesh as a whole are shown in Table 4.10. Although the time period is short it is clear that wages in Comilla did not rise faster than those in Bangladesh as a whole. In fact the real wages in both followed an inverted U-shaped path over this period. For Bangladesh as a whole we know that this constitutes a short-term fluctuation along a long-term trend of decline.[34] From the evidence of Table 4.10 there is no reason to suppose that the situation in Comilla District was any different. It has also been found that the amplitude of seasonal fluctuations in the wage rate in Comilla has been no less than for Bangladesh as a whole over the period under consideration.[35]

Table 4.10 Agricultural Wage Rate in Comilla District and Bangladesh as a whole

	Comilla District		*Bangladesh*	
Year	*Money Wage* (taka *per day*)	*Real Wage* (taka *per day at 1967 price*)	*Money Wage* (taka *per day*)	*Real Wage* (taka *per day at 1967 price*)
1967	2.71	2.71	2.60	2.60
1968	2.64	2.65	2.75	2.76
1969	3.07	2.96	3.12	3.01
1970	3.32	3.38	2.98	3.03
1971	3.19	—	3.15	—
1972	3.91	2.15	3.93	2.16

Source Money wage for Comilla from S. R. Bose, 'The Comilla Co-operative Approach and the Prospects for Broad-based Green Revolution in Bangladesh' in *World Development*, August 1974. Money wage for Bangladesh and the cost of living index (for low income groups) from ILO, *Poverty and Landlessness in Rural Asia* (Geneva 1977) Chapter 7. The cost of living index (originally with 1963/64 weights) has been converted to 1967=100.

To summarise, the main facts of the change in income distribution were as follows:

1. Among the members of the co-operatives the larger farmers obtained greater increases in output and income.

2. Overall, perhaps more among those who were kept out of the 'protective' fold of the co-operative, landlessness and near proletarianisation increased.
3. Over the longer time period real wages declined.
4. All this happened over a period of relatively successful 'Green Revolution' which brought about significant increases in output and income.

(d) Capital Accumulation and Credit

As the founder member of the Comilla experiment put it:

> We laid great emphasis on thrift and savings. We kept on repeating like a litany, 'You are being crushed by the power of capital. The same power will redeem you if you learn to possess and control it. Therefore, every week, everyone of you must make a thrift deposit, however small. Gradually your co-operative will accumulate a substantial amount. It will also get cheap loans for you. Ultimately, through your co-operative, you will become your own financier.'[36]

We have information on the savings deposited by the members with the co-operative societies in the form of savings deposits and shares but not on possible saving in the form of: (i) direct purchase of capital goods; (ii) direct capital construction; (iii) accumulation of cash balance outside the co-operative; and (iv) accumulation of other financial assets. We cannot, therefore, judge the overall savings performance of the co-operative members or compare it with the performance of non-members. Similarly, our information on credit relates almost exclusively to the credit made available by the co-operative society. We have little information about borrowing from other, more traditional sources. However, the available information will enable us to analyse whether the programme has achieved reasonable success in generating high savings through the co-operatives and in organising the credit programme in an efficient and equitable way.

The impetus for capital accumulation came from a number of sources. First, the initial purchase of a share (for Tk.10) and another one in each subsequent year and regular deposits at given intervals were made pre-conditions of membership. Secondly, these shares and deposits earned dividends and interest. We have no estimate of the rate at which dividends were paid although it has been claimed that investments are carefully chosen to ensure a high return.[37] The interest on deposits was about the same as that offered by most banking institutions and, in *real* terms, was negligible or even negative. Finally, the most important incentive for saving probably derived from the consideration that the maximum limits on credit were set as fixed multiples of the accumulated shares and savings.[38]

It is, therefore, not surprising to see shares and savings rise steadily with the expansion of the credit programme. Indeed, it would make sense for the prudent primary societies and individuals to borrow a bit more than would be necessary in a given period to build up shares and savings deposits so that the ability to borrow in the subsequent periods expanded by a multiple of the increment in their quantities.

Table 4.11 contains some of the details of the accumulation of savings and shares and borrowing made by all the primary co-operative societies under the Agricultural Co-operative Federation (ACF) in Comilla Thana. Per member annual capital formation in the form of the acquisition of savings and shares shows no clear trend: the peak was reached in 1967/68 and thereafter a decline began. In real terms the decline in the volume of annual savings was more pronounced than is indicated by the figures as the rate of inflation accelerated sharply in the 1970s. As a percentage of the current value of per capita agricultural production or income in the area the value of annual savings must have declined sharply in the late 1960s and early 1970s.

In no period, except in 1975/76 when special circumstances prevailed, was aggregate accumulated saving greater than total borrowing from the co-operative.[39] Indeed, per capita net indebtedness (i.e. debt less accumulated shares and savings) increased most rapidly in the period in which the accumulation of savings grew steadily – an association which lends support to the view that the primary motivation behind such accumulation was to qualify for still greater loans.

This also suggests an explanation why accumulation slowed down to a negligible rate in the late 1960s and early 1970s. During this period the credit programme was in serious difficulty due to a high and increasing rate of default (see column 9 of Table 4.11). According to the rules the defaulters could not be given new loans. Thus, by the early 1970s the proportion of members receiving loans dropped to 10 per cent from over 80 per cent in the mid-1960s (see

Table 4.11 Capital Accumulation and Credit in Comilla Agricultural Co-operatives (all values, except in columns 5 and 8, are in thousand *taka*)

Year	Accumulated Saving (Closing Balance)			Annual Capital Formation (shares and savings)	Total Members (thousand) End Year	Per Capita Annual Capital Formation (Tk.)	Total Outstanding Loan (closing balance)	Net Indebtedness outstanding loan less accumulated savings and shares	Per Capita Net Indebtedness	Total Overdue Loan (closing balance)
	Savings (1)	Shares (2)	Total (3)	(4)	(4)	(5)	(6)	(7)	(8)	(9)
1961/62	28	11	39	—	1.86	—	176	137	74	—
1962/63	82	41	123	84	3.16	27	345	222	70	—
1963/64	151	69	220	97	3.83	25	578	358	93	—
1964/65	170	135	305	85	4.91	17	751	446	91	—
1965/66	202	214	416	111	5.16	22	887	471	91	—
1966/67	287	264	551	135	8.46	16	1586	1035	122	—
1967/68	554	549	1103	552	11.52	48	4029	2926	254	—
1968/69	831	780	1611	508	11.67	44	4604	2993	256	—
1969/70	975	874	1849	238	11.15	21	4542	2693	242	276
1970/71	1192	1021	2213	364	11.26	32	4981	2768	246	1328
1971/72	1387	1036	2423	210	11.79	18	4931	2508	213	2358
1972/73	1611	1083	2694	271	13.39	20	4983	2289	171	2968
1973/74	1702	1117	2819	125	13.74	9	4384	1565	114	3632
1974/75	1721	1154	2875	56	14.24	4	3389	514	36	2878
1975/76	1973	1180	3153	278	14.72	19	2833	−320	−22	2531

Source 1968/69, 1969/70, 1973/74 and 1975/76 *Annual Reports* of BARD.
Note Columns 5 and 8 show values in *taka*. All other values are in thousand *taka*. Figures for total overdue loan for the years up to 1968/69 are not available. These are believed to have been very small.

Table 4.12). Due to this extraneous limit on credit the need to build up qualifying deposits and shares was sharply reduced.[40]

Supervised credit was the cornerstone of the co-operatives. The programme was not one of mainly channelling the savings of the members who were able to save to those members who were in need of credit. Most of the fund for lending was made available to the apex societies (the ACF and the KTCCA) by the government and the public sector banks at a very low rate of interest. To obtain credit, individual members had to apply through their respective co-operative associations which scrutinised such applications before endorsing them for the sanctioning of credit by the Agricultural Co-operative Federation (ACF). The loan request had to be for a specific 'investment' purpose. Nearly 80 per cent of loans were short-term, that is repayable within one year. This category of loans consisted mainly of the so-called crop loans meant to finance the working capital requirement for specific crops. About 20 per cent of the loans were medium-term, that is repayable within three years, and were given mainly to finance the lease, re-lease or purchase of land. Purchase of cattle accounted for only 2.5 per cent of total loans granted and the purchase of farm machinery, a mere 1.45 per cent. No long-term credit was made available by the ACF.[41]

Table 4.12 Loan Received by the Members of Agricultural Co-operative Societies in Comilla Thana

Year	*Percentage of Members Receiving Loan*	*Average Loan Per Recipient in* Taka
1965/66	85	—
1966/67	83	235
1967/68	62	591
1968/69	68	368
1969/70	34	450
1970/71	33	412
1971/72	10	402
1972/73	10	403
1973/74	12	417

Source P. K. Chowdhury, *Loan Default Under the Agricultural Co-operative Federation of Comilla Kotwali Thana* (mimeo), (Comilla: BARD, 1976).

A member's loan request could not exceed 75 per cent of the value of the land that he mortgaged to the ACF as security for loan. Even after the sanctioning of a loan the ACF inspector exercised control by phasing the release in such a way as to coincide with the progress of the activity for which the loan was granted. The purpose of all this supervision was to ensure proper utilisation of a loan and to adhere strictly to the repayment schedule outlined in the original loan proposal.[42]

Some of the main features of the evolution of the credit programme may be analysed with reference to Table 4.11 and additional information that has been generated by two surveys – Ali Akhter Khan's 1969 Survey of the performance of 30 societies in the administration of credit and P. K. Chowdhury's 1975 Survey of defaulters in 31 societies.

The first fact to note is that the dependence on credit continued to grow through the first decade of the co-operative programme. Too frequently this has been justified by arguing that with the spread of the HYV techniques the credit required to finance the application of new inputs rose. This is a false defence because all capital inputs for the HYV (e.g. tubewells, pumps and tractors) were provided by the apex societies with state assistance and none of the capital cost of these inputs was recovered. Current inputs such as fertilisers, pesticides and new seeds were provided at prices which contained massive subsidies.[43] The adoption of HYV resulted in a rapid increase in output and income. In this situation it should have been easy to switch gradually toward self-financing of current inputs which, in any case, were seriously underpriced. To suggest that the spread of HYVs made the farmers increasingly less able to self-finance current inputs is to contradict all the accumulated evidence which indicates that net income per acre increased very substantially after the adoption of such varieties, even after allowing for the cost of such inputs. This is further strengthened by the evidence that much of the credit went to the bigger farmers.

This brings us to the next important question of who received credit and who were the major beneficiaries from the credit programme. In deciding about the allocation of credit to the members of the societies the members of the managing committees played a crucial role. This may, therefore, be the place to look at the composition of the members of the managing committees – the chairmen, the managers and the directors. Table 4.13 shows the farm size distribution for the ordinary members and managers of

Table 4.13 Farm Size Distribution of Ordinary Members and Managers of the Agricultural Co-operative Societies in Comilla Thana

Size of Farm (above lower limit and up to upper limit in acres)	*Percentage of Ordinary Members*	*Percentage of Managers*
Below 0.50	7	—
0.50–1.00	28	9
1.00–3.5	53	52
3.50 and above	12	39

Source P. K. Chowdhury, 'Some Experiences of Co-operative Credit under IRDP in the district of Comilla', *Journal of the BARD*, July 1976.

the Comilla co-operative societies. While 35 per cent of the ordinary members belonged to the category of farm size of up to one acre, only 9 per cent of the managers belonged to this size group. At the other extreme, those with a farm size of above 3.5 acres represented only 12 per cent of ordinary members but supplied about 40 per cent of the managers. While we do not have similar information on the farm size distribution of the chairmen – an even more respectable breed – and directors, it would not be unreasonable to suppose that Table 4.13 provides a reasonable picture of the divergence between the class origins of the ordinary members (who, as we may remind ourselves, were already much above average farmers in size) and those who mattered in the making of decisions. It is, therefore, reasonable to expect that much of the credit, and other resources, would be allocated in favour of their own classes.

In Ali Akhter Khan's 1969 Survey a smaller sub-sample of 278 borrowers in six co-operatives showed that the managing committee members who represented 15 per cent of the total members were indebted to the extent of 36 per cent of the total outstanding loans. The average debt of the managing committee members was more than three times that for the ordinary members. The managing committee members also accounted for a disproportionately large share of the overdue loans.[44]

Table 4.14 shows per capita indebtedness of the various size groups to the co-operative societies: the concept of need that underlies the distribution of credit seems to coincide exactly with the size of landholding and, plausibly, other assets. While the rich and the powerful obtained cheap credit from the government

Table 4.14 Size of Farm and Extent of Per Capita Indebtedness in Comilla Co-operatives

Size of Farms (in acres)	*Per Capita Indebtedness (in taka)*
Small (up to 1)	85
Medium (1–3)	1391
Large (3 and above)	3112

Source Ali Akhter Khan, op cit, p. 27. The data, based on a small sample of 30, refer to the year 1969. The figures are reproduced from the last column of Khan's Table 6. There seems to be some inconsistency between that column and the other columns (showing the number of borrowers and total debt for each group), but the per capita borrowing by the largest group is at least about three times as great as for the smaller groups, however one may choose to calculate the figures.

through the co-operative, the poor and the powerless were condemned to borrowing from private money lenders at astronomical rates. About 17 per cent of the members borrowed from private sources according to the 1969 Survey. Their average debt was Tk.1340

> which is quite a considerable amount. The members having less than one acre of land and non-agricultural occupation as their major source of income contracted this kind of loan. These loans are of diverse nature and have the ramification of borrowing for interest in kind . . . , usufructuary land mortgage etc.[45]

The rich and the powerful not only allocated much of the institutional credit to themselves, but they also accounted for a disproportionately large number of loan defaults. In 1969 more than two-thirds of the overdue loans belonged to members of the managing committees.[46] A 1975 survey of 73 defaulters in 31 societies in Comilla Thana established the following facts.[47]

1. Fifty-two per cent of the defaulters were members of the managing committee and they accounted for 17 per cent of the membership of the co-operatives. Since the members of the managing committee borrowed much more than did the others on average, it is likely that the proportion of overdue loans that belonged to them in 1975 was even greater than that in 1969 (when their share was two-thirds).

2. The average size of the holding of the defaulters was much higher than that of ordinary peasant households. The size of landholding and annual income of the defaulters were as follows:

Category of Defaulters	*Average Size of Holding (acres)*	*Average Annual Income (*taka*)*
Managing committee members	3.05	17476
Ordinary members	2.37	13517

The average size of holding in Comilla Thana in the same period was just over an acre and average family income was probably no more than Tk.4000. thus it was not so much the inability to repay as the probability of getting away with it that determined the degree of default. The complete size distribution of the defaulters was as follows:

Size of Holding (acres)	*Percentage of Defaulters*
No land	—
up to 0.5	1
0.51–1.00	14
1.01–2.00	36
2.01–3.00	22
3.01 and above	27

Considering that the amount of the loan varied more than proportionately with the size of holding, the group belonging to up to one acre in size would appear to account for a negligible portion of the total amount defaulted, while the group having more than three acres would probably account for more than half.

Our conclusions are obvious: the programme of promoting savings through the co-operative did not achieve the success that was hoped. In the three years ending in June 1976 the annual average accumulation of savings and shares per member amounted to Tk.10.67, which was not much in excess of one-tenth of one per cent of the annual income per member household. The underpriced credit programme, funded by the government, mainly went into the hands of the rich and the powerful. The managing committees, firmly controlled by the large farmers, not only allocated much of the credit in favour of their own class but also hampered the entire credit operation by defaulting on repayment on a massive scale. The modest expectation of the founder of the movement that the penny capitalists would respond to the virtues of co-operative capitalism remained unfulfilled.

It is, however, unlikely that the beneficiaries of the movement, particularly the richer ones, did not save. Together with the relatively vast sums of co-operative credit that they appropriated, they must have amassed considerable volumes of surplus. What did they do with this surplus? As has been mentioned, the policies of the co-operatives and the government made it senseless for them to invest in capital equipment such as pumps and tubewells. The services of such capital equipment were supplied through the co-operative organisation at a fraction of the cost that would be incurred if the rich farmers invested in them privately (even after allowing for the sale of their surplus services to other farmers). It seems inevitable that they would turn their attention to the acquisition of more land. An interesting fact that emerged from the 1969 Survey of Credit is that the average size of landholding of members of the high and medium loanee societies increased respectively by 11 and 32 per cent over what they were before becoming members of the co-operatives, only a few years previously. Such estimates conceal the distribution of gain between the various size groups of members of the same category of co-operatives. But the evidence leads one to hypothesise that the richer members made even greater gains in terms of landholding and that this was largely made possible by the asymmetrical distribution of benefits from the co-operatives.

(e) The success and failure of the Comilla experiment: a summary

In Comilla the co-operative experiment was an organisation of farmers to facilitate the absorption of inputs supplied by the government from outside. Co-operation did not extend to the mobilisation of the latent productive resources (e.g. unused and under-used labour) or to any kind of pooling of private productive resources for joint production activities. Farming was carried out strictly on the basis of private ownership, management and planning. Even the most rudimentary type of co-operative effort, such as the exchange of seasonally surplus family labour or the joint construction of common capital on the basis of self-financing, is known not to have been organised on a significant scale. The founder of the programme was quite right in suggesting that it was an experiment in co-operative capitalism.

The basis of the programme was a massive, by Bangladesh standards, concentration of modern inputs (e.g. pumps, tubewells, tractors, improved seed, fertilisers, pesticides and training) which were very heavily subsidised. Above all, the presence of the Bangladesh Academy of Rural Development and its remarkable leader, Akhter Hamid Khan, ensured the availability of a particularly unique resource for the implementation of the programme. The co-operative organisation was intended to serve as the vehicle for the reasonably equitable and efficient absorption of these massively subsidised inputs.

Given such attractive subsidies and organisational talents, it is somewhat surprising that the co-operatives failed to attract the majority of rural households. The proportion of enrolment is itself not clearly documented but would appear to have reached a maximum of something like a third of the rural households.[48] The reasons for such a low enrolment are unclear: it is probable that the apparently modest conditions of membership were too difficult for the poor peasants to fulfil. It is also probable that the poor peasants recognised that the structure was such that the benefits would,at best, be proportional to the initial landholding. They may also have been afraid of an unequal partnership with the large farmers. Finally, the programme had virtually nothing to offer to the landless labourers.

There was a significant, even impressive, gain in production. Output per acre increased rapidly for the crops incorporating the HYV technique. But this gain seems to have owed little to the co-operative organisation itself. The credit is more appropriately due to the concentration of resources made available by the government. Indeed, the co-operative members seem to have lagged a little behind the non-members in terms of *growth* in production per acre.

The management of the co-operatives was dominated by farmers who, by Comilla standards, were large. The benefits were largely concentrated in the hands of the rich and the powerful. The medium farmers also appear to have made some gains. But relatively little gain accrued to the small peasants and none did directly to the landless. The naive view that benefits would trickle down to the poorest does not seem to have been correct. Landlessness and land concentration increased. The course of real wages was no different from what it was elsewhere in Bangladesh.

The sacrifice of equity was not the only cost. The co-operatives failed to distribute the subsidised resources efficiently. Although the efficiency of utilising the pumps and tubewells was greater than elsewhere in Bangladesh, one could not claim that a desirable absolute standard was achieved. The credit programme virtually broke down by the middle of the 1970s as a result of default mainly on the part of the rich and the powerful. The co-operatives not only failed to prevent this breakdown but, being controlled by those who were powerful enough to get away with non-repayment, actually brought about such a breakdown.

The whole exercise constitutes a lesson in the futility of 'co-operation' in a situation of inequality. The programme accepted the unequal distribution of land as an initial condition and hoped to work around the big farmers not by arranging that 'they should be excluded from the new co-operatives', but by hoping that they would 'not be allowed to dominate'.[49] The lesson of the Comilla experiment is that it is impossible to prevent the big farmers from dominating co-operatives if they are allowed to enter them. In fact, it is impossible to prevent them from dominating if they are allowed to exist.

4.3 The expansion of the Comilla-type of co-operative in the rest of Bangladesh

The decision to adopt the Comilla type of co-operative under the banner of the Integrated Rural Development Programme (IRDP) as the organisational framework for rural development in Bangladesh was made before independence. Since independence the programme has gradually expanded. The main features of the Comilla co-operatives which have been adopted in the IRDP framework are the two-tier co-operative structure and the integration of multiple services, e.g. credit, irrigation, training and inputs for HYV. The progress in terms of the gradual adoption of the 434 administrative thanas of Bangladesh has been as follows:

Year	*Cumulative Total of Thanas*
1971/72	33
1972/73	86
1973/74	151
1974/75	161
1975/76	162
1976/77	200

Although the programme has expanded quite rapidly, to 46 per cent of the administrative thanas within a period of six years, the spread has been rather thin compared to Comilla. By the middle of 1977 the following enrolments had been achieved:

Total no. of KSS	21874
Total membership of all KSS	649088
KSS per thana	109
Members per thana	3245

The average number of *rural* households in a thana is about 29,000, so that in those thanas which have been adopted by the IRDP no more than 11 per cent of the households have, on average, joined the co-operatives. Thus the rate of participation has not been more than a small fraction of that in Comilla at the comparable historical period.

The general approach is the same as in Comilla – to concentrate the supply of heavily subsidised inputs to induce the adoption of HYVs. The First Five-Year Plan reported the following rates of subsidy on pump and tubewell irrigation in 1973/74:

Type	*Subsidy in Taka per Acre of Irrigation*	*Subsidy as Percentage of Total Cost*
Low-lift pump	107	68
Deep tubewell	166	77

These rates are generally higher than those in Comilla. The subsidies on fertiliser, at the rates shown below for the year 1977/78, are much higher than the corresponding rates in Comilla, as elsewhere in Bangladesh, in the 1960s:[50]

Type of Fertiliser	*Subsidy as Percentage of Cost*
Urea	48
Triple super phosphate	67
Potash	60

Credit is made available on an annual interest and service cost of 17.5 per cent. This in nominal terms is slightly higher than the 15 per cent that used to be charged in Comilla during the 1960s, but the difference is more than compensated by the higher rate of inflation and generally higher structure of nominal interest rates in the 1970s, as compared with those in the 1960s. From time to time the government talks about reducing the subsidies on inputs. It is clear, however, that to date it has been very difficult to initiate a major change in the situation.

Although the subsidies are high and the general orientation is the same, the IRDP programme has not yet succeeded in achieving the same degree of concentration of inputs as in Comilla. We do not have separate estimates for the 200 IRDP thanas. For Bangladesh as a whole, irrigation from all public sources amounted to only 1.84 million acres in 1977/78 or about 8 per cent of the cultivated acreage.[51] Even if we assume that irrigation facilities in the IRDP thanas were far above that elsewhere, the expansion of irrigation in these thanas would appear to have been far less than that in Comilla (where, by the early 1970s, about a third of the acreage was under pump and tubewell irrigation). This seriously limits the spread of HYVs and other modern inputs in these areas to a level far below that of Comilla Thana. Credit is probably the only programme under the IRDP which has even remotely approached the performance in Comilla: by 1977 loan distribution per member in the IRDP co-operatives was Tk.407. In real terms (i.e. deflated by the price index of those goods and services which are acquired by funds obtained through credit) this is probably no more than a third to a half of the peak rate of credit in Comilla Thana.[52]

That the rapid adoption of 200 thanas by IRDP has meant a rather thin spread of resources has come to be recognised. Perhaps this realisation has led to what may be called a second wave of organisation of co-operatives in some of the IRDP areas. In 1976 the project RD-1 was launched on the basis of joint planning and financing by the Government of Bangladesh and IDA of the World Bank (which financed 69 per cent of the cost).[53] The objective was to intensify rural development within the IRDP framework in seven thanas that had been adopted by the IRDP programme, three in Mymensingh District and four in Bogra District. While the chief purpose was to intensify the IRDP programme (thereby approaching the levels attained in Comilla Thana), it was claimed that the following additional features have been incorporated. First, the compilation of land and cattle registers has been provided for in order to develop a clearer idea of the input requirements and to guard against the possibility of fraud on the part of members.

Secondly, the older co-operative organisations which survived in the RD-1 areas have been reorganised to be integrated gradually into a new comprehensive co-operative framework. Thirdly, loans in kind have been envisaged, presumably as a safeguard against the use of cash loans for consumption purposes. Finally, in some RD-1 projects the reclamation of the *publicly owned* derelict tanks by employing landless labourers has been included in order to find a place for the latter in the programme.

Even with these additional features the programme is hardly an improvement over the Comilla co-operatives. Land and cattle registration will merely provide some information for better planning. In a comparison with Comilla this must be balanced against the large action research programme of the BARD. Similarly, the incorporation of the old, ineffective co-operative organisations will add little to the programme. The loans in kind probably are an unmitigated restraint on the already dubious credit programme. This is analagous to the tying of international credit which does no good to the borrower. At the least it is highly paternalistic and fraught with the danger that the bureaucratic planners will recommend an inappropriate composition of inputs. Almost certainly it raises the cost of credit, especially to the poor and the weak, by subjecting the borrower to greater bureaucratic control. The notion that loans in kind prevent the channelling of funds into consumption is false in that borrowers can allocate a greater proportion of their own resources to consumption. After all, credit alone does not account for all financing of production activities. Most farmers resort to significant self-financing. Finally, the reclamation of derelict tanks is not fundamentally different from the Works Programme which was a prominent feature of the Comilla project. Since the project is limited to the publicly owned derelict tanks, its quantitative impact must be limited. It also appears that this component is considered to be somewhat 'optional' to the RD-1 programme. The landless continue to be just as peripheral under the RD-1 as they have been in Comilla and in IRDP.

In several ways the IRDP and the RD-1 programmes are distinctly inferior to the Comilla experiment. A mere glance at these programmes is enough to show how overwhelming is the official bureaucratic structure.

Many resources are devoted to the building of infrastructure (e.g. buildings, houses and vehicles) for use by the government officials who head these programmes at the thana level. In Comilla it is unlikely that bureaucracy had such a great share of resources. Compared to Comilla the capital investment programme (especially in tubewells) is more resource intensive, based on more sophisticated techniques and, as a consequence, more costly.

It is not yet possible to evaluate the IRDP and the RD-1 projects in the same way as we have analysed the Comilla experiment. Separate data about production, distribution and accumulation in the IRDP thanas are not available and the RD-1 has only recently been launched. If, however, the IRDP had transformed yields in the same way as did the Comilla experiment the result would have been reflected in the national average (since these thanas constitute 46 per cent of the total). This, clearly, has not yet happened and the reason seems to lie in the fact that the distribution of inputs and the degree of effort have not been above the critical minimum level.

As and when the intensity of the IRDP programme reaches a level comparable to that in Comilla there is little reason to expect that the outcome will be fundamentally different. Production will increase unless the lack of effective demand applies a brake.[54] The distributional consequences are unlikely to be any better than those in Comilla. There is nothing in the IRDP or the RD projects that will reduce the control of the rich and the powerful over the co-operative organisation, which will continue to serve as the vehicle for the absorption of subsidised inputs. If anything, the cost will be greater than in Comilla due to greater bureaucratic control and dependence on more conventional methods.

One can interpret the official strategy as due to ignorance on the part of policy-makers about the true consequences of the Comilla programme. Alternatively, one might attribute it to the decision on the part of policy-makers that an increase in production is the overwhelmingly important target and that the programme should aim at achieving that target even if the cost is the further enrichment of the rich and continued impoverishment of the poor.

This is a cause of extreme concern in view of the trends revealed by the data in Tables 4.15 and 4.16. Table 4.15 shows that landlessness has been increasing – in absolute numbers, as a proportion of the farm labour force and as a proportion of rural households. The increase has been particularly rapid in the last decade. Since the increase in landlessness has been greater than the increase in the agricultural population, the trend cannot be explained by demographic factors; the distribution has been changing sharply in an unfavourable direction.

Table 4.15 Landlessness in Rural Bangladesh

Year	*Landless Labourers as a Percentage of Total Farming Labour Force (1)*	*Number of Landless Agricultural Labourers in Millions (2)*	*Percentage of Rural Households which are completely Landless (3)*
1951	14.3	1.51	—
1961	17.5	2.47	—
1963/64	17.8	2.71	—
1964/65	17.5	2.75	—
1966	—	—	6.97
1967/68	19.8	3.40	—
1977	26.1	4.54	11.07

Note For the estimates of columns (1) and (2) up to 1967/68 see A. R. Khan, *Poverty and Inequality in Rural Bangladesh,* ILO (WEP) Working Paper, 1976. The 1977 estimate is based on the Land Occupancy Survey (LOS) of that year reported in F. T. Zannuzi and J. T. Peach, *Report of the Hierarchy of Interests in Land in Bangladesh* (Washington: USAID, September 1977). The LOS estimate refers to the number of rural households (and their total population) which are completely landless. The 1974 population census shows that there were 0.243 members of the agricultural labour force (i.e. *excluding* non-agricultural rural labour force who constituted about 15 per cent of the total rural labour force) per rural population. Applying this to the number of households and persons reported by LOS the *agricultural* labour force in landless and total rural households have been found. For columns 1 and 2 the LOS definition of Landless-2 (i.e., all those who have no land or only have homestead land) has been used as this seems to be the same as the quantity measured in preceding years. Column 3 shows the percentage of rural *households* that are absolutely landless. The source for 1966 is Registrar of Co-operative Societies. *Agricultural Credit in East Pakistan, January 1966 – A Survey Report* (Dacca 1967) and the source for 1977 is the LOS. For this purpose the LOS definition of Landless-1 (i.e., households that claim no ownership of land either homestead or other land) has been used. The main reason the column 1 proportions are so much larger is that many landless labourers may have a homestead but no cultivable land.

Table 4.16 shows the distribution of operational landholdings in 1967/68 and the distribution of land ownership in 1977. The two are not comparable: in general the ownership distribution is likely to be more unequal.[55] It is, however, somewhat unlikely that the vastly worse distribution in 1977, as compared with that in 1967/68, is due entirely to this factor. Given the grouping of the published

Table 4.16 The Distribution of Landholding in Rural Bangladesh in 1967/68 and 1977

	The Distribution of Operational Holdings: 1967/68			*The Distribution of Ownership in 1977*		
Size	*Percentage of Landholding Households*	*Percentage of Land*	*Average Size of Holding in Acres*	*Percentage of Landholding Households*	*Percentage of Land*	*Average Size of Ownership Acres*
Up to 1 acre	25.0	4.3	0.53	53.4	9.3	0.23
1–3 acres	39.2	23.8	1.90	28.5	27.6	1.78
3–5 acres	18.7	23.3	3.91	9.6	20.1	3.84
5–12.5 acres	14.4	33.3	7.24	7.1	28.1	7.27
Above 12.5 acres	2.7	15.5	18.58	1.3	14.9	19.62

Note The 1967/68 distribution is from Bangladesh Bureau of Stastistics, *Master Survey of Agriculture*, Seventh Round, Second Phase, Dacca 1972. The 1977 distribution is from the LOS reported in Zannuzi and Peach, op cit. The 12–13 acre group in the latter has been equally distributed to the two highest groups in the above table which means a slight understatement of the land share and a slight overstatement of family share for the highest size group. Homestead land is included in landholdings and land owned. Average farm size in 1967/68 was 3.1 acres and the average size of an ownership unit in 1977 was 1.84 acres. In calculating the percentages of households only the aggregate of landholding (landowning) households has been used.

data, we can only make approximate comparisons. In 1967/68 the lowest 56.6 per cent of the landholding households had 21.4 per cent of the land while in 1977 the bottom 53.4 per cent of the landowning households had only 9.3 per cent. At the other end, in 1967/68 the top 7.9 per cent of the landholding households had 31 per cent of land, while in 1977 the top 8.4 per cent had 43 per cent. In 1967/68, at the upper end of the distribution, 20 per cent of land was held by 4 per cent of the biggest landowners. In 1977 the same proportion of land was owned by 2.2 per cent of the richest landowning households.[56] A part of the difference probably is explained by the fact that the ownership distribution is more equal than the distribution of operational holdings. But it is also highly probable that the distribution itself has worsened.

The intolerable inequality of land ownership is clearly revealed by the 1977 Land Occupancy Survey data. The top 10 per cent probably own close to 50 per cent of the land. The bottom 50 per cent of the households own less than 9 per cent of the land. The average ownership of the top 8.4 per cent was 40 times the average ownership of the bottom 53.4 per cent!

The consequence of intensifying the implementation of the IRDP programme in a situation of such initial inequality will be to perpetuate, more likely to aggravate, such inequality. In the official discussions of rural development programmes and policies there is little awareness of the gravity of the situation. It seems that the decision has been made to march towards greater inequality in the belief that, given the existing structure of the distribution of power in the rural areas, nothing else is feasible and in the hope that this strategy will provide some, possibly high, growth.

Notes

1. The term 'co-operative capitalism' as a description of the Comilla experiment has been borrowed from an article written by Akhter Hamid Khan, the founder of the Comilla experiment. For the complete quotation and reference see Section 4.2 (c).
2. See Chapter 7 of ILO, *Poverty and Landlessness in Rural Asia* (Geneva: ILO, 1977).
3. R. D. Stevens, H. Alavi and P. Bertocci (eds.), *Rural Development in Bangladesh and Pakistan* (The University Press of Hawaii, 1976), pp. 96–7. Among the less favourable accounts of the Comilla experiment are: Harry Blair, 'Rural Development, Class Structure and Bureaucracy in Bangladesh', *World Development*, VI, No. 1 (1978), and his earlier

work, *The Elusiveness of Equity: Institutional Approaches to Rural Development in Bangladesh* (Cornell University, 1974).

4. A thana in Bangladesh is a lower level administrative unit with an average area of 125 square miles and an average population of about 180,000. Traditionally, the distinctive feature of a thana was the presence of a police station (the word for which in Bengali is thana). The Comilla Kotwali Thana is smaller than average in size (107 square miles) but larger than average in population (about 311,000 according to the 1974 census). The administrative level above the thana is called a sub-division (of which there are 55 in Bangladesh) and that above the sub-division is called a District (of which there are 19).
5. In Comilla the Kotwali Thana Central Co-operative Association (KTCCA) Ltd. serves as the apex organisation. Besides the ACF its constituents include the Special Co-operative Societies Federation (the SCFC, which itself is a federation of the non-agricultural primary co-operative societies, e.g. those of automobile drivers, mechanics, small traders and rickshaw-pullers) and six small societies of special categories. In 1973/74 the membership of all the KSS accounted for 72 per cent of all the members of all societies belonging to the KTCCA. In this paper we shall be concerned with the ACF and its constituent KSS and not with the non-agricultural and special co-operative societies.
6. Akhter Hamid Khan, *The Comilla Projects – A Personal Account*, Paper presented at the Development From Below Workshop, October 1973, Addis Ababa. Despite his own statements to the contrary, most observers of the Comilla experiment emphasise the crucial role played by Mr A. H. Khan in making it what it became. It is clear that this personality factor was important in shaping the Comilla project. This is a factor of which the programme of replicating the Comilla model all over Bangladesh was deprived.
7. Ibid.
8. Ibid. A. H. Khan's figures are illustrative orders of magnitude, not exact estimates.
9. Ibid.
10. Ibid.
11. Ibid. The by-laws of the KSS generally did not exclude the landless from membership. But agricultural labourers could not obtain credit (for which land had to be mortgaged). Nor did they have any use for subsidised water, fertiliser or pesticide. Thus there was no incentive for them to become members for which strict obligations had to be undertaken.
12. In 1960/61, 1963/64 and 1966/67 the shares of rice in total cropped acreage in Comilla thana were respectively 98.9 per cent, 96.5 per cent and 94.3 per cent. There is some evidence that the declining trend continued through the early 1970s, though the share is probably greater than 90 per cent even today.
13. According to the FAO *Production Yearbook*, 1976, the yield of rice (in *paddy* form) per acre would be 2.37 tons in Japan, 2.02 tons in North Korea and 2.08 tons in South Korea in 1974. In terms of rice, *paddy* yield per acre in Comilla was 1.75 tons.

14. Trend rates of growth (g) have been found by fitting Log $X = a + gt$ (from $X = Ae^{gt}$) where X is yield per acre and t is time (= 0 for 1963/64, 1 for 1964/65, . . . 11 for 1973/74). The fitted functions are as follows:

Aus

Members: $\log X = 2.9119 + 0.0363t$ $R^2 = 0.33$
(0.0173)

Non-members: $\log X = 2.6302 + 0.0541t$ $R^2 = 0.48$
(0.0186)

Boro

Members: $\log X = 3.0440 + 0.0903t$ $R^2 = 0.64$
(0.0258)

Non-members: $\log X = 2.6745 + 0.1148t$ $R^2 = 0.63$
(0.0336)

For *Amon* crop g is not significantly different from zero at any reasonable level of confidence and R^2 is approximately zero for both groups. For *Boro* and *Aus* the 95 per cent confidence intervals for estimated gs for members and non-members overlap widely. Thus the estimated gs for non-members are higher than those for members only at a very low level of probability.

15. There are problems in comparing the trend rates of growth of the two groups over time. As is well-known, over time members increased in number. If one accepts that initially the farmers with higher yield became members, then their ranks in the later years might have been swollen by the influx of those who had relatively a lower initial yield. This would bias the estimated trend for the members downward. But, if one further accepts that among the non-members those with a higher initial yield joined the co-operatives then, over the years, the depletion in their ranks must have come about by the desertion of those with relatively higher yields. This would bias the estimated trend for the non-members downward. Given the state of information one cannot hope to resolve these hypotheses. One must also note the relatively small size of samples and plots of crop cutting surveys before taking the results too literally. All that the present writer wants to do is to show that there is little statistical basis for the claim that the members attained a rate of growth higher than the non-members.

16. Until a well-documented and complete explanation is available one should perhaps view the extremely high cropping intensity estimates for 1966/67 with some scepticism.

17. All the data quoted on the costs and payments for Comilla tubewells have been taken from M. Ahsanullah, *Economics and Management of Tubewell Irrigation in Comilla Kotwali Thana*, (BARD, April 1972). The publication puts together a good deal of information but, unfortunately, makes no systematic use of it. Due to some unfortunate gaps in information it is impossible to put together the complete picture. But there is enough information to enable us to demonstrate the validity of our basic arguments.

18. The 15 per cent rate of interest is essentially an arbitrary figure but is

probably something like the lower limit of the estimated social value of capital. The farmers paid 15 per cent for credit from the co-operatives – which was by far the cheapest source of credit. A general equilibrium exercise carried out by the present writer for the 1960s shows that the accounting rate of interest for Bangladesh was about 15 per cent under favourable assumptions. See A. R. Khan, *The Economy of Bangladesh* (London: MacMillan 1972), Chapter 10.

19. This is because, among other factors, the capital cost was almost certainly underpriced due to the artificially low value in domestic currency of the imported tubewell equipment because of the overvalued rate of exchange. Also, repair and maintenance is probably underestimated and allowance has not been made for the fact that each year a number of tubewells remained non-operational.
20. For example, these were the changes levied in the village Joypur. See Md. Ahsanullah, op cit, p. 41.
21. The cost to the KTCCA was: Tk.3585 of annual interest on capital, Tk.500 of pay to the driver and Tk.478 of estimated repair and maintenance. The total was Tk.4563 which, spread over 48 acres, gives a per acre cost of Tk.95. Adding Tk.34 of cost per acre borne by the primary society, we arrive at the total of Tk.129 per acre. Once again, one must emphasise that this was almost certainly an underestimate of the true cost to the society. Also note that our cost estimates refer to that of tubewell water only and not to the whole process of tubewell irrigation (which involved additional costs at the field level).
22. For Bangladesh as a whole, the subsidy on tubewell water in 1973/74 had grown to TK.166 per acre which was 77 per cent of the annual cost per acre. For low-lift pump the subsidy per acre for Bangladesh as a whole was estimated to have been lower, Tk.107 or 68 per cent of the cost. See Planning Commission, *The First Five-Year Plan* (Dacca, 1973), pp. 150–1.
23. One should note that even if one could prove that the co-operative members had a higher or more rapid rate of adoption, one could not argue that membership constituted a higher level of innovativeness. It would still remain possible that more disadvantaged peasants were excluded from membership which bestowed special privileges which in turn enabled a more rapid rate of adoption of new practices.
24. The *Annual Report* of the BARD, 1967 (pp. 46–7) shows that in 1966/67 in Comilla 2274 members and 2145 non-members of tubewell agricultural societies had adopted irrigation. The average size of irrigated area was much lower for the non-members. This probably reflects nothing more than the lower average size of holding for the non-members. More on this later.
25. Data from BARD source quoted in Stevens, Alavi and Bertocci (eds.) op cit, p. 134.
26. The Commilla experiment was to serve as the laboratory for action research for the BARD. Whatever the success of the programme in other areas its performance in monitoring and documenting the effects of its own actions must be rated at best as mediocre. Scores of surveys have been carried out since the programme was launched. But these surveys

have been ill-designed, lacking a framework for a reasonably comprehensive coverage of the basic aspects on which the performance of the programme, on its own terms, was to be judged. Perhaps the worst failure relates to the quantitative measurement of change in the distribution of income. Since one of the objectives of the programme was to improve the distribution, it would have been natural for its action researchers to monitor systematically any change in inequality by collecting reasonably comprehensive information. Unfortunately, this expectation has remained unfulfilled. It, therefore, is necessary for us to depend on incomplete and fragmentary evidence.

27. A 1973 Survey (M. Solaiman, *Landholding and Co-operatives in Five Comilla Villages*, Comilla, BARD, 1974) shows that only 23 per cent of the landless had access to land on the basis of sharecropping, leasing or mortgaging. The landless represented only 14 per cent of all those who obtained land on this basis.

28. The requirements for becoming a member were as follows:

> Each member of the village society has to agree to fulfil the following obligations: he has to attend weekly meetings, he has to make regular deposits with the society, he has to buy at least one share at the initial stage and one share in each subsequent year, he has to repay loans timely and pay interest at 15 per cent per year if he takes a loan from the society.

M. Nurul Haq, *Comilla Co-operative Experiment*, (Comilla: BARD, 1966).

29. In later years the participation of the landless and small peasants has grown somewhat. In 1975, according to a survey of 3702 families in 24 villages in Comilla (A. Rahim, *Amon Crop Survey in Comilla Kotwali Thana*, Comilla: BARD, 1977) 9.1 per cent of the landless (presumably a proportion of those who rented in land from others) had become members. The same survey does not indicate the membership among those who had less than one acre. But a smaller survey in 1973, based on 558 households in 5 villages (M. Solaiman, *op cit*) claimed that 41 per cent of those who had up to 0.5 acres and 53 per cent of those who had more than 0.5 acres and up to one acre had joined. Indeed, these 1975 data, based on only 5 villages, may represent overestimates of the participation of these groups for Comilla as a whole. Thus this survey seems to overstate the overall participation rate for Comilla and, rather implausibly, claims that 32 per cent of the landless had joined while at the same time admitting that only 23 per cent of the landless had any access to land and fulfilled the criterion of being cultivators. In a private communication with the present writer Solaiman gave the following information:

> In the mid-sixties a special drive was given at Comilla to increase the membership and volume of loan disbursement. Many landless families at that time joined the co-operatives. Some of them were working as farm labourers. Many of them have left later, and wherever available, joined non-agricultural co-operatives like rickshaw pulling. Still some remained with the agricultural co-operatives, but not very actively, who are not regularly involved in farm work through renting in land.

30. A. H. Khan, op cit.
31. According to Solaiman's 1973 survey, the average size of a farm was 1.16 acres. In the late 1960s it was probably a little higher.
32. Unfortunately, we have little explanation of the data. In 1963/64 there was very little difference between per capita incomes of the small and large farmers. If one is to take the income data seriously, then the size must be taken to refer to the operational rather than ownership units. One could then argue that the large owners were content to rent out a part of their land (and thus become medium operators) while at the same time they pursued non-agricultural activities yielding high income.
33. In the section on productivity increase we argued that the members had a lower trend rate of growth in output per acre than the non-members. Is the present argument inconsistent with that? We do not think so. A higher yield per acre does not necessarily mean a higher *net income* per acre. In particular, it normally has a very tenuous connection with net income *per worker* or *per capita* since higher yield is frequently achieved by the more intensive application of family labour.

    In the village of South Rampur of Comilla Kotwali Thana all sales and purchases of land over the 10 years ending in December 1973 have been compiled (M. Rezaul Karim, 'Land Transaction in a Comilla Village' in *Journal of BARD*, July 1976) and the size distribution of 59 sellers and 73 buyers were as follows:

Landholding per Family (acres)	*No. of Sellers*	*No. of Buyers*
up to 1.00	27	7
1.01–2.00	18	16
2.01–3.00	4	14
3.01–4.00	4	8
4.01–5.00	6	12
above 5.00	—	16

    Clearly the smallest size group holding up to one acre was mainly losing land which was being bought up by the larger size groups. Unfortunately, the study does not show how many in each category belonged to the co-operatives.
34. See ILO, *Poverty and Landlessness in Rural Asia*, op cit, Chapter 7, for a time series on real wages for agricultural labour over the period 1949–75.
35. See S. R. Bose, op cit.
36. A. H. Khan, op cit.
37. See, for example, M. Ahsanullah, *Capital Accumulation Through Co-operatives*, (Comilla: BARD, 1973) p. 5.
38. 'A' class societies could borrow up to five times total deposits and shares; 'B' and 'C' class societies up to four and three times respectively. The classification was based on total deposit size, regularity of repayment and general co-operative behaviour (e.g. attendance at weekly meetings). See Ali Akhter Khan, *Rural Credit Programme of Agricultural Co-operative Federation* (Comilla: BARD, 1971).
39. By 1975/76 the credit programme was virtually completely clogged up

by a high rate of default. As a result total outstanding debt, for the first time, was below accumulated shares and savings.

40. There were probably other factors causing a reduced incentive for institutional saving. Thus, the sharp acceleration in the rate of inflation in the 1970s must have rendered the return on these assets highly negative.
41. These figures are to be found in Ali Akhter Khan, op cit. His estimates are based on a 1969 Survey of the borrowing profile of 30 co-operative societies in Comilla Thana.
42. For a detailed account of the procedure for application for and sanctioning of loans see Ali Akhter Khan, op cit.
43. Pesticides contained a 100 per cent subsidy and fertiliser a minimum of 25 per cent. In recent years the subsidy on fertiliser has been higher. More data will be presented in the last section.
44. Ali Akhter Khan, op cit, p. 28. The same report contains accounts of the malpractices on the part of the managers and the resentment of the ordinary members against such malpractices.
45. Ali Akhter Khan, op cit, p. 29. The reference to interest in kind applies to cases of extraordinarily high implicit monetary rates of interest.
46. Ali Akhter Khan, op cit.
47. P. K. Chowdhury, *Loan Default Under the Agricultural Co-operative Federation of Comilla Kotwali Thana*, (mimeo) (Comilla: BARD, 1976).
48. See the note to Table 4.9.
49. Akhter Hamid Khan, op cit.
50. Ministry of Agriculture and Forest, Government of Bangladesh, *The Revised Two Year Plan (1978–80) of the Agricultural Sector* (mimeo, Dacca, April 1978), Table IX. Today, the farmers in Comilla receive the same subsidies as do the farmers elsewhere in Bangladesh. It may be useful to reiterate that these subsidies are based on the difference between price and actual cost and not on the evaluation of social cost.
51. Ibid., Table XII.
52. This and the above figures about membership etc. of the IRDP are from Government of the People's Republic of Bangladesh (Ministry of Land Administration, Local Government, Rural Development and Co-operatives), *Rural Development Expansion Programme in Bangladesh* (Dacca 1977). It is not clear if the distribution of credit shown in Annex II refers to accumulated outstanding credit or only to the credit distributed in 1976/77. We assume that it refers to outstanding credit. In Comilla the peak outstanding credit was Tk.442 in 1970/71. In real terms the difference would of course be much greater.
53. Similar programmes have been launched elsewhere in Bangladesh with the assistance of bilateral donors. A more ambitious RD-2 programme, covering 29 thanas, has been formulated along these lines and World Bank support is expected to make its implementation possible.
54. This was not a problem in Comilla since it was a small island in a vast famine prone country. This may yet become a problem if the big push on an all Bangladesh scale becomes a reality. If demand from abroad, with or without a change in the cropping pattern, can resolve the problem then one may witness the coexistance of surplus production and domestic hunger.

55. For evidence see A. R. Khan, *Poverty and Inequality in Rural Bangladesh*, ILO (WEP) Working Paper, 1976.
56. These figures are based on the more detailed data than those presented in Table 4.16. See the references in Table 4.16 for the actual data. These proportions are based on the exclusion of landless households from the denominator. The last comparison is based partly on linear interpolation within size groups of the distribution. This has led to slight inaccuracies.

5 The State and Agrarian Change: A Case Study of Egypt, 1952–77

Samir Radwan and Eddy Lee

5.1 Introduction

The purpose of this study is to trace the changes in Egypt's agrarian system over the last quarter of a century in an attempt to assess the extent to which the successive agrarian reforms introduced in Egypt in 1960, 1962 and 1969 have succeeded in resolving the agrarian question, that is the creation of an alternative agrarian system more conducive to egalitarian rural development. After briefly outlining the pre-reform agrarian system, we shall try to assess the effect of the reforms on production and productivity, distribution of land and income in general and the problem of rural poverty in particular, and provide an appraisal of the 'supervised co-operatives' – the most important institution created by the Egyptian agrarian reform. Finally, we shall conclude with a note on the nature of the present-day agrarian system.

The study of the Egyptian case is interesting at both the general and the specific levels. Egypt is a good example of that group of Third World countries that have tried to change their agrarian systems through a fairly radical land reform and the introduction of new forms of organisation, usually co-operatives, but without altering the basic characteristic of the pre-reform system based on private land-ownership. The Egyptian experience is, however, particularly instructive due to several elements specific to it. One is the centrality of its political system and the active role of the state in controlling irrigation and hence agriculture. This provides a unique historical example for studying the contrasting role of the state in development both under the colonial and post-colonial conditions. Moreover, Egypt is a typical case of 'arrested development' where spurts of growth are followed by periods of stagnation, and it would be extremely instructive to observe the pattern of agrarian change in the context of a stagnant economy.

5.2 The evolution of Egypt's agrarian system before land reform

A central episode in the economic and social history of Egypt has been the introduction of cotton cultivation in the 1820s and the consequent transformation of a poor subsistence agriculture into a commercialised sector producing for exports. The agrarian system that evolved over the following one and a half centuries was strongly conditioned by this episode.

One of the most distinguishing features of this transition was the impressive growth of agricultural output. Available data indicate that Egypt has experienced something like an 'agrarian revolution' during the course of the nineteenth century.[1] Over the fifty years between 1820 and 1870, total agricultural output, in real terms, increased just over 12 times, while per capita production rose nearly six times. This impressive performance could be attributed mainly to two factors: the extension of the cultivated area through the massive investment by the state in the development of one of the most complex systems of irrigation in the world; and the reallocation of land in favour of more profitable cash crops, especially cotton. From 1870 onwards, the growth of agricultural output began to slow down. The average annual rate of growth of agricultural output was 3.5 per cent during the period 1872/78–1895/99 compared to more than 5 per cent over the previous 50 years. The main explanation for this decline in the growth rate can be found in the fact that by the 1880s Egypt had reached that stage which Ishikawa calls 'the virtual disappearance of the arable land frontier' where a new round of 'basic investment' in irrigation becomes an essential precondition for any further addition to the extensive margin of land and increase in agricultural output.[2] The history of agriculture has been mainly the history of attempts to add either to the extensive margin of land through investment in irrigation, or to the intensive margin through multiple cropping and the heavy use of fertilisers. But by the 1930s a stage was reached where any addition to the arable land was almost impossible and the only increase in productivity would have to come from investment in 'new inputs'. However, one of the striking features of Egyptian agriculture has been the lack of investment in technical innovations beyond fertilisers. Small peasants have been unable to afford the price of these new inputs and large landowners (who mostly rented out their land) had no incentive to invest in new machinery due to the existence of cheap labour.

Rapid growth of population in the context of fixed supply of land and the absence of technical change has resulted in the early appearance of diminishing returns to the abundant factor of production: labour. The average land to labour ratio deteriorated from 100 at the turn of the century to 68 in the 1950s and the average output per worker also decreased from 100 to 73 during the same period (Table 5.1). The turning point was the 1930s when it became clear that Egyptian agriculture could no longer absorb the increasing population and hence large-scale migration to the urban centres began.

This pattern of production and accumulation took place within the framework of, and was also perpetuated by, a land system based on private ownership and a production system characterised by extreme duality between large estates and small peasant farms.

Egypt's agrarian system of the 1950s had its historical roots in the system of private landownership, established in the second half of the nineteenth century and the concomitant emergence of a powerfull class of large landowners.[3] The most striking feature which characterised the evolution of this system was the heavy concentration of landownership on the one hand, and the continued fragmentation of small landholdings on the other. An initial situation of extreme inequality in the distribution of land had actually worsened with the passage of time. Thus we find that at the turn of the century small farmers (owning less than five *feddans*) who represented 80 per cent of the total number of landowners, possessed only 20 per cent of the land, while large proprietors (owning over 50 *feddans*) representing only 1.5 per cent of landowners, held 44 per cent of the land.[4] The process of differentiation continued. A striking feature of this process was the way in which medium landowners were able to consolidate their position. Between 1910 and 1952, the average size of medium properties as well as their share in the total cultivated area remained virtually constant. By contrast, the number of small owners more than trebled in 50 years, and although their share of the total cultivated area increased from 22 per cent to 35 per cent between 1900 and 1952, the average size of their properties fell from 1.46 to 0.80 *feddans*, i.e. by 45 per cent. Furthermore, a finer breakdown shows that most of the increase in

Table 5.1 Indices of Agricultural Growth, 1895–99 to 1950–54

	Total Output	*Rural Population*	*Agricultural Labour Force*	*Cropped Area*	*Cultivated Area*	*Net Fixed Capital Stock*	*Input of Chemical Fertilisers*	*Output/ Worker*	*Output/ Unit of Land*	*Land/ Labour Ratio*	*Cropping Intensity*	*Capital/ Unit of Land*	*Capital/ Worker*
1895–9	100	100	100	100	100	100		100	100	100	100	100	100
1900–4	110	108	108	109	108	120		102	101	101	101	111	111
1905–9	116	117	118	113	109	160		98	103	96	104	147	136
1910–14	121	125	126	114	107	176		96	106	90	106	164	140
1915–19	103	131	136	115	107	181	100	76	90	85	107	169	133
1920–4	113	138	153	120	107	194	144	74	94	78	112	181	127
1925–9	133	143	169	127	113	213	387	79	105	75	112	188	126
1930–4	135	150	186	136	111	267	474	73	107	68	114	241	145
1935–9	135	156	207	123	107	343	910	74	124	59	115	321	167
1940–4	153	162	205	133	108	377	397	64	98	65	123	349	184
1945–9	131	173	204	136	116	410	600	69	103	66	117	353	201
1950–4	140	188	206	140	115	420	1246	73	107	68	122	365	204

Source Samir Radwan, *Capital Formation in Egyptian Industry and Agriculture, 1882–1967*, Ithaca Press (London: 1974) pp. 123 and ff.

small properties took place in the bracket 'less than one *feddan*'. Between 1910 and 1952 the number of owners in this bracket increased from 783,000 to 2,018,000 (compared to an increase from 464,000 to 624,000 in the 'one to less than five *feddans*' bracket) and the average size of property declined from 0.47 to 0.38 *feddans*.[5] This development was largely due to a rapid increase in rural population combined with the Muslim laws of inheritance which required the breaking up of properties among heirs upon the death of the owner. Despite the operation of these laws, both the number of large- and medium-sized properties and their area showed little change over time. This is partly explained by the fact that newly reclaimed land and land sold by the state from public domains from time to time was generally acquired, in the first instance, by the bigger landlords, a practice which largely offset the negative effect of the breaking up of their properties on the total area owned by them.[6] Thus, through the parallel processes of fragmentation of small properties and consolidation of medium and large properties, the distribution of landownership, extremely uneven from the beginning, became even more unequal. The Gini coefficient for the distribution of landownership increased from 0.696 in 1896 to 0.785 in 1950. Consequently, we find that on the eve of the First Agrarian Reform of 1952, large landlords, those owning over 50 *feddans*, numbered less than half of 1 per cent of the total number of proprietors but held 35 per cent of the land, while, at the other end of the spectrum, small farmers, owning less than one *feddan*, represented 72 per cent of total landowners but held only 13 per cent of the land. It is not surprising, therefore, that Egypt was then described as the 'half per cent society'. The two million families who owned less than one *feddan* each constituted the poorest section of the landowning class; some, having leased their land to other peasants, lived in towns or joined the army of casual labourers in the villages; others rented land to supplement their holdings wherever possible.[7]

Outside the spectrum of landowners we find the landless peasants; the poorest of the poor. With no access to land either through ownership or rent, the number of landless families as well as their proportion to total rural population have been growing considerably. In 1929 there were 697,000 landless families, or 37 per cent of the rural households. The ratio increased to 53 per cent in 1939, and by 1950 there were some 1.5 million families, representing 60 per cent of rural population, who neither owned nor rented land at all and who had to rely exclusively on their labour for earning a living.

The lopsided development of private landownership sketched above gave birth to an agrarian system that can best be described as *capitalist* in structure and *feudal* in its social relations.[8] The introduction of cash crops and the consequent integration of Egypt as a cotton producer into the world economic system; the development of a complex transport network linking the farms to the market place; the creation of a sophisticated trade and financial structure, have all contributed to the transformation of Egyptian agriculture from a subsistence into a highly commercialised and profit-maximising activity. The dual structure of landownership – large estates and small peasant holdings – was also reflected in the mode of land exploitation. Egyptian agriculture was dominated by a large sector of small family farms totally dependent on family labour and using primitive tools and implements. By contrast, medium and large estates depended mainly on hired labour, permanent or casual, and employed modern machinery. The organisational structure of factor markets (land, labour and capital) and the land tenure relationship functioned under a *laissez-faire* system.[9] Wages, in a labour-surplus economy, were always pushed to subsistence levels and below. The credit market was controlled by foreign banks that were mainly concerned with financing the cotton trade. Their lending policies favoured large landowners and small farmers had to rely on private money-lenders who charged exorbitant rates of interest. The extremely unfavourable man to land ratio was associated with an active land market. An increased demand for an almost perfectly inelastic supply of land contributed to a tremendous inflation of land prices without any actual increase in productive investment. Moreover, where ownership of land was concentrated in a few hands, leasing was the only way in which the land 'hunger' could be satisfied. Absentee landlords leased their land either directly or through intermediaries under a wide range of tenancy contracts: cash rent, rent in kind and sharecropping. It was estimated that by 1950 some 60 per cent of Egypt's cultivated land was tenanted (as against 17 per cent in 1939).[10] Acute competition and exploitation by intermediaries sent rents rocketing. Average rents increased five times between 1938 and 1950. In many cases it was more advantageous for the landlord to lease out his land than to farm it.

This defective land tenure system not only resulted in an extremely unequal distribution of income as we will see later, but also enforced a pattern of social relations that could be described as feudal. The village scene was dominated by the large landlord usually referred to as the *Pasha.* His manor house, which overshadowed the mud houses of the *fellahin,* was run by a small army of permanent workers and their families. Invariably, he, as the *Umda* (or village mayor), was the representative of the state. In short, he represented the political and social power in the village, and he certainly used this power to ensure his mastery. Tales of social subjection, illegal abuses, repression and human degradation of the *fellahin* were a pale description of the grim reality of Egyptian rural life. It is not surprising, therefore, that by the early 1950s, the so-called passivity of the Egyptian peasantry was wearing thin and signs of unrest were appearing in various parts of the countryside.[11]

It is against this background of mounting agrarian crisis that a number of writers, social reformers and politicians called for the reform of the agrarian system.[12] Various elaborate schemes were proposed but all went unheeded by a government dominated by large landowners. The new régime brought in by the 1952 Revolution recognised the urgent need to tackle the agrarian question. One of their first actions was the promulgation of the Land Reform Law of September 1952 which represented a frontal attack on the power of the entrenched class of big landowners and an attempt at a radical reform of an ailing agrarian system.

5.3 The post-reform agrarian system

In this section we shall delineate the main features of the agrarian system that resulted from the agrarian reform. The first, and perhaps most important point to note was the Land Reform of 1952 and subsequent measures in 1961 and 1969 did not result in the total upheaval of the old system. In fact, this was not one of the aims of the reform, the major objective of which was to break the power of large landowners and raise the standard of living of the peasantry through the enforcement of complementary measures such as tenancy reforms and minimum wage legislation within the framework of private landownership. Thus, as regards expropriation of large landowners, we see that the ceilings of landownership were extremely high in relation to the average size of landownership; the ceiling under the 1952 legislation was 300 *feddans* compared to the average ownership size of 2.1 *feddans* in the same year. Although this ceiling has been lowered in subsequent legislation, it still remains very high in relation to average farm size and as a consequence by 1970 only 13 per cent of total farming land was expropriated and redistributed to only 9 per cent of the farming population. As a point of reference we note that the land reforms in Japan, Taiwan and South Korea were far more drastic in scope; in the case of South Korea, the ceiling on ownership was only three hectares and two-thirds of the farming households received land under the land reforms.

The net redistributive impact arising out of the land that was actually transferred was also relatively weak. First, the level of compensation, set at ten times the annual rental value of the land, was relatively generous. Secondly, the full burden of compensation was imposed directly on the beneficiaries. They were to pay this compensation by instalments over a 30 year period together with an additional 15 per cent in charges. There was also an annual interest of 3 per cent that had to be paid on bonds issued to the expropriated landlords. Thus, not only was the transfer of land limited in extent, its redistributive impact was also limited by the negligible extent of the transfer of wealth between the expropriated and the small group of beneficiaries.

The tenancy and minimum wage reforms had a potentially wider redistributive impact. At the time of land reform 60 per cent of the total land area was tenant-cultivated and an estimated 60 per cent of rural families were landless and hence dependent, to varying degrees, on income from agricultural wages. The number of potential beneficiaries from these measures was thus very great. The 1952 Land Reform Law fixed rents at seven times the land tax in case of cash tenancy and, in the case of sharecropping, the tenants' share in the crop at 50 per cent, with equal participation in costs between owner and tenant. Security of tenure was also guaranteed for three years. The impact of these measures, fully applied, could have lowered the level of cash rents by 30–42 per cent[13] and brought about an increase of one-third in the share of total output retained by sharecroppers. Similarly, the minimum wage for an eight hour day was set at 18 *piastres* for men and 10 *piastres* for women compared with the prevailing rates of 10–15 *piastres* and 6–7 *piastres* respectively.

Undoubtedly, there was a significant initial redistribution as a result of these measures but the enforcement of rent control and minimum wage legislation was patchy and evasion was widespread. The strong demand for land and the domination of land reform institutions by large landowners made it easy to continue squeezing peasants. In addition, under conditions of growing population pressure, many landlords had succeeded in evading the minimum wage stipulations and, as we show in the next section, there has been no sustained improvement in real wages.

Thus, neither the redistribution of land itself nor the accompanying tenancy and wage reforms had any radical effect in changing the agrarian system. There were some initial, once-and-for-all gains in asset and income redistribution but these were, as we shall show, eroded away as the system reverted to a new inegalitarian equilibrium. The post-reform system was still based on the private ownership of land and the distribution of land remained highly unequal; the top 5.6 per cent of landowning households still owned 53.4 per cent of the total cultivated area after the 1952 reform. The vast majority of the landowning population (94.4 per cent) owned only 46.6 per cent of total cultivated land and owned an average of one *feddan* per owner; in contrast, 20 per cent of the total area was in ownership units of more than 50 *feddans* and another 14 per cent were in units of between 20–50 *feddans*. Thus, the range in the size of ownership units still remained enormous in spite of the elimination of the top of the landowning pyramid through the 1952 reform.

Furthermore, a large pool of landless families continued to exist. No less than 40 per cent of agricultural families were still estimated to be landless after the land reform.[14]

Because land continued to be highly unequally distributed and a large pool of landless peasants continued to exist, it is clear that the structural conditions for the perpetuation and deepening of rural poverty remained untouched. Unequal landownership confers monopsonistic power to the landowners in labour markets against a large and powerless rural labour reserve.[15] Market power also extends to credit markets and economic control by the rich of the poor through usurious relationships. Moreover, political power, as a result of the land reform, had merely shifted from the very rich to the rich; the political position of the mass of the rural poor remained unchanged.

Another distinctive feature of the post-reform agrarian system, and one that is perhaps unique to the Egyptian case (as well as the case of Mexican Eijidos), was the introduction of the system of 'supervised co-operatives'. The First Agrarian Reform of 1952 made it obligatory for all land reform beneficiaries in any one village to form a co-operative society among themselves. The co-operative was to replace the former landowner in the organisation of cultivation, provision of credit and other inputs, and the marketing of produce. With the government concentrating its services through co-operatives, the system was extended beyond the boundaries of agrarian reform areas until it covered almost all of rural Egypt by the mid-1960s. We can distinguish three types of agricultural co-operatives: (i) the agrarian reform and the land reclamation co-operatives, created in land reform and newly reclaimed areas to cater exclusively for the needs of land distribution beneficiaries; (ii) the general or multi-purpose co-operatives (also known as the credit co-operatives) which have as their members all the farmers on the 'old' land, i.e. outside land reform areas; and (iii) the specialised co-operatives, with a membership of producers engaged in agriculture-related activities such as animal husbandry, fishing and food processing. Of these three categories, the general, or multi-purpose is the most important: these co-operatives service some 87 per cent of the country's cultivated land, account for 83 per cent of the total membership of co-operatives and control 70 per cent of their capital.[16]

The organisational structure of these co-operatives is pyramid shaped with the local or village society at its base, the co-operative union at the top and three middle strata corresponding to the country's administrative hierarchy: the District (*Markaz*), the Governorate (*Muhafaza*), and the country as a whole. The local co-operative, which is the cornerstone of this structure, is created in each village with a minimum membership of 20. It is run by an elected council of five to seven members assisted by a supervisor (*Mushrif*) who is usually an agricultural engineer appointed by the Ministry of Agriculture. In many co-operatives, the government also provides a clerk and an accountant.

This elaborate system of agricultural co-operatives has two basic functions: to be the primary channel through which government services reach the small farmers and to serve as an efficient instrument of government policies concerning the improvement of production and the mobilisation of agricultural surplus. The wide network of co-operatives has enabled the government, perhaps for the first time, to reach the remotest areas of rural Egypt and to influence in a direct way the course of rural development. Through the co-operatives, the government was able to re-organise production by introducing the system of land consolidation and the arrangement of crop rotation. An elaborate marketing and pricing policy also enabled the government to manipulate the agricultural surplus.

The organisation of supervised co-operatives was thus a systematic attempt by the state to control the functioning of the agricultural economy. In this sense it can be seen as the manifestation of the general drive towards state control and planning in the Egyptian economy and an extension of the bureaucratic apparatus to the rural areas. There were three basic instruments of control in the system. First, there was the control through the agricultural engineer attached as a supervisor to the elected co-operative council. His writ extended to the enforcement of crop-rotation schedules, water control and the organisation of key farming operations such as mechanised ploughing, crop fumigation and pest control. Secondly, the co-operatives had a monopoly over the supply of key inputs such as chemical fertilisers, insecticides, seeds and institutionally-provided credit. Thirdly, the co-operatives exercised control over the marketing of the most important crops and were also the instrument for enforcing the direct procurement of agricultural produce.

It is worth noting that such an extent of state control has few parallels in non-socialist developing countries. In many cases the network of co-operatives either represent localised experimental schemes or cover only a minority of the rural population. Secondly, the enforcement of centralised crop-rotation schedules and direct control over other aspects of agricultural operations over the entire countryside is probably unique to Egypt.

5.4 The impact of agrarian reform

Having outlined the main features of the post-reform agrarian system, we shall now draw on available data to assess the effect of agrarian reform on rural Egypt. Here, we shall concentrate on the performance of the system in three areas: output growth, income distribution, and poverty and the impact of the co-operative system.

(a) Growth performance

One of the interesting features of Egypt's co-operative system was the attempt to increase agricultural production through the introduction of the system of land and crop consolidation – thus reversing the effects of rapid fragmentation – and through the supply of improved inputs – seeds, fertilisers, pesticides, technical advice and, above all, credit. Available data suggest, however, that the period 1952–77, taken as a whole, was one of slow growth for Egyptian agriculture. The value of four of the major crops, cotton, wheat, maize and rice, increased from £E182 million in 1950 to £E232 million in 1975 (measured in constant 1950 prices); an average annual rate of growth of 1.2 per cent. But the growth of output was not even over this period. We can in fact distinguish between two major sub-periods in terms of growth performance. Table 5.2 shows that the average annual rate of growth of value added in agriculture at constant prices amounted to some 4.0 per cent during the 1950s and mid 1960s, then declined to 2.0 per cent between the mid 1960s and the 1970s. The improved performance during the first period must reflect, at least partially, the impact of improved co-operative services and the effect of the new rotation system. This is confirmed by the observed increase in productivity – as measured in terms of yields of major crops – during the same period. In 1965, indices of yields per *feddan* for the seven major crops (average 1948–51=100) amounted to: cotton 124; wheat 143; maize 167; millet 142; barley 134; rice 132; and sugar cane 130.[17] This initial improvement did not continue during the second period. The decade 1965–75 witnessed a negligible increase in the yields of these crops. Taking 1965 as a base, the yield indices in 1975 were: cotton 105; wheat 122; maize 102; millet 102; barley 111; rice 101; and sugar cane even declined to 92. This reflects the combined effects of a number of factors most important of which were the increased waterlogging and salinity, the lack of continued investment in agricultural infrastructure and the fact that Egyptian agriculture had reached a stage where any increase in productivity would be almost impossible given the techniques of production used. There is no doubt that Egyptian crop yields are high compared with world averages,[18] but not unduly so considering the country's warm climate and the fact that agriculture is fully irrigated. Had the co-operatives continued the initial push by introducing high-yield varieties and improved production techniques, and had the government carried out the planned investment in covered drainage, the impressive performance of the 1950s and 1960s could have continued.

A more disturbing aspect of the peformance of Egyptian agriculture over the period studied is the decline in output per capita. Table 5.2 shows that after a modest increase in the index of output

Table 5.2 Growth Performance of Egyptian Agriculture

	1960	*1965*	*1969*	*1970*	*1971*	*1972*	*1973*	*1974*
Value of output £E (1)	559	797	966	1040	1123	1223	1392	
Costs of inputs £E (1)	147	190	241	262	306	318	373	
Value added £E (1)	412	606	725	778	817	905	1019	
Index of total output (2)	84	107	123	123	127	129	130	132
Index of output per capita (2)	88	102	105	102	102	101	99	97

	1956/61	1961/66	1966/71	1972/77
Average annual growth rates (%): (3)	3.5	4.0	2.0	2.0

Source Central Agency for Public Mobilisation and Statistics (CAPMS), *Estimates of National Income from the Agricultural Sector*, various issues and FAO, *Production Yearbook*, various issues.

Notes 1. At current prices. 2. FAO Index (1961–5=100) based on a weighted average of physical crop production. 3. Based on value added at constant prices.

per capita during the 1960s, a slow but steady decline began by the early 1970s. This reflects the effect of a slow rate of growth of total output (2 per cent per year) combined with a rate of population growth of 2.5 per cent per annum. The deterioration in land to man ratio which began as early as the 1930s had reached its critical limit by the 1970s. Neither the marginal additions to the arable land as a result of the High Dam, nor the modest increase in productivity were sufficient to counterbalance the effects of this increase in rural population. These developments have serious implications not only for the future growth of agriculture, but also for the economy as a whole. With agriculture being unable to support the increase in population, rural-urban migration has been increasing at unprecedented rates. The 1976 Population Census reported that the proportion of urban to total population in Egypt was 44 per cent compared to 40.5 per cent in 1966 and 37.4 per cent in 1960.[19] Moreover, Egypt's agriculture is becoming less and less capable of supplying the growing population with their needs of food with the consequent increase in the country's dependence on imports and the rapid inflation in food prices. Agricultural imports which accounted for 34 per cent of total imports in the mid 1960s have increased to 60 per cent since 1974.[20] Wheat and flour alone account for about half of these imports. Moreover, the official wholesale price index of 'Foodstuffs and Beverages' has almost doubled during the same period.[21]

The growth performance of the Egyptian agriculture over the last 25 years outlined above demonstrates both the potential and the limitations of the agrarian reform. While improved co-operative services and better crop rotation resulted in a better performance, the decline of the co-operative system since the late 1960s was partially responsible for the ensuing deterioration. Egyptian agriculture is at cross-roads with the deterioration in land to man ratio and stagnation in productivity. Future growth will have to depend on improvement in input supplies and production techniques. In an agriculture where the overwhelming majority are very small holders (95 per cent of landowners are holders of less than five *feddans*) these innovations can be provided by the co-operative system only if it is restructured radically.

(b) Income distribution and poverty

We have seen in the previous section that aggregate figures on production and yields show a moderate upward trend. The same has not, however, been true of the trend in average income per household in rural areas over this period. The picture that is revealed by the four surveys of household income and expenditure that are available is one of stagnation in real incomes between 1958/59 and 1977, where gains in money incomes have been wiped away by a high rate of inflation. These surveys are the official Family Budget Surveys carried out in 1958/59, 1964/65, and 1974/75, and the ILO survey on rural poverty carried out in 1977. The figures on rural household expenditure from these surveys are as follows:

	Average Expenditure Per Rural Household	*Rural Consumer Price Index*	*Real Expenditure Per Household*
1958/59	£147.7	100	£147
1964/65	£224.2	143	£157
1974/75	£375.5	237	£158
1977	£383.0	257	£149

Source CAPMS, *Family Budget Surveys*, 1958/59, 1964/65, 1974/75; and for 1977 Samir Radwan and Eddy Lee, *The Anatomy of Rural Poverty: Egypt, 1977* (ILO, forthcoming).

It will be seen that average expenditure per household in money terms more than doubled between 1958/59 and 1977 but that in real terms there has been only a negligible increase and even this was short-lived as figures for subsequent years show.

It should be noted that there are statistical objections to the procedure of inferring a trend from scattered observations from sample surveys but there are some factors which tend to weaken the strength of these objections in the Egyptian case. The first three surveys were all carried out by the same statistical agency (CAPMS) and form part of a regular sample survey national programme. This ensures consistency in definitions, concepts, sampling size and procedures etc. which removes many of the objections to the making of intertemporal comparisons on the basis of these data. The fourth survey was also carried out in consultation with CAPMS and consistency with the previous surveys in these matters was ensured. Moreover, all four surveys showed such consistency in the average household size of the sample, that objections about the distorting effect of changes in demographic structure cannot be seriously entertained.

Be these objections as they may, a broad generalisation that average rural incomes have stagnated between 1958 and 1977 seems incontestible, especially since such a trend is consistent with other indicators of economic change in the rural economy. Firstly, as we have seen earlier, the trend in physical agricultural output per man has been downward, although yield per hectare has risen. Secondly, landlessness has been increasing and more importantly, the proportion of the rural population engaged in the low income, low productivity, non-agricultural activities has increased sharply. As we shall argue later, this can be interpreted as an indication of 'distress adaptation' within the countryside to mounting demographic pressure on limited land. Thirdly, there has been a clear trend of increased differentiation among the peasantry,[22] a factor which explains how average incomes have remained constant while the incomes of some have increased significantly. Fourthly, the data on per capita real income over this period is broadly consistent with the trend from the survey data (see Table 5.11). Fifthly, real wages in agriculture have shown no increase. Finally, the data on the size-distribution of rural incomes and on the proportion of households falling below a poverty line from the four surveys we mentioned point to growing inequality and absolute poverty, a trend which is consistent with all the above indications of stagnation in average incomes.

We shall proceed by documenting the changes in income distribution and poverty that are revealed by the survey data. The first point to note is that inequality has increased; the Gini coefficient for household consumption expenditure for the four years were:

	Gini coefficient
1958/59	0.370
1964/65	0.353
1974/75	0.392
1977	0.393

Although there was a reduction in the degree of inequality between 1964/65 and 1958/59, by the mid-seventies this trend had been more than reversed. It should be noted that these measures related to money incomes and probably understate the extent of increase in inequality. As already mentioned, the rate of inflation was high during this period and there is evidence that the price of 'wage goods' increased faster than the overall index.[23] This would suggest that measures of inequality based on real incomes obtained by applying the appropriate group-specific cost-of-living index would show a greater increase in inequality.

A corollary to the increase of inequality around a constant level of income is the increase in absolute poverty. This can be established by using the Poverty Line approach. Our results are summarised in Table 5.3.

It should be emphasised here that these estimates should be treated as rough indicators of the orders of magnitude rather than accurate measures of the problem of rural poverty. Nevertheless, some broad conclusions can be drawn from these data. The most important perhaps is that despite a noticeable decrease in the number of the rural poor, both in absolute and relative terms, between 1958/59 and 1964/65, the problem of rural poverty remained unsolved. Thus, by the mid-1960s, more than one-quarter (27 per cent) of the rural families comprising over 3 million people were living below the poverty line. Much more serious is the dramatic increase in rural poverty over the following decade both in absolute and relative terms. Thus, by 1974/75, 44 per cent of the families in rural Egypt, comprising some 5.8 million people, were living below the poverty line. The results of our survey suggest that the trend towards increased impoverishment has continued. In 1977, there were 35.3 per cent of the rural households whose per

Table 5.3 An Estimate of the Rural Poor in Egypt, 1958/59, 1964/65 and 1974/75

	1958/59	*1964/65*	*1974/75*
1. Household income corresponding to poverty line (Egyptian pounds)	93	125	270
2. Total population ('000s)	25832	30139	36417
3. Rural population ('000s)	15968	17754	20830
4. No. of rural families ('000s)	3224	3345	4166
5. Families below poverty line:			
% of rural families	35.0	26.8	44.0
No.	1160640	903150	1833000
6. Population below poverty line:			
% of rural population	22.5	17.0	28.0
No.	3592800	3018180	5832400

Source For details on the data and methods used see S. Radwan, *Agrarian Reform and Rural Poverty, Egypt, 1952–1975*, ILO (Geneva, 1977) pp. 40–50.

capita expenditure fell below the poverty line. Moreover, an additional 20 per cent of the rural households could be classified as 'marginal' in the sense that they were able to satisfy their food requirements, but their incomes fell below the poverty line, thus indicating their vulnerability.[24]

Our recent survey throws some light on the characteristics and causes of poverty in rural Egypt. First of all, we found a substantial correlation between the various manifestations of poverty. Thus, households with incomes below the poverty line suffered a shortfall of their calorie intake from their energy requirements; their consumption pattern was dominated by expenditure on food and their diet was cereal-based and protein-deficient; and their access to such basic needs as education and housing fell short of minimum requirements. In an attempt to understand the causal factors in the process of poverty generation, we identified these basic attributes: the ownership of assets, employment and demographic characteristics and sources of income. The picture that emerges is extremely revealing. First, we found out that the disparity in the distribution of land and other productive assets had serious implications for income distribution. Assets were highly concentrated; Gini coefficient for the size distribution of assets was as high as 0.725. Moreover, the results of the regression between household income and assets shows that a substantial proportion (about 30 per cent) of the variation in household incomes can be explained by differences in asset holdings. Assuming *Y* to be the household income and *K* assets, the regression results were as follows:

$$\log Y = 1.3343 + 0.5354 \log K \quad R^2 = 0.286$$
$$(0.07156)$$

The following table sums up the situation: the strong positive correlation between the value of assets and both household and per capita incomes.

Household Groups	*Average Value of Assets £E*	*Average Income Household £E*	*Average Income Per Capita £E*	*Household Size*
Poor	1016	327	46	7.1
Marginal	764	267	50	5.3
Non-poor	2125	486	93	5.2
Total	1501	384	65	5.9

But asset ownership alone cannot explain all these variations in income. In fact our data show that employment was perhaps equally important. A striking feature of the distribution of income by source is the remarkably high proportion of earnings (51 per cent) generated from wage employment in and outside the village. This is hardly surprising in view of the fact that half the labour force in the villages surveyed were engaged in non-agricultural activities. What is more significant as a differentiating factor between poor and non-poor households is the difference in the sources of income. While the non-poor derive 53 per cent of their earnings from ownership of assets and 47 per cent from wage employment and remittances, the order is almost reversed in the case of the poor and marginal households; 43 per cent from assets, 53 per cent from wages and 4 per cent from remittances.

The picture that emerges from the preceding analysis has extremely interesting implications for our understanding of the process of poverty generation. It is evident that in a country like Egypt, with a rapidly growing population and a fixed supply of arable land, employment becomes an important source of income. It appears that a process of 'tertiarisation' has been taking place in rural Egypt where the poor who face limited access to land and have little prospect of being absorbed in the urban economy try to earn their living by engaging in such low productivity activities as handicrafts and petty trade. The expansion of this tertiary sector reflects the crisis of development in this country: the difficulty to increase productive agricultural employment given the present structure of land ownership and level of technology, and the inability of the industrial sector to create employment at a rate fast enough to absorb the surplus agricultural labour.

The dramatic increase in rural poverty can also be analysed in terms of our earlier conclusions about the inability of the agrarian reform policies to alter fundamentally the unequal distribution of land and income in rural Egypt. The marginal redistribution of land and income which resulted from the major reforms of the 1950s and early 1960s, has no doubt led to some improvement. But these improvements were soon to be overcome by the dynamics of the situation which reproduced inequality. Another important factor in explaining this increase in poverty is inflation. It is obvious from our estimate of the poverty line, that the cost of the minimum diet per capita was more than doubled between 1964/65 and 1974/75, and increased by about 25 per cent over two years, 1975/77. The spurt of inflation which occurred in the beginning of the 1970s (especially since 1973) has inevitably accelerated the process of marginalisation of the lower income groups who previously survived just above the poverty line by virtue of a price control and subsidy system.

It is clear, therefore, that the agrarian reform has not arrested the process of impoverishment in rural Egypt. It is of course possible that in the absence of agrarian reform this process might have occurred faster or with greater intensity. Indeed, the evidence of the temporary improvements in distribution until the mid 1960s would be consistent with such an interpretation. However, such a question is not central to our analysis. Regardless of its temporary impact, it remains true that the reform did not change the system and that the fundamental forces which perpetuate poverty in rural Egypt have continued to operate. The outcome that we have observed in terms of increasing poverty is, to us, evidence of the ease with which these forces re-assert themselves after a partial agrarian reform. These forces can be basically classified as those arising from diminishing access to land and the process of growing differentiation among those with access to land as a result of the operation of market forces. In addition, the inability of the industrial sector to provide employment opportunities for the growing rural labour force made it difficult to take the pressure off a stagnant agricultural sector. These pressures would have been reduced had the rate of industrial expansion been higher, as was the case in South Korea and Japan. The industrialisation drive of the 1960s has no doubt increased the contribution of industry to GDP growth. Industry contributed 25 per cent of the increase in real GDP between 1952/53 and 1959/60, and this contribution increased to 27 per cent between 1959/60 and 1969/70. Nevertheless,

> Hopes that industrialisation will solve the country's employment problem, caused by the inability of a land-scarce and labour-intensive agriculture to absorb the ever increasing number of new entrants to the labour force, are not yet fulfilled. Manufacturing created new jobs for 18 per cent of the increase in labour force between 1937 and 1960; and, despite the employment drive, for only 16 per cent of the increase between 1960 and 1970.[25]

In the post-agrarian reform period the rate of increase in rural population has remained at almost 2.5 per cent per annum in spite of rapid increase in urbanisation.[26] Between 1960 and 1970, the total rural population increased by 19 per cent whereas the number of landholdings, an indicator of access to land, increased by only 12 per cent over a similar period (1961–72).[27] This implies growing landlessness and a recent estimate shows that the proportion of landless families had increased from 40 to 45 per cent of those engaged in agriculture between 1961 and 1972.[28] Thus, apart from the once-for-all gain in terms of improved access to land (the extent of which we believe to be less than that given in previous estimates), the process of growing landlessness has continued. Continued growth in rural population and a highly unequal distribution of landownership ensured this outcome.

Our recent survey[29] has yielded some interesting information on the nature and extent of landlessness. Forty-nine per cent of households in our sample operated a farm. By this criterion, 51 per cent of households could be considered 'landless' but this would be an inaccurate depiction of the situation. The detailed information available from the survey shows that there is no simple way of defining landlessness because of the high incidence of multiple economic activities among rural households. Thus if we classify households according to the occupation of the household head we find that only 36.7 per cent of households have been 'operating a family farm' as their main economic activity and that only 13 per cent of households are headed by 'landless agricultural labourers'. Of the remaining 50 per cent of households, however, about one-quarter operate a farm even though the household head is in a non-agricultural occupation. Any estimate of landlessness is thus highly sensitive to the definition and classification system that is used. Thus, from our data we see that the non-agricultural population amounts to a staggeringly high 51 per cent if we use the occupation of the household head as the classificatory principle, but it drops to a more reasonable figure of 38 per cent if we define the non-agricultural population as those households not operating a farm and not engaged in agricultural wage labour.

A crucial point to note, however, is that, whatever the definition used, the proportion of the non-agricultural population is extremely high. Although these proportions are higher than those revealed by recent census figures, it should be noted that they were already high then and the trend had been upward. For example, Mabro estimates, on the basis of the 1969/70 figures that only 67 per cent of employment in the 'provinces' was agricultural while 'services' accounted for 25 per cent. Moreover, 'nearly half of all new jobs in the provinces were created in the tertiary sector'.[30]

Two interrelated characteristics of Egypt provide the essential explanation. The first is the blurred distinction between town and country in the Egyptian context. 'The rural habitat is not dispersed, with isolated houses in the fields, but concentrated in small, densely populated settlements.'[31] Large villages (up to a population of 30,000) in this context would thus have many urban attributes and would be considered towns in other countries. Furthermore, the 'countryside' in Egypt is, for the most part, a thin ribbon of land on either bank of the Nile. Settlements are thus linear and 'market' towns almost equidistant from each other (30–40 km.) form a long string along the valley and a tight network in the Delta .[32] The physical economic distance between village and town is thus very small and the 'commuting rural-urban migrant' can exist as an economic category.

Given these spatial parameters, it has become possible for some of the adaptation to population pressure on the very limited arable land to take place in the countryside itself. Whereas in most countries, mounting population pressure and/or growing differentials in income and economic opportunity between town and country must necessarily be reflected in an exodus from the countryside, in Egypt this has not been necessarily so. Undoubtedly, rural-urban migration has been important in Egypt too, but scope has also existed for a transfer to a non-agricultural livelihood without a physical move out of the village. We also know from other data in the survey that 6.5 per cent of income earners' earnings come from 'non-agricultural productive assets', which refers to a variety of handicraft, petty commodity production and trading activities, and that another 10.7 per cent are engaged in non-farm wage-employment within the village. It would appear that a low productivity tertiary sector of significant proportions has become established within the rural economy through the process of improvised adaptation as the limits of agricultural involution have been reached. The sector could thus be seen as the refuge of the rural poor who face limited access to land and have no immediate prospect of being absorbed in the urban economy. In this way it is a reflection of the limited capacity of a non-egalitarian agrarian structure to cope with population increase and is a manifestation of increasing landlessness and marginalisation.

We have seen thus far that the land reform has not conferred continuing access to land. It is also important, however, to look at the changes in the structure of landownership and of landholdings that have occurred among those households which have access to land. Here the data situation is highly unsatisfactory; the latest figures from an Agricultural Census that are available relate to 1961. Comparing these figures with those from the 1950 Census would show that there has been an improvement in both the ownership and holding distribution as a result of the land reform but developments since 1961 are difficult to document with any degree of rigour.

Two reservations should, however, be noted concerning the improvement in land distribution that has been mentioned. First, is that the figures for ownership distribution are not sufficiently disaggregated; there is no breakdown for the ownership category of 'less than 5 *feddans*' although 95 per cent of all ownerships fall into this category. Changes within this group are thus very significant for the overall distribution but are not recorded. Again data from our recent survey throw additional light on this question. The figures for land ownership show that there is an overwhelming preponderance of very small ownerships; 55.6 per cent of all ownerships are less than 1 *feddan* in size and another 19 per cent are between 1 and 2 *feddans* in size. The group of under 5 *feddans* accounts for 94.3 per cent of all landownerships and it is clear from the previous figures that the distribution within this group is highly skewed. In fact the value of the Gini coefficient of 0.556 for landownership is higher than the figure of 0.432 estimated from the 1961 Census figures.[33] We cannot, of course, on the basis of these figures, rush to the conclusion that the ownership distribution has worsened. The figure for 1961 is biased downwards because of the lack of disaggregation we complained of and our figure is based on a small sample survey whereas the earlier one was from a census. What we can say, however, is that land distribution in 1977 remains highly skewed and that our figures would not be inconsistent with the continued operation of disequalising tendencies since the land reform. It also shows that inequalities within the 'small farmer' category are extremely wide.

The second reservation about the figures showing an improvement in landownership between 1950 and 1961 relates to the first. The claim is of an improved distribution made on the basis of the observation that the share of the total area owned in the 'under 5 *feddans*' category increased from 35 to 52 per cent while there was a corresponding decrease in the share of large ownerships. It is important to note, however, that we cannot infer from these figures that there has been a general improvement in the landownership position of the 95 per cent of small landowners. In the first place under land reform the land was distributed in units of close to the upper limit of this group of landholdings (5 *feddans* was the general norm). We also know that the number of beneficiaries was a small minority (13 per cent) of the number of landowners within the 'under 5 *feddans*' category. It follows therefore that the increase in the share of total land owned within this category due to the land reform resulted in an increase in inequality in land ownership within this group.

It will be recalled, however, that not all of the increase in the share of total land owned in this category was due to direct distribution under the land reform. It was also due to pre-emptive private sales and this raises the question of whether or not we can consider the increase in the share of small ownerships due to such transactions as being equivalent to an improvement in the welfare of small peasants in general. *A priori*, we would not expect small landowners in this category with their indebtedness and poverty to have had the capacity to buy land from large landowners. It seems more likely, therefore, that the purchases were made by absentee landowners who do not belong to the small farmer category. Thus, although the land sold by large landowners was sold in sub-divided plots of 5 *feddans* or less, there is no reason to suppose that they were entirely bought by small peasants. The increase in the share of total land owned in ownerships due to this cause cannot therefore be considered as an unambiguous indicator of an improvement in the distribution of landownership in the sense of a general increase in the welfare of small farmers.

An important piece of evidence in support of the contention that disequalising forces continue to operate in the Egyptian agrarian system is the data on land movements from our survey. Information was obtained on the details of changes in the landownership position of households over the lifetime of the household head and the data show that land transactions have played a significant role in determining the current pattern of landownership.[34] Furthermore, the effect of these transactions has been highly disequalising. As shown in Table 5.4, the incidence on land loss has been overwhelmingly concentrated on the small landowners. Forty-three per cent of all households who reported some loss of land are now completely landless and since the average amount of land lost was small (only 16 *kirats*), it is safe to conclude that most of these households had previously belonged to the 'less than 24 *kirats*' category. Another 27 per cent of households who lost land now own less than 24 *kirats*. By contrast, there was only one case of land loss coming from the 'more than 120 *kirats*' (i.e. 5 *feddans*) category. In terms of the total amount of land lost, we find that 76 per cent of the total area lost came from households in the 'less than 24 *kirats* category' (including those who are now landless) whereas only 17.3 per cent of total current landownership falls within this ownership category. Thus we see that land loss has been overwhelmingly concentrated on very small farmers and has therefore contributed to a worsening of the distribution of landownership.

In principle, of course, land loss has to be balanced against land gained and it is only after we have determined the net change that we can judge the overall impact of land distribution. The data on land purchases, however, sweeps away this doubt; they show an inverse relationship to that for land loss. In column (4) of Table 5.4B we show the relative incidence of land purchases of each landownership class (i.e. the number of households reporting a purchase divided by the number of households in that landownership class). It is clear that the incidence rises sharply with the size of land ownership. Only 7 per cent of households in the 'less than 24 *kirats*' class had ever purchased land, whereas the incidence is substantially higher for the other groups. Another way of measuring the impact of land sales on current land distribution would be to compare the proportion of land bought by each income class (column 6) with their share in current landownership (column 8). Here we see that 45 per cent of total area purchased went to those with large landownerships (240–480 *kirats*) and that this is far larger than this group's share of 16.5 per cent in current landownership. In contrast to this, the 'less than 24 *kirats*' category accounted for only 4.5 per cent of total area purchased, although it accounted for 17 per cent of current landownership. Land purchases have thus contributed to a greater concentration of landownership.

It was not only limited access to land and continued inequality in land distribution that conditioned the neutralisation of the gains from the agrarian reform. A related process was the lack of countervailing power to protect their real incomes, let alone increase them, in the face of rapid inflation in the 1960s and 1970s. The two groups we shall consider in this connection are the beneficiaries of the land reform (i.e. those who received land) and the landless labourers.

It has been estimated that beneficiaries of the land distribution programmes had money incomes '50 per cent above the pre-reform level'.[35] This arose from the fact that repayments for land received were lower than the rent they had previously paid and that yields had sometimes increased. The increase in money incomes was

Table 5.4 Land Movements
A. Land Loss by Size of Landownership

Land Owned (kirats)	*Sale*			*Forfeiture*			*Others*		*Total*		
	No.	*Area*	*Av.*	*No.*	*Area*	*Av.*			*No.*	*Area*	*Av.*
0	16	256	16.0	—	—	—	—	—	16	256	16.0
24	5	31	6.2	4	31	7.8	1	24	10	86	8.6
24–48	1	12	12.0	2	18	9.0	—	—	3	30	10.0
48–72	2	12	6.0	2	17	8.5	1	9	5	38	7.6
72–96				1	14	14.0			1	14	14.0
96–120				1	2	2.0			1	2	2.0
				—					—		
				—					—		
168–192				1	24	24.0			1	24	24.0
Total	24	311		11	106		2	33	37	450	

Table 5.4 Land Movements (*continued*)
B. Land Purchases by Size of Landownership

Land Ownership (kirats)	(1) Total No. of owners	(2) %	(3) No.	(4) % Incidence (3÷1)	(5) Area Bought (kirates)	(6) % Area Bought	(7) Present Total Area	(8) % Present Land Owned
<24	174	0.556	12	0.069	120	0.045	2253	0.173
24–48	60	0.192	11	0.183	354	0.132	2467	0.189
48–72	33	0.105	8	0.242	284	0.106	2106	0.162
72–96	21	0.067	6	0.286	293	0.109	1854	0.143
96–120	7	0.022	1	0.143	60	0.022	794	0.061
120–144	7	0.022	2	0.286	120	0.045	977	0.075
144–168	2	0.006	0	—			—	
168–192	1	0.003	1	1.000	130	0.048	173	0.013
192–240	1	0.003	1	1.000	111	0.041	231	0.018
240–360	4	0.012	3	0.750	636	0.237	1350	0.104
360–480	2	0.006	2	1.000	576	0.215	796	0.061
>480	1	0.003	0	—			—	
Total	313		47	0.150	2684	1.000	13001	

sustained between 1952 and the mid-sixties by improved yields, the recovery in agricultural prices from the depressed levels after the Korean War and by the successive reductions in the annual instalments fixed for repayment on the land received. Since then, however, these gains have been washed away by a combination of sharp increases in the rural cost of living and in the costs of agricultural production, especially contribution to co-operative expenses. It has been claimed that the latter charges have lately become so exorbitant that they outweigh the peasants' revenue, leaving agrarian reform beneficiaries in permanent debt.[36]

The landless agricultural labourers have fared no better. The average daily money wage in 1960 was almost the same as in 1951, the year preceding agrarian reform. (Table 5.5). In fact, it was not until the mid-sixties that the minimum statutory wage level was reached and actually exceeded in subsequent years. There was an upward trend between 1955 and 1966, the years of 'socialist transformation', the second agrarian reform and the government drive against the 'remaining pockets of feudalism'. However, there was a decline thereafter due mainly to rapid inflation – with the real wage index in 1974 declining to the 1948 level. It is therefore our conclusion that, with the exception of a brief period in the late 1960s, the standard of living of agricultural labourers (essentially landless peasants and very small farmers) has more or less remained unchanged over the last 25 years and, in fact, the decline which began in the late 1960s has continued uninterrupted until the present day.

(c) The impact of the system of supervised co-operatives

In a previous section we described the introduction of a system of supervised co-operatives as one of the major institutional changes associated with the agrarian reform. In this section we shall complete our evaluation of the performance of the Egyptian agrarian system by considering how this system of co-operatives fared in relation to its two principal objectives – the rationalisation of agricultural production and the wider diffusion of modern inputs to the peasantry.

Evidence suggests that beyond the establishment of co-operatives and the enforcement of the crop rotation system, the land and crop consolidation programme had only limited success. First of all, it has failed to solve the problem of fragmentation which, many writers believe, has actually worsened during the last few

Table 5.5 Movement in Real Wages in Rural Egypt, 1938–74 (1938=100)

	Average daily					*Average daily*			
	Money Wage		*Cost-of-living*	*Real Wage*		*Money Wage*		*Cost-of-living*	*Real Wage*
Year	*PT*	*Index*	*Index*	*Index*	*Year*	*PT*	*Index*	*Index*	*Index*
1938	3.0	100	100	100	1959	12.5	417	334	124
1939	3.5	117	101	116	1960	12.5	417	337	123
1941	3.6	120	132	90	1961	12.3	410	358	113
1942	5.0	167	198	83	1962	14.0	450	367	122
1943	6.3	210	238	87	1963	15.0	480	377	127
1944	9.3	310	262	117	1964	19.0	609	438	138
1945	9.3	310	262	117	1965	22.0	704	519	135
1946	9.5	317	297	107	1966	25.0	801	468	170
1948	10.0	333	271	123	1967	24.5	784	479	162
1949	10.0	333	259	130	1968	24.5	784	499	156
1950	11.6	387	264	147	1969	25.5	817	536	151
1951	12.6	420	263	160	1970	25.0	801	576	138
1952	12.0	400	265	151	1971	25.5	817	580	140
1953	12.0	400	269	150	1972	27.5	880	613	143
1955*	7.6	253	294	87	1973	29.2	930	661	140
1956	10.0	333	342	97	1974	32.2	1001	792	125

Source Radwan, op cit, Appendix I.

Note *We suspect this sharp decline but in the absence of other data we decided to leave it unchanged since it will not affect the trend in any case.

years.[37] Ironically enough, the Agrarian Reform Law, which prohibited the division of landholdings below a minimum of five *feddans*, has itself provided for the distribution of land to agrarian reform beneficiaries in parcels of two to five *feddans* (Article 9 of Law No. 178 of 1952). More important perhaps was the adverse effect of the consolidation system on poor and small landowners who found it difficult to diversify their products. Their small plots of land would inevitably fall within a single block and they would be obliged to grow a single crop. For example, if, according to the crop rotation, their plot falls in the cotton block, they will have to buy their needs of wheat and corn and clover for their animals on the open market. Here, the large landowners have an advantageous position: their holdings are large enough to enable them to diversify their production and have a surplus to sell. A study of ten villages has shown that this situation has resulted in the creation of an active black market, especially in cereals, rice and fodder, where large landowners sell their surplus to small farmers at exorbitant prices. Moreover, the practice of small farmers hiring land for specific seasons to grow a specific crop at rents much higher than the official rate became widespread.[38] It is not, therefore, surprising that the system was strongly resisted by small farmers.

From the above discussion we can conclude that the ideas of land and crop consolidation had a great potential in creating a new form of communal land system. While the co-operatives were, to a large extent, instrumental in preventing further exhaustion of the soil by introducing the new triennial rotation, they have certainly not succeeded in providing a viable alternative to the system of peasant agriculture except, perhaps, on land reform areas. For the new system of consolidation to succeed a more radical change in land distribution would have been necessary. Moreover, the lack of enthusiasm on the part of bureaucrats in charge of the project – who, in most cases, were in collusion with large landlords who remained the dominant power in the village – stifled the experiment from the start. To sum up, the Egyptian co-operative system has failed to introduce an alternative form of organisation for agricultural production. All it did was to patch up some of the obvious deficiencies of the existing system mainly through crop consolidation. A negative aspect of this system, however, was its adverse effect on poor and small peasants since it has resulted in a redistribution in favour of larger farmers.

As regards the diffusion of inputs the system of co-operatives has been only partially successful.

The total value of inputs supplied by the co-operatives increased considerably during the 1950s, and by the mid 1960s they became the sole supplier of most inputs. The government had a monopoly of supply for fertilisers, certain selected seeds and pesticides, and used the co-operative as a distribution channel.

Increases in the volume and changes in the terms of credit were among the most important complementary measures to land reform. By 1962, the co-operatives had become the sole channel through which rural credit facilities could be extended to farmers, and loans were advanced against the security of crops instead of land, as was the case before. In 1964, there was a major reorganisation of the rural credit system. The most notable feature of this new scheme was its emphasis on short-term loans in kind – such as seeds, fertilisers and insecticides – which were made to each farmer according to the area of his farm and the type of crop planted. During the 1960s the volume of credit (at current prices) was more than doubled. However, the significance of this increase in the supply of credit should not be exaggerated. The *real* growth of rural credit was not that impressive. The volume of credit at constant prices increased from 41 million Egyptian pounds per year on average during the period 1960–4 to 65 million Egyptian pounds for 1965–70, and then dropped to 56 million during 1970–5. Moreover, it seems that, despite this growth in the volume of credit, co-operative credit was far from satisfying the farmers' needs. The ratio of co-operative credit to the cost of agricultural production increased from 24 per cent in the 1950s to 28 per cent in the 1960s, but declined to 24 per cent again in the 1970s.[39] A detailed breakdown of loans by purpose and kind reveals a typical pattern: the dominance of short-term loans which provide finance for current agricultural operations (seeds, fertilisers, pesticides and hiring of wage labour). Medium-term loans for development purposes (purchase of livestock and machinery, land reclamation and development of fruit orchards) constitute a very small proportion of the total.

The impact of improved supply of credit on the process of differentiation among the peasantry is an important consideration. Fragmentary evidence suggests that the system may have favoured large farmers rather than small and poorer peasants. Table 5.6 provides a breakdown of co-operative loans by size of landownership of debtors for the year 1963/64.

A number of interesting features emerge from the above table. Despite an apparent equality in the distribution of credit by a unit of

Table 5.6 Co-operative Credit by Size of Landownership

	1963/64[1]				
Size of Holding (feddans)	*Loans Advanced (E£m)*	*%*	*Number of Debtors ('000s)*	*%*	*Average Debt/ Person*[2] *(E£)*
<5	25.4	50	1177	85	22
5–25	15.6	31	187	13	83
>25	9.7	19	26	2	373
Total	50.7	100	1390	100	36

Source Sami Abu el-Ezz and Ahmed Abu el-Ghar, *Co-operative Financing*, Dar El-Taawon (Cairo, 1971) p. 485.

Notes 1. Percentages and averages rounded.
2. Average debt per unit of land was available only for this year. It amounted to E£10.597, 10.573 and 10.597 per *feddan* for the three ownership classes respectively.

land (as evidenced by the data available only for one year, 1963/64), the co-operative credit policy did not fully succeed in serving the primary target group: the small peasants. Thus we find that small peasants, who represent the majority of debtors (85 per cent) and the most needy, get half the credit advanced by the co-operative system, while medium and large landowners get the other half. Various writers have suggested that medium and large landowners found in the credit facilities advanced by the co-operatives a cheap source of finance not only for their agricultural operations, but to enhance their wealth by buying more land, machinery, livestock and sometimes to finance some commercial business.[40] When the government tried to check this trend by imposing a rate of interest of 4 per cent on loans advanced to holders of ten *feddans* and more, many large landowners reacted by dividing their holdings into plots of less than ten *feddans* to benefit from the exemption from interest.[41]

Unequal distribution of credit was not the only means by which large landowners appropriated the greater part of co-operative finance. A common practice has been to raise substantial amounts of capital through continuous accumulation of co-operative debts. They were encouraged by the lenient penalty rate of interest on arrears, 6 per cent, which was much lower than the current market rate. In fact, there were cases where some large landowners as-

Table 5.7 Accumulated Debts by Size of Landownership as at 31 December 1966

Size of Holding (feddans)	*Value of Debts (E£m)*	*%*	*Number of Debtors ('000s)*	*%*	*Average Debt/ Person (E£m)*
<5	18.7	45	1470	80	13
5–25	16.7	41	351	19	48
>25	5.7	14	22	1	260
Total	41.1	100	1843	100	22

Source Abu el-Ezz and Abu el-Ghar, op cit, pp. 340–1.

sumed the role of the private moneylender, using the credit they obtained from the co-operatives. Table 5.7 provides an idea about the position of accumulated debts by the mid-1960s, the height of co-operative activity.

Finally, many privileges were enjoyed exclusively by rich farmers by virtue of certain provisions of the agrarian reform laws. For instance, in the system of livestock insurance, only the owner of at least three head of cattle is formally eligible to insure his livestock, and subsequently to obtain a ration of 150 kg. of fodder at the subsidised price fixed by the state. The poor peasant who cannot insure his livestock is deprived of such privileges; in most cases he has to buy fodder on the black market.[42] Moreover, only owners of more than 15 *feddans* were eligible to buy selected seeds at subsidised prices.[43] It is not, therefore, difficult to understand the often-expressed view among small and poorer peasants that co-operatives have done very little to relieve their monumental burden of indebtedness.

5.5 Sectoral balance and the mobilisation of agricultural surplus

(a) Inter-sectoral terms of trade

A crucial issue in the development literature is the extent to which the agricultural surplus is used to finance the growth of other sectors of the economy. The experience of many developing countries shows that the growth of industry has been financed to a great extent by 'squeezing' the agricultural surplus either through taxation or the manipulation of inter-sectoral terms of trade. Here, we shall try to assess the extent to which this was true of Egypt and what were its implications for the rural population.

As mentioned earlier, the wide network of co-operatives established in Egypt over the last quarter of a century provided the government with an effective institutional set-up through which it was able to carry out its policies. One of the most important of these policies was that of 'achieving an effective maximisation and mobilisation of agricultural surplus'. Two major policy instruments were used:

1. The control of the market through the gradual extension of the system of co-operative marketing to cover the most important crops necessary for the manufacturing and export sectors; and
2. The system of 'compulsory deliveries' where farmers were obliged to deliver a certain part of their crops to the government at 'administered prices' in order to ensure a continuous flow of grains (in the case of wheat) to the local market, and (in the case of rice and onions) to expand the exports of some cash crops. By the end of the 1960s, co-operative marketing covered 100 per cent of the cotton crop, 80 per cent of sesame, 60 per cent of flax, 60 per cent of groundnuts, 50 per cent of rice, 30 per cent of onions and 20 per cent of potatoes. Meanwhile the quotas for procurement amounted to 60 per cent in the case of rice, 57 per cent for onions and 27 per cent for wheat.

Evidence suggests that the government, through the procurement system and co-operative marketing was able to regulate the domestic terms of trade primarily by purchasing agricultural products at lower prices than those for export or for local producers and by selling agricultural inputs at prices higher than their import costs. Price differentials on procured crops amounted to 125 per cent in the case of wheat, and were as high as 145 per cent and 200 per cent in the case of onions and rice respectively. Moreover, the difference between producers' prices and *fob* export prices in the case of cotton amounted to some 70–80 per cent over the period 1960–70. The differential between producers' prices and the selling prices of cotton to domestic mills ranged between 115–25 per cent over the same period. Similarly, in the case of rice the differential between government buying prices and export prices ranged between 161–86 per cent over the period 1965/66–1969/70. Other crops were no exception.[44]

We have attempted to measure the effects of such policies on the movement in terms of trade for the agricultural sector over the period 1960–75.[45] Table 5.8 provides a summary view of our calculations. We have calculated three sets of indices:

1. An overall index of the terms of trade between agricultural output and all manufactured commodities;
2. An index of the terms of trade between agricultural output and manufactured consumer goods; and
3. An index of the terms of trade between agricultural output and manufactured inputs.

These three indices were calculated for poor farmers, rich farmers and all farmers. A number of inferences can be drawn from the data. First of all, looking at the period 1960–75 as a whole, we find that the overall agricultural terms of trade for all farmers (Index A-1) have remained more or less constant. A closer look at this index shows that the terms of trade actually deteriorated following the cotton crisis of 1961, began to recover in 1965, remained constant until 1971 when they slightly deteriorated again, and it was not until 1975 that a very slight improvement took place. The constancy in the terms of trade reflect the impact of two opposing movements: an improvement in the prices of agricultural products vis-à-vis prices of manufactured goods used as inputs to agriculture (Index A-3), which was counterbalanced by a deterioration in agricultural prices vis-à-vis the prices of consumer goods consumed by the rural population (Index A-2). It appears that farmers in general have benefited from the government policy to stabilise the prices of manufactured inputs to agriculture (especially fertilisers), but this gain was wiped out by the rapid increase in the prices of commodities entering into the consumption basket of the rural population (especially foodstuffs and textiles). We can therefore conclude that, on the whole, the time series for the overall agricultural terms of trade over the period 1960–75 indicate no improvement in the relative prices of agricultural output vis-à-vis agricultural inputs and consumer goods.

But the impact of government price policies was not even for different groups of farmers. A comparison of the two sets of terms of trade indices (for 'poor farmers' and for 'rich farmers') indicates

Table 5.8 Indices of Terms of Trade for the Agricultural Sector, 1960–75 (1960=100)

	A. All Farmers			B. Poor Farmers			C. Rich Farmers		
Year	*(1)*	*(2)*	*(3)*	*(1)*	*(2)*	*(3)*	*(1)*	*(2)*	*(3)*
1960	100	100	100	100	100	100	100	100	100
1961	99	99	100	99	99	101	100	100	101
1962	88	96	68	88	96	68	90	99	70
1963	86	94	71	87	94	70	90	97	73
1964	94	100	78	94	100	78	100	107	83
1965	85	88	79	86	88	79	92	95	86
1966	93	94	90	93	94	89	105	106	101
1967	99	96	112	99	95	115	109	107	121
1968	97	93	116	98	92	120	106	102	124
1969	99	95	118	100	94	123	108	104	126
1970	98	98	101	99	98	104	111	111	111
1971	97	96	100	98	96	103	109	109	111
1972	97	95	105	98	95	109	113	111	120
1973	95	91	106	95	91	109	114	111	125
1974	97	93	112	97	93	116	118	115	132
1975	102	97	120	102	96	124	127	122	146

Notes (1) Overall index of terms of trade between agricultural output and all manufactured commodities.
(2) Terms of trade between agricultural output and manufactured consumer goods.
(3) Terms of trade between agricultural output and manufactured inputs.

that while the former has remained constant, if not slightly deteriorated, the latter has definitely improved. This conclusion has serious implications concerning the process of differentiation in rural Egypt: government policies, especially the co-operative marketing and pricing policies, are much more favourable to rich farmers than to the poor. The last point seems to be confirmed also by the resistance of small peasants to co-operative marketing in general and the compulsory deliveries system in particular. The latter seems to produce regressive results. After delivering their quotas to the co-operative, small farmers usually have very little left of their crops, and thus they have to obtain their needs of grain and fodder for their animals at considerably greater cost on the black market where larger farmers dispose of their surplus.

The crucial question is how effective were these governmental policies in mobilising the agricultural surplus for transfer to finance accumulation in the other sectors of the economy. While various

Table 5.9 Financial Contribution of Agriculture to the Public Treasury (selected years)

Financial year	(1) Total Financial Contribution of Agriculture to Public Treasury* (E£m)	(2) Agricultural Income at Current Prices (E£m)	(3) (1)/(2) %	(4) Total Government Capital Expenditure Connected with Agriculture (E£m)	(5) Net Financial Contribution of Agriculture to Public Treasury (E£m)	(6) (5)/(2) %
1950/51	33.5	340.0	10.0	10.0	+23.5	7.0
1960/61	55.0	402.7	13.7	33.2	+21.8	5.4
1965/66	95.5	608.5	15.7	86.1	+ 9.4	1.5
1967/68	106.2	644.4	16.5	61.5	+44.7	7.0
1968/69	115.4	688.3	16.7	68.1	+47.3	6.9
1969/70	128.7	771.9	16.6	62.0	+66.7	8.6

Source Mahmond Abdel Raouf and H. Saad, *Vertical Agricultural Development in Egypt, 1952/53–1969/70,* Institute of National Planning (Cairo, 1973) Table 26.

Note *Including the yields of land tax, export duties on farm products, and revenues accruing to the public treasury as a result of marketing agricultural products in the home market as well as income from the lands of the public domain and land reform estates.

writers agree that the net resource transfer from agriculture has been substantial, they differ in estimating the magnitude of this transfer. According to Fadil '. . . the net transfers of financial surplus out of agriculture would have amounted only to about 5–7 per cent of the total agricultural income during the period 1965–70.'[46] The World Bank, on the other hand, estimates that the net transfer was of the order of 12–22 per cent of agricultural output.[47] We believe that a realistic estimate lies somewhere between these two figures. An attempt was made to measure the magnitude of this transfer as the net outflow from agriculture to the Treasury as measured by the difference between the total contribution of agriculture to state revenues and total government expenditure on agriculture. The results are summarised in Table 5.9. It is immediately clear that the total resource outflow from agriculture has increased from 10 per cent of agricultural income in 1950/51 to 16 per cent by the end of the 1960s. If we allow for the resource inflow to agriculture – government investment and subsidies – the net outflow would amount to 7.0 per cent of agricultural income in 1950/51 and 8.6 per cent in 1969/70. But these estimates do not include government gains from price differences between export and producers' prices for leading cash crops. It was estimated that the government gains from such price differentials for cotton and rice alone amounted to some £E43 million on average per year over the period 1960–71.[48] This represents about 7 per cent of agricultural value-added over the same period. It is reasonable to assume, therefore, that the net contribution of agriculture was in the order of 15 per cent of agricultural income. It appears that this trend has continued into the 1970s.

A recent study by the World Bank concluded that intersectoral transfers of income have been considerable, especially the transfers out of the agricultural sector.[49] The average subsidy ratio on full costs of agricultural production has been about 18 per cent, whereas the average tax ratio on output of major crops has been about 30–40 per cent. The study also shows that over the period 1973–6, the transfers to the Treasury from the cotton crop alone more than covered the full cost of *all* subsidies, direct and indirect (Table 5.10).

To sum up, the system of co-operative marketing together with the pricing policy attached to it, must have resulted in an improvement on the pre-reform methods which depended mainly on intermediaries who appropriated substantial amounts of the agricultural surplus for themselves. The government was certainly successful in

Table 5.10 Net Effect of Price Transfers (E£m)

	1973	*1974*	*1976*
Transfers to Treasury of cotton organisation	64.8	136.8	92.4
Exchange rate gains	114.0	177.0	100.0
Total transfers out	178.8	313.8	192.4
Direct subsidies	15.8	12.7	56.8
All public sector investments in agriculture	51.0	54.0	49.0
Current expenditure of ministry of agriculture	16.4	19.8	26.0
Current expenditure of ministry of irrigation	18.4	19.9	28.8
Total transfers in	101.6	106.4	160.6
(1) Net flow	−77.2	−207.4	−31.8
(2) Agricultural income	1062.0	1280.0	1553.0
1÷2	7.3%	16.2%	2.1%

Source World Bank Staff paper, *Egypt – Agricultural Prices, Taxes and Subsidies*, World Bank (Washington, 1978).

using the new system to mobilise and transfer the agricultural surplus to finance industrial accumulation and redress the balance of payments deficit. Nevertheless, the new marketing and pricing policies seem to have contributed to the aggravation of social differentiation in rural Egypt by favouring the large farmers and discriminating against the poor and small peasants for whom the agrarian reform was promulgated in the first place.

(b) Rural-urban disparities

The impact of government policies on rural incomes in general, and rural-urban balance in particular were also important. In Table 5.11 we have tried to estimate the development of real rural and national income per capita over the last 25 years. A number of conclusions can be drawn from our data. The *first*, is that the rural per capita income has improved slightly for the period as a whole: there was a notable improvement between the mid-1950s and 1960s, followed by more or less stagnation. *Second*, perhaps the most significant trend revealed by these statistics is the widening urban–rural differentials in real per capita income. The ratio of per capita rural income to per capita income of the total population has gone down from 51 per cent in 1952/53 to about 45 per cent in the 1960s and

Table 5.11 Per Capita Real Income of Total and Rural Population (all at 1959/60 constant prices)

	(1) Per Capita Income of Total Population £E	*(2) Per Capita Income of Rural Population £E*	*(3) Per Capita Income Differential (2)÷(1) %*
1952/53	45.8	23.3	51
1953/54	44.7	20.9	47
1954/55	44.8	16.5	37
1955/56	45.4	18.3	40
1956/57	45.3	19.6	43
1957/58	46.8	19.8	42
1958/59	48.4	24.8	51
1959/60	50.2	25.6	51
Average annual growth rate %	1.2	1.2	45
1960/61	51.7	24.9	48
1961/62	52.0	22.8	44
1962/63	55.1	25.6	46
1963/64	58.3	26.8	46
1964/65	59.8	27.3	46
1965/66	60.7	27.8	46
1966/67	59.9	25.7	43
1967/68	59.0	25.6	43
1968/69	60.9	25.7	42
1969/70	63.4	27.0	43
Average annual growth rate %	2.1	0.8	45
1970/71	64.9	26.7	41
1971/72	66.1	26.9	41
1972/73	65.9	27.3	41
1973/74	66.1	26.9	41
1974/75	66.8	26.5	40
1975/76	71.8	26.8	37
Average annual growth rate %	1.7	0.0	40

Notes and Sources (1) Based on data from D. Mead, *Growth and Structural Change in the Egyptian Economy*, Richard D. Irwin, Inc. (Illinois, 1967) for the years 1952/53 – 1962/63; the *Statistical Indicators* of the UAR for the years 1962/63 – 1969/70; and the Ministry of Planning, *Follow-up Reports* for subsequent years.

(2) Value-added in agriculture divided by rural population. The latter was calculated by interpolating census figures for inter-census years. Data for the years 1952/53 – 1969/70 from M. Abdel Raouf, *Vertical Agricultural Development in Egypt*, Institute of National Planning (Cairo, 1973); and *Follow-up Reports* for subsequent years.

40 per cent in the 1970s. The real discrepency between per capita rural and urban income must, therefore, be of a greater magnitude.

A number of reasons account for the increase in rural-urban disparities. The first, of course, is the strict constraint on the expansion of agricultural output from a fixed area of cultivated land and the difficulties to increase productivity given the present techniques used in agriculture. Second is the relative neglect of the Egyptian village by the government. There is no doubt that there has been an impressive increase in the supply of investment and social services to the rural areas since 1952. But the gap between the share of urban areas and the countryside is so wide that one can speak of a case of 'urban bias' characterising government policies which favour the cities. In Table 5.12 we have tried to piece together some indicators of this type of disparity between rural and urban areas in terms of the distribution of some basic social services. It is clear that with the exception of primary education, urban areas have more than their proportionate share of social facilities.

Tables 5.12 Some indicators of Rural-urban Distribution of Social Facilities 1966/67 (percentages)

	Urban Governorates	*Provinces*
Population	21.8	78.2
Students in primary education	27.6	72.4
Students in preparatory education	35.7	54.3
Students in secondary education	42.4	57.6
Hospital beds	41.4	58.6
Social centres	13.2	86.8
Local administration revenues	29.7	70.3
Local administration expenditures	31.3	68.7

Source Ministry of Planning, *Indicators of Regional Development 1964/65–1966/67*, unpublished memorandum (Cairo, 1968) pp. 148–57.

5.6 Conclusion

In the preceding appraisal of the Egyptian experience of agrarian change, we have tried to avoid the narrow 'technocratic' approach in favour of a wider and dynamic one in which recent developments were placed in historical perspective. The main purpose of the study was to assess the extent to which the various measures of the last quarter of a century, triggered off by the 1952 land reform, have changed the agrarian system and in which direction.

Our assessment centred around three main issues. *First*, the redistributive effect. Here we find that the distribution of some 13 per cent of the cultivated area to 9 per cent of the rural population has perhaps eliminated the very large ownerships, slightly improved the lot of small peasants, but certainly consolidated the position of the medium stratum which represents the economic and political power now in rural Egypt. Thus, the initial inequality in the distribution of landownership was not changed fundamentally, and the problem of landlessness has but increased. Changes in income distribution reflected this basic inequality and, as a result, poverty has been increasing. *Second*, the growth performance of Egyptian agriculture was more than outweighed by a rapid population growth causing output per capita to decline. *Third*, we have assessed the system of 'supervised co-operatives' as an alternative organisation for rural development. Despite its potential, the new system did not achieve its stated objectives of increasing agricultural production through improved supply of inputs and the improved production system. Moreover, the co-operatives favoured the medium and rich farmers to small peasants. *Finally*, we found that although the government policies were effective in mobilising a substantial part of the agricultural surplus to finance other sectors of the economy, these policies had the negative effect of widening the gap of rural-urban disparities and, by favouring rich farmers, aggravated inequalities in the rural society.

The crucial question is that of finding an appropriate framework for understanding these changes in the agrarian system: their extent and their limitations. Various writers have offered different interpretations. Anouar Abdel Malik, for instance, argues with Wittfogel, that the key to the analysis of these changes is the central role of the state which has, for so long, controlled the irrigation system, and, therefore, agriculture.[50] For M. Abdel Fadil agrarian reform can be seen as having paved the way for the virtual transition of Egypt's agrarian system from semi-feudalism into capitalism by eliminating the power of the very large landowners, and as a result of organising the labour market.[51] In contrast, Robert Mabro stresses the growing pressures on the agricultural sector arising from the unfavourable imbalance between land and people,[52] while Bent Hansen concentrates on the absence of allocative efficiency.[53] There is no doubt that each of these writers has pointed to an important element of the process of agrarian change. But none answers fully our question: how is it that such a significant agrarian reform has failed to alter the agrarian system in any fundamental way?

An interesting framework for interpreting the developments in Egypt's agrarian system over the last twenty-five years would be Michal Kalecki's analysis of 'intermediate regimes'.[54] In fact, Kalecki based his conceptualisation on the findings of an empirical study of Nasser's Egypt.[55] According to Kalecki, the basic element of specificity in underdeveloped countries with a mixed economic system which distinguish them from both the developed capitalist and socialist economies is the role of the state and the interests it represents. The main problem in developing mixed economies is not one of generating effective demand but 'the expansion of productive capacity indispensable for the rapid growth of the national income'. This role is performed mainly by the state through public sector investments and the control of private investment, trade and prices – thus, the increase in the power of the state. The results of state action would depend on the interests they represent. In Kalecki's scheme, the state in most developing mixed economies would represent the alliance of lower middle classes and the rich peasantry. To gain a power base, the state would, in the initial stage, carry out progressive policies such as massive nationalisations of big business and agrarian reforms. In so doing, the régime would contribute to development and benefit the poor. But because of its very nature, as an alliance of state bureaucracy and rich peasants, the régime would soon reach a stage where a contradiction arises between its interests and the requirements of egalitarian development. Invariably, this contradiction would be solved in favour of this alliance at the expense of the rest of the population, and the regressive phase of 'state capitalism begins'.

The parallel of Kalecki's analysis with the Egyptian case is very striking: the state has been in command of the mobilisation and allocation of the economic surplus; there has been an initial stage in the 1950s and 1960s where the state launched a relatively radical and comprehensive reform which hurt the large landowners and benefited the landless and the tenants; the co-operatives provided the state with an effective institution for control; but because of the marginal nature of the reform and the growing influence of the bureaucracy-rich peasants' alliance, these 'gains' started to be eroded in the late 1960s, and in fact have almost reversed since the launching of the new 'open-door policy' in the 1970s.

It is thus the class character of the state that generated pressures to bring about an agrarian reform and yet constrained these reforms to narrow limits. The old system had to change but an intermediate régime could only bring about an 'intermediate' transition which did not represent a lasting change.

Kalecki's scheme thus provides an elegant general framework within which our analysis of the specific elements of change in the Egyptian agrarian system can be fitted. It should be noted, however, that this framework does not constitute a comprehensive explanation; it merely offers an illuminating categorisation of the type of changes that occurred in Egypt. The specifics of each case analysed within such a general framework thus remain very important. In the case of Egypt, regardless of how one chooses to categorise the nature and role of the state, certain elements remain crucial in explaining the outcomes we have described. Thus the extreme land scarcity, the stagnation in both agriculture and industry and the inherited structure of the agrarian system remain vital explanatory variables.

Notes

1. For details see Samir Radwan, *Capital Formation in Egyptian Industry and Agriculture, 1882–1967*, Ithaca Press (London, 1974), Chapter Five.
2. S. Ishikawa, *Economic Development in Asian Perspective*, Kino Kumya Bookstore Co., Ltd. (Tokyo, 1967) pp. 57–83.
3. The intricate history of landownership in Egypt has been the subject of several excellent studies. See in particular: Ibrahim Amer, *Land and Peasant: the Agrarian Question in Egypt*, Publisher unknown (Cairo, 1958); Gabriel Baer, *A History of Landownership in Modern Egypt, 1800–1950*, Oxford University Press (London, 1962); Rauf Abbas Hamid, *The Social System of Egypt under Large Landowners* (1837–1914), Dar El-Fikr al-Hadith (Cairo, 1973); and Asem A. El-Dessouky, *Large Landowners and Their Role in the Egyptian Society, 1914–1952*, Dar al-Thakafa al-Jadida (Cairo, 1975).
4. 1 *feddan*=1.038 acres.
5. R. Mabro, *The Egyptian Economy 1952–72*, Clarendon Press (Oxford, 1974).
6. E. Eshag and M. A. Kamal, 'Agrarian Reform in the UAR (Egypt)', *Bulletin of the Oxford University Institute of Economics and Statistics*, XXX, 2 (May 1968) p. 75.
7. Ibid., pp. 77–8.
8. This statement is a gross simplification of the outcome of the rich debate on the nature of Egypt's agrarian system which took place in the 1950s. For a good summary see Anouar Abdel-Malek, *Egypt: Military Society*, Vintage Books (New York, 1968) pp. 50–7.
9. M. Riad El-Ghonemy, 'Economic and Institutional Organisation of Egyptian Agriculture since 1952', in P. J. Vatikiotis, *Egypt since the Revolution*, George Allen and Unwin (London, 1968) p. 68.

10. Gabriel Saab, *The Egyptian Agrarian Reform, 1952–62,* Oxford University Press (London, 1967) p. 11.
11. The early 1950s witnessed violent peasant uprisings on large estates: Saab (op cit, p. 13) reported that peasants in Bashmur, Daqhaliya, occupied the Averoff estate twice in the space of three years; that agitation continued after the sale of the holding by its foreign owners, and only brutal intervention by security forces reduced the squatters to submission. Baer (op cit, p. 221) also mentioned three rebellions in 1951 on the Buhut estate of the Badrawi Ashur family and the Kufur Nigm estate of the Crown Prince as unprecedented events in modern Egyptian history.
12. The subject of land reform became most controversial during the 1940s. Various projects were discussed in the press or submitted to Parliament, notably the bill of Mr M. Khattab to limit the size of landownership to 50 *feddans,* the call by Mureit Ghali (in his book *The Agrarian Reform*) to raise the ceiling to 100 *feddans,* and the argument by Sadek Saad (in his book *The Problem of the Fellah*) that the state should expropriate land above 50 *feddans* and run it as collectives. For interesting details see Abdel-Azim Ramadan, 'The Revolution and Agrarian Reform', *Al-Kateb* (Cairo, August, 1971).
13. Samir Radwan, *Agrarian Reform and Rural Poverty: Egypt 1952–1975* International Labour Office (Geneva, 1977) p. 29.
14. Radwan, op cit, p. 23.
15. See K. B. Griffin, *The Political Economy of Agrarian Change,* Macmillan (London, 1973) for a full statement of these arguments.
16. These figures refer to 1969; see Central Agency for Public Mobilisation and Statistics (CAPMS), *Annual Bulletin of Co-operative Activities in the Agricultural Sector, 1969* (Cairo, 1971) p. 8.
17. Mabro, op cit, p. 81.
18. IBRD, *Egyptian Agriculture Development Problems, Constraints and Alternatives,* World Bank (Washington 1976) p. 13.
19. CAPMS, *1976 Population Census: Preliminary Results* (Cairo, 1977) p. 9.
20. UN, *Yearbook of International Trade Statistics,* various issues.
21. CAPMS, *Monthly Bulletin of Consumer Price Indices,* various issues. The index amounted to 190.3 per cent in 1975 with base year 1965/66=100.
22. M. Abdel-Fadil, *Development, Income Distribution and Social Change in Rural Egypt (1952–1970),* Cambridge University Press (Cambridge, 1975) Chapter 2.
23. Radwan, op cit, pp. 64–9.
24. Radwan and Lee, op cit, Chapter 5.
25. For details see Robert Mabro and Samir Radwan, *The Industrialisation of Egypt: 1939–1973,* Oxford University Press (Oxford, 1976).
26. Abdel-Fadil, op cit, Chapter 6.
27. Radwan, op cit, Table 2.3.
28. Ibid.
29. Radwan and Lee, op cit.
30. R. Mabro, *The Egyptian Economy,* op cit, pp. 206–7.

31. Ibid., p. 43.
32. Ibid., p. 44.
33. Radwan, op cit, p. 20.
34. For details see Radwan and Lee, op cit, Chapter 4.
35. Doreen Warriner 'Land Reform and Community Development in the U.A.R.' (UN, May 1971) p. 25.
36. Literature in current press abounds with descriptive material on this point. For a vivid description of the condition of peasants in the 1970s, see 'The Worries and Troubles of an Agrarian Reform Peasant' *Al-Taliaa* (July 1976).
37. Ahmed Hassan Ibrahim, 'Co-operative Farms: A System We Should Try to Eliminate – Fragmentation', *Al-Taliaa* (October 1972) p. 35.
38. Ahmed Hassan, *A Field Study on Crop Rotation in Ten Villages* (Institute of National Planning, Cairo, 1974).
39. In the absence of data on total credit supply to farmers, we used the cost of agricultural inputs as a proxy of total credit needs. Detailed figures from M. A. El-Shahaat and S. Z. Nassar, 'An Economic Analysis of State Farm Credit in Egypt', *L'Egypte contemporaine* (Cairo, July 1973) p. 96.
40. See for instance Adel Ghoneim, 'Notes on the Evolution of Economic and Class Relations in the Egyptian Countryside', *Al-Taliaa* (September 1966).
41. Abu el-Ezz and Abu elGhar, op cit, p. 341.
42. M. A. Fadil, op cit, p. 149.
43. Ibid.
44. S. Radwan, Agrarian Reform, op cit, pp. 72–3.
45. For details see Ibid., pp. 60–77.
46. Abdel Fadil, op cit, p. 120.
47. IBRD, *Arab Republic of Egypt: Economic Management in a Period of Transition 1978*, vol. III, World Bank (Washington, 1978) p. 37.
48. Essam Montasser, *Agricultural Prices, Growth and Sectoral Terms of Trade in Egypt*, FAO (Rome, 1975) Table 12.
49. IBRD, *Egypt – Agricultural Prices, Taxes and Subsidies*, Staff Paper, World Bank (Washington, 1978).
50. Anouar Abdel Malek, *Egypt: Military Society*, Vintage Books (New York, 1968) pp. 353 and ff.
51. M. Abdel Fadil, op cit.
52. R. Mabro, op cit.
53. B. Hansen and K. Nashashibi, *Foreign Trade Regimes and Economic Development: Egypt*, National Bureau of Economic Research, Columbia University Press (New York, 1975).
54. Michal Kalecki, 'Intermediate Regimes', in *Essays on Developing Economies*, The Harvester Press (London, 1976).
55. Ignacy Sachs, 'Kalecki and Development Planning', *Oxford Bulletin of Economies and Statistics*, XXXIX, 1 (1977) p. 48.

6 Agrarian Change in a Plantation Economy: the Case of Guyana[1]

Clive Thomas

6.1 Introduction

Sugar production was first introduced into Guyana by the Dutch in the late seventeenth century during the high season of European colonial expansion. Rivalry among the colonial powers caused the territory to change hands several times until finally in 1814 it became a possession of the British who ruled until independence in 1966. Throughout the entire period of colonisation sugar production was a major activity, and over the past century and a half, sugar has been the dominant crop – and agricultural production has been the principal means of livelihood – of the population. Following nationalisation, state farms have emerged as the dominant sugar producing structures. In this essay we analyse change within the sugar producing industry, tracing its larger impact on the entire agrarian system, and *vice versa*. Because of the importance of sugar in the national economy, the dynamic interaction between sugar and the entire society is also central to our presentation.

The long period of domination by plantation institutions did not mean that the structures of production in the sugar producing industry were static. These structures were organically linked to the historically dominant mode of production and to the definite forms of the social relations under which surplus was produced and appropriated in Guyana. Since these have always been changing, by way of either their extended social reproduction or qualitative transformations, they have inevitably produced great changes in the plantation as the dominant sugar producing institution. On this basis we can identify three major periods of change in the sugar economy of Guyana during the period of plantation dominance.

First, the period in which the plantation was founded on colonial slavery, that is, until 1838. Second, a subsequent period of modified colonial slavery in which the plantation was based on immigrant labour and indentured contracts, that is, until 1917. Third, the period of transformation of the plantation into the branch operation of a multinational corporation (MNC), that is, until 1975/76. These three periods together with the period after the nationalisation of the estates, that is, post-1976, will constitute the four basic units of our analysis. The earlier periods will be discussed only in so far as they have a direct bearing on contemporary issues of agrarian change in Guyana.

6.2 Colonialism, slavery and the plantation system

When the Dutch first settled the territory, sugar was produced on the basins of the numerous rivers which flow north into the Atlantic. By the middle of the eighteenth century rapid soil depletion had forced them to cultivate on the coastal strip where sugar production (and indeed 90 per cent of economic activity) continues to be located until today. The coastal strip is below sea-level. This, together with difficulties in the clay soils, heavy seasonal rainfall and a permanent swampy belt to the south of this strip required a complex system of water control before coastal farming could take place. Moreover, during this period European settlers were reluctant to occupy the territory because of the presence of the indigenous Amerindian inhabitants. Soon, however, Dutch special know-how, European capital and African slave labour were being combined by the planters to establish a network of drainage and irrigation canals along much of the coastal strip.

In the course of these developments it became obvious that only a large-scale and highly capitalised system of agricultural organisation would be successful. Several factors accounted for this. First, the provision of drainage and irrigation works required huge capital outlays and hence production on a large scale to be profitable. Second, at the prevailing levels of technology, a large amount of unskilled labour was needed to dig the canals. The slave trade provided this and the labour force with which subsequently to produce sugar. Again, this required large outlays of capital. Third, certain features of sugar production facilitated the bias in favour of large-scale agricultural organisation. One was that sugar cane cultivation at that time did not really require highly sophisticated farming techniques and, provided there was adequate supervision, could be run efficiently by using a large, unskilled labour force. It did, however, involve the movement of huge quantities of material. The ratio of plant to sugar produced was then of the order of 12:1. Another feature of the sugar cane was that after it was manually cut, it had to be processed within hours or lost its sucrose content. This necessitated the placement of the factories near the cultivation areas and a complex transportation system for the movement of materials and fuel. Again a premium was placed on large-scale operations. The skilled personnel and the technology used up in these operations could at that time only be found in Europe, and their procurement also required significant expenditures. Finally, if there were to be large-scale producing units, large markets had to be found. The local market was obviously too small and so from its inception production had to be geared towards satisfying the needs of consumers in far-off Europe. Producing for such a market placed a further premium in favour of large scale operations.

The plantation soon emerged as *the* large-scale agricultural organisation in Guyana. Several aspects of its early institutional form are important for understanding the contemporary agrarian system in Guyana. First, it grew out of the process of the overseas expansion of European capital. This was concretely expressed in two forms. One was that its output was marketed in Europe. The other was that the local planter class was organically linked to, and derived its initial capital from, the propertied classes in Britain. Second, Guyana was a colony and the dominant position of the plantation was directly supported by the colonial state. Third, the planter class linked sugar directly to the plantation and pursued mono-cultural specialisation with an amazing zeal. Fourth, the dominant mode of production in Guyana at that time was slave labour and the plantations were organised on this basis. This created a highly stratified society. The planters were the dominant economic, social, and political force. The direct producers, as slaves, had no economic resources or political status. Since the slaves were black, this added a racial stratification onto the basic differentiation between owners and managers and the direct producers. In this system of class and racial differentiation, ideology and culture played important roles in rationalising the *status quo*.

Finally, although the planters placed heavy investments in water-control systems and in acquiring slaves, their outlook on profit making was essentially speculative. This derived from two basic considerations. One was the undeveloped nature of capital at the time and their immediate origins in merchant capital and the landed classes of Britain. The other was their uncertainty of the future price of sugar in an era when mercantilism was giving way to free trade based on the supremacy of the British industrial capitalists. The planters' outlook was to speculate in the years of 'bonanza' prices, when a 'killing' could be made, and to keep the plantation turning over in the lean years. This speculative outlook contradicts the normal expectations of a more fully developed capitalism where investors associate long-term investments with low and steady yields. Such an outlook combined with an approach to accumulation which further highlighted the influence of pre-capitalist features on the plantation. Thus the approach of the planters was towards extensive exploitation of labour as a means of generating profit. Instead of seeking to raise productivity per man hour through technical change, the emphasis was on acquiring more slaves and extending the working day – often to the point of physical collapse. The land was also exploited extensively. Expansion of new acreage using existing techniques, rather than newer methods and better farming practices, accounted for most of the increases in output. Throughout this period, technological innovation and the application of new technology to sugar production in Guyana were not significant features of the rural economy. When soil depletion made itself manifest, the planters opened up new virgin lands. Because of the extremely low man to land ratio, this was an easy option.

The system of extensive exploitation of slave labour and land meant that most of the benefits of sugar production flowed to the owners of the estates. Frequently this was used to enhance their social status within Britain where many of these owners returned to live. To the planter class capital did not have the driving, self-expanding value that it had for the class of capitalist manufacturers which was to emerge and dominate the British economy by the middle of the nineteenth century. This practice of returning 'home' soon led to the creation of an absentee landlord structure on the plantations and the rise of a class of local attorneys and managers. This tended to reduce the patriarchal tendencies within the plantation and so made easier the transition at a later date to a more capitalist structure.

6.3 Change in the plantation system: immigration and indenture

The iniquitous system of colonial slavery collapsed under the strains of slave resistance, rebellion and insurrection, and the development of capitalism in Britain. The slaves were legally freed in 1838. But from then until 1917 a modified form of colonial slavery, as enshrined in local law, persisted. During this period four developments in the agrarian system are of profound significance. First, despite a post-slavery apprenticeship system the slaves abandoned the estates, and wages rose immediately after the abolition of slavery. At the same time the dissolution of the mercantilist system had led Britain to abandon protection for the colonies which resulted in a drastic fall in sugar prices. The result of higher wages and lower export prices was to plunge the sugar economy into deep crisis. To solve this the planters obtained a state supported system of immigration. Immigration was based on a system of indenture in which the immigrants entered into binding long-term forward contracts for their labour.[2] The main period of immigration was between 1837 and 1919. Immigrants came from Germany, Malta, Brazil, China, Madeira, Mauritius, Europe, Africa, other Caribbean territories and India. The last were found to be the most suitable and became the largest source of immigration – today the descendents of these immigrants account for approximately 53 per cent of the total population.

Indentured immigration, like slavery, was a legally controlled system of surplus extraction. Like slavery it was also aimed at limiting the development of a labour market. But unlike slavery, indenture coexisted with a class of 'free' men; free, that is, to sell their labour power on the market. As such indenture did not altogether contain the spread of commodity relations in the economy, time progressed, and the labour market expanded more and more, and it was also fuelled by those indentured immigrants whose contracts had expired, but who did not opt for their return passages to India. The pattern of immigration did, however, yield two consequences of lasting significance. One was that the slaves after moving off the plantations, when forced eventually to re-enter as wage labourers, avoided field operations and concentrated on factory work. The newer immigrants from India were concentrated in the field. This led to a second development of great consequence, namely a racially compartmentalised labour force.

The second development of this period was the rise of a peasant mode of production in the rural economy. Initially the slaves bought lands either as individuals or on a communal basis to form new villages. This 'village movement' met with great success initially, but a combination of punitive taxes, inadequate drainage and irrigation, a narrow domestic market, poor communications and hostile attitudes by planters and state officials who saw the village movement as a direct threat to the plantation, led to the collapse in the dynamism of this movement. The planters were particularly opposed to the communal village structure. Yet this was the only structure in which the peasants could afford the overheads of a water control system for independent farming. Without such a structure individual peasants would be heavily dependent on plantation drainage and irrigation.

The result of these changes was that the peasant sector remained weak and the plantation dominant. The small farmers concentrated on foodstuffs for the domestic market as they did not have the factory or marketing capacity to produce sugar. The peasant economy was therefore decidedly not founded on sugar. Given the limitations of the local market, the competition of imported staples (e.g. salted fish, flour), the low level of peasant farmer capitalisation, etc., returns were too low to secure an adequate rate of expansion of peasant operations. The peasant economy therefore remained essentially a subsistence economy within this dualistic structure, with periodic surpluses sold in the local market. This development was reinforced by one other aspect of the policy of the planters to the immigrants. To deter immigrants leaving the estates after the expiry of their contracts, particularly those desirous of returning to India (which would have been costly to the planters), small parcels of land adjoining the estates were given out. These were just enough to contribute to the feeding of the family without deterring them from working on the estates during the reaping and grinding seasons.

The third development of note was the fact that during this period the landed proprietors became more and more absentees. Later this paved the way for the rise of the limited liability company within the industry and its eventual transformation into a subsidiary within an MNC structure. This process was accompanied by a marked consolidation of land and capital in the industry. Thus, whereas at the time of emancipation there were 308 sugar estates (out of a total of 400 separate estates in Guyana), by 1853 the numbers of sugar estates were reduced to 173; by 1885 there were 105; by 1890, 84;

and by 1904 this was reduced to 46. When the industry was nationalised there were two MNC subsidiaries, one owning just over 80 per cent and the other just over 10 per cent of sugar industry assets.

The final development of note was the introduction of policies to contain the peasantry and to promote immigration.

In order to contain the communal peasant purchases a law was passed in 1853 prohibiting the joint purchase of land by more than 20 persons. The law further required that if more than 10 persons held land, it had to be partitioned and levied for maintenance rates. The aim of this taxation was to move the peasants into the cash economy and prevent the desertion of the plantations by freed African slaves. Similarly, immigration was for the most part financed through the general exchequer. In addition, labour regulations were passed which progressively limited the freedom of movement and association of indentured workers.

Two consequences of present importance followed from the deliberate use of such policies. First, the present system of confrontationist industrial relations descends directly from this period of managing slave and half-free labour. Second, the break up of the communal land systems led to the fragmentation of peasant holdings and fostered an aggressive individualism among peasants. Given the ecology of the coastal area, with its special water-control requirements, this forced immense obstacles in the way of the growth of the peasant economy.

6.4 Sugar as an MNC activity

(a) Transition

With the abolition of indenture in 1917, the labour process based on legal coercion was ended. Thereafter the sugar plantations operated through a market for labour and the extraction of surplus from the direct producers had to take place as part of the normal functioning of an economy dominated by market relations. This change produced huge convolutions in the economy eventually resulting in the plantations being incorporated into MNC subsidiary operations. Certain characteristics of the labour market were clear from the early stages. First, the work system was based principally on piece work, with time work largely confined to factory, clerical and other such operations. Second, employment was offered on a seasonal basis with no obligations by the planters for off-season work. Third, a significant proportion of the labour force was women and children. Fourth, the physical arduousness of field labour was not reduced, since the field operations were not mechanised. Finally, the planters were opposed to the rise of worker organisations.

Except for the reprieve based on shortages created by World War I, the first half of the twentieth century was one of depression and severe economic crisis in the Guyana sugar industry. Falling export prices and rising costs of labour after 1917 forced many plantations into bankruptcy. Their assets were frequently taken over by the more successful plantations. This tendency towards the centralisation of plantation ownership was accompanied by expansion in the scale of operation on each estate as the benefits of large-scale economies and technical change became more evident. In order to survive the larger units introduced a wide range of new technology: fertilisers, steam pumps, evaporators, centrifugal driers, juice clarifiers, chemical controls in the factory, changes in the lay-out of the sugar fields, etc. All these were aimed at increasing productivity and reducing unit costs of operations. The crisis therefore produced a strong association between tendencies towards concentration in the industry and technical innovation.

The process of increasing consolidation of land and capital and technical change was accompanied by two changes in ownership structure of the plantations. First, many of them became public companies, with their shares traded on stock markets. Second, the absentee landlords either gave way to, or became more organically linked to, industrial interests in Britain. Indeed it was those plantations where these links were the closest which survived the best. Changes along these lines induced the shift in outlook away from speculation, and low capital-specific technology to a greater concern with costs, reserve accumulation for hard times, and an orderly system of investments and expansion. But these same developments gave rise to diversification along two lines. First, the plantations moved into other areas of local activity, often related to their activities in sugar – shipping, warehousing, retail, insurance, etc. Second, their new owners overseas started to diversify internationally, to cushion against effects in Guyana. Thus, by the time of the nationalisations in 1975/76, the two companies which controlled the industry had world-wide investments in Canada, Africa, other parts of the Caribbean, Latin America and Britain as well. As it turned out, sugar profits made in Guyana were an important source of investment funds.[3]

(b) Sugar economy: post-World War II structure

Table 6.1 indicates the main GDP aggregates for Guyana (1960–76). These data indicate the wide fluctuations in the levels of activity in sugar and agriculture – a feature characteristic of primary production. However, between 1960 and the years 1971–3, there is unmistakable evidence of a decline in the contribution of the agricultural sector to the national economy. The great commodity boom of 1974/75 reversed this tendency substantially, but this can be treated as unusual. Sugar is by far the largest contributor to agricultural GDP giving a clear indication of its domination of the rural economy. Rice is the second largest agricultural sector but this is substantially below sugar in importance. The other agricultural sectors combined (livestock and other agriculture – primarily domestic food crops and fruits) are also small in relation to sugar. The dominant position of sugar in national production has been challenged only by the development of the bauxite–alumina mining industries. During 1964–73, mining was the leading sector in national production. Since then, mainly on account of the boom in sugar prices and the post-1975 depression in the aluminium industry, sugar has regained the position of leadership. Both these sectors, however, are subject to wide fluctuations in their levels of economic activity.

Ninety per cent of the acreage under sugar was controlled by the two MNC plantations until their nationalisation in 1975/76. This acreage represented the best drained and irrigated, the most fertile, and the most accessible agricultural land in Guyana. Over the past two and a half decades, the acreage of sugar cane harvested, although fluctuating, has nearly doubled to a present total of approximately 120,000 acres. Yet sugar output on the estates in 1976 was below the level of 1961/62, and at 304,000 tons was below output capacity already achieved in the 1960s.[4] Yield of sugar cane per acre has decreased and the conversion rate of sugar cane to sugar worsened during this period. Over the years 1972–6, the yield of sugar cane per acre was only 30.2 tons (compared with 37.4 tons in 1956–9) and 11.85 tons of sugar cane was required to produce a ton of sugar during the same period (compared with 10.94 in 1956–9). The overall yield of sugar per acre between the two periods declined from 3.42 tons to 2.55 tons. This was a remarkable development since the post-war years witnessed a number of vast technical changes. Many operations were mechanised and farming methods reorganised, resulting in a decrease of

Table 6.1 Sectoral Contribution to GDP at Factor Cost 1960–76 (percentage)

Product	*1960*	*1961*	*1962*	*1963*	*1964*	*1965*	*1966*	*1967*	*1968*	*1969*	*1970*	*1971*	*1972*	*1973*	*1974*	*1975*[2]	*1976*[2]
1. Sugar	17.6	17.2	17.0	22.6	14.7	14.7	12.5	13.9	11.8	13.8	12.4	14.8	14.5	11.7	29.0	30.6	20.4
2. Rice	5.3	5.4	5.0	4.8	6.7	6.3	5.7	4.2	4.2	3.0	3.7	2.9	2.1	2.8	3.4	3.8	2.9
3. Livestock	1.3	2.4	3.2	2.5	2.8	2.6	2.2	2.3	2.3	2.3	2.2	2.3	2.2	2.3	1.7	1.8	2.1
4. Other agriculture	3.3	3.2	3.2	2.7	3.3	2.9	2.9	2.9	2.8	3.1	2.9	2.9	2.9	3.1	2.2	2.0	2.4
5. Agricultural sector[1] (1+2+3+4)	27.5	28.2	28.4	32.6	27.5	26.5	23.3	23.3	21.1	22.2	21.2	23.0	21.7	19.9	36.3	38.2	27.8
6. Fishing	1.6	1.4	1.4	1.3	1.6	1.4	1.4	1.4	1.2	1.1	1.1	1.1	1.1	1.2	1.2	0.9	0.9
7. Forestry	2.5	2.4	2.1	1.6	1.6	1.6	1.8	1.4	1.4	1.3	1.1	1.0	1.1	1.0	0.9	0.8	0.9
8. Manufacturing[1]	5.2	5.2	5.4	5.8	6.5	7.7	7.7	7.6	8.0	7.8	8.2	7.9	7.9	7.6	5.8	6.3	7.4
9. Mining & quarrying	11.0	12.9	16.3	13.0	17.6	16.6	17.2	17.8	19.6	19.6	20.4	18.3	17.0	14.0	13.5	13.0	13.7
10. Transport and communication	7.5	7.5	7.2	6.6	6.5	6.6	6.8	6.6	6.4	6.2	5.9	6.0	6.2	6.4	5.1	4.6	5.7
11. Engineering and construction	10.5	7.2	6.8	5.0	5.0	5.3	6.4	6.8	7.4	8.1	7.9	7.8	8.1	8.1	6.2	6.2	7.6
12. Distribution	14.1	13.5	11.9	12.2	12.9	12.1	12.3	11.9	12.7	11.9	11.5	11.0	11.1	11.2	8.9	8.7	9.9
13. Services and rent	11.3	10.6	10.2	10.9	9.9	9.9	10.4	9.8	9.5	9.3	9.6	9.9	9.6	9.7	7.4	6.8	8.0
14. Government	9.7	10.5	9.9	10.2	10.7	12.4	12.8	13.4	12.8	12.6	13.2	14.1	16.3	21.0	14.9	14.6	18.0
15. Change in total product on previous year	—	+9.7	+6.0	−10.4	+10.0	+8.3	+5.2	+9.6	+8.1	+8.1	+6.7	+5.9	+6.9	+8.9	+50.1	+26.8	−3.6

Note 1. The agricultural sector is given as the sum of rows 1–4. In the Guyana national accounts data sugar and rice processing are placed in the manufacturing sector. For our purposes, because a very rudimentary stage of processing is involved they are treated as agriculture. The manufacturing sector (row 8) includes manufacturing other than rice and sugar processing.

2. The data for 1975 and 1976 are provisional.

Source *Statistical Digest Various Issues* and *Bank of Guyana Annual Reports.*

one-third in the labour force. New factories were constructed and old ones modernised. New bulk loading and distributive facilities were built. Plant selection and breeding, pesticide control (aerial spraying) and fertilisers were all introduced on a tremendous scale. Many of these changes occasioned serious conflicts on the estates and a deterioration of labour relations is perhaps the single most important factor accounting for the eventual decline in yields.

All these modernisations and expansion were financed from resources within the industry, either through levies on sugar sales for such purposes or ploughed back profits. Despite this, real wages did not show significant increases. Between 1965 and 1975 real earnings in the factories increased by only one per cent and in the field by nine per cent. Yet labour productivity between the years 1956/57 and 1974–6 had increased by approximately 37 per cent in the field and 45 per cent in the factory.

Throughout this period, sugar continued to be a predominantly export crop. During the period 1960–75, 92 per cent of the volume of sugar and 96 per cent of the value of its sales were made overseas. The bulk of the sugar sales was made under preferential quotas and agreements, principally the Commonwealth Sugar Agreement (the US market until 1974) and later the Lomé Convention of the EEC. These agreements were vital to the industry as the plantations had claimed that their costs were uncompetitive on the world market. Since these negotiated concessions required reciprocal preferential access of US and UK manufactures to the Guyana market, a strong element of national subsidy was entailed. Comparable support was not given to other types of local agriculture.

The MNC structure of the industry gave rise to considerable scepticism in some quarters over the frequent and almost ritualistic declarations of impending insolvency in the industry. It has been claimed that transfer pricing, interlocking directorates, the development of allied activities (e.g. shipping, alcohol, distribution, wholesale and retail outlets, foundries, and motor vehicle repair outlets), which served sugar as part of the same company complex combined with an ancient local company law (going back to 1929) which did not require local incorporation, led to fears of 'hidden' surpluses being transferred outside the economy and that the MNC plantations were operating as a drain on national resources. The failure of labour incomes to grow and the fact that the companies, despite their vast size, were contributing less than five per cent of total government current revenue between 1966 and

1973 strengthened this feeling. It was not until the sugar boom of 1974/75 that the government introduced a variable levy on sugar which led to more than a one hundred times increase in sugar revenue between 1973 and 1974 and a further increase of more than 50 per cent the following year![5]

As with most areas of primary tropical production, the level of processing and the degree of internal linkages in the sugar industry are low. After initial processing, refining takes place abroad. The main local by-products are: bagasse, molasses and alcohol. The first is used as fuel in the factories. Molasses was exported through an MNC subsidiary, and the alcohol produced (except for that retained for local consumption) was exported in bulk to be bottled and marketed by another subsidiary of the same MNC. Most of the major material inputs: fuel, lubricants, fertilisers, machinery and equipment, industrial chemicals, transport equipment, storage facilities, agricultural tools and implements, etc. are imported. It was in order to exploit the need for these inputs that the larger of the two MNCs (Booker McConnel Ltd, controlling 80 per cent of the industry) developed its own shipping, engineering and agricultural divisions. The linkages were however created abroad and not within Guyana. All the activities of the MNC, except the most technologically specific were conducted overseas.

We have already referred to the displacement of labour which accompanied technological improvements in the MNC plantation structure. Despite this the industry still remains highly dependent on labour in the crucial areas of planting and reaping. The lay-out of the sugar cane fields in Guyana, together with the slushy nature of the clay soils when wet, has so far defied all efforts to mechanise these operations. The sugar industry thus remains highly dependent on labour. At the time of nationalisation sugar was the largest single employer of labour, using approximately 10 per cent of the total labour force. During the period of mechanisation, women and children were retrenched in order to meet the many social criticisms levied at the use of these types of labour. While most of the work on the estates is very seasonal, the bulk of the labour force had no guaranteed system of out-of-crop work. This generated great insecurity and a high turn-over of the labour force. When combined with the physical arduousness of work on the plantations, it is not surprising to find a definite aversion to estate labour and strong social stigmas attached to it. Thus it has frequently happened that despite high unemployment levels in the country, of the order of 35 per cent, sugar estates could still have found it difficult to find labourers to reap the crop!

The post-World War II period witnessed the final consolidation of the dominance of the MNC structure in the sugar economy. By then, however, the dependence of the economy on sugar was only a part, albeit one of vital importance, of the wider dependence of the economy on foreign capital inflows and primary export production. This dependence gave rise to growing inequalities accompanied by widespread distress in the rural areas. The problems posed by this sugar producing structure were indissolubly linked to the larger and more fundamental problems of social change in Guyana.

6.5 Peasant economy and sugar: the other side of dualism

(a) Broad characteristics

From the abolition of slavery right up to the nationalisation of the plantations in 1975/76, the peasant mode of production struggled to assert itself. We earlier noted the opposition of the planters, after the break-up of the colonial slave mode of production, to the movement of the slaves to build an independent peasant economy. Would-be peasants had to struggle for access to land, drainage and irrigation and markets. Later this struggle was merged into the nationalist movement which resulted in independence and later nationalisation of the plantations. It is significant that both the major political parties of the post-war period have espoused socialism for Guyana. Even more significant is the fact that the rural areas are overwhelmingly in support of the opposition Marxist-led political party. It is true that ethnic divisions contribute to the distribution of political support among the two major parties: the rural areas are predominantly Indian and the urban areas are predominantly African. But it is also equally true that the extreme impoverishment of the peasant economy together with the bad conditions of labour on the estates have produced a long and deep discontent among the rural inhabitants. This discontent was so deep that the present ruling party undertook the nationalisation of the plantations.

Peasant cultivation has developed along the lines of four major areas of economic activity in Guyana. First, there is the rice industry which was introduced and developed under special state support schemes, such as the land settlement programme of the late 1950s and early 1960s. Rice was developed as a peasant crop complementary to plantation sugar. Its growing schedule did not cause competition with the plantations for labour. State support was principally by way of the provision of drainage and irrigation facilities, access roads and credit. The objective of the rice programme was to minimise peasant land hunger. Currently rice contributes about 3 per cent to the GDP, although in the 1960s its contribution was twice this. The second line of activity is domestic foodstuffs, i.e. ground provisions, vegetables, fruits and so on. This production is aimed almost entirely at the domestic market and it has contributed about 2.5 per cent to the GDP over the past five years. The third line of activity is livestock and dairying which is also aimed at the local market. This accounts for about 2 per cent of the GDP. The final line of production is sugar. Peasant production of this crop now averages about one-tenth of the total production of sugar cane.

About two-thirds of the population live on the coastal rural areas where approximately 0.5 million acres are under crop cultivation and 0.4 million under developed pasturage. There are two main categories of peasant lands. The first is within the estate empoldered areas, and are usually small plots often given by the estates to the farmers to encourage their settlement there and hence contribution to the work force of the estates during the sugar crop season. The other type of settlement is outside the estate empoldered areas. One half of the acreage under crop cultivation is used to produce rice, which is the main provider of staple food for the bulk of the population. One-third of the acreage is under sugar (at any point of time about one-third of this amount is reserve lands). This is the bulk of the best drained and irrigated and the most accessible lands. In the peasant economy, 68 per cent of the total number of farms are under 10 acres.[6] These farms account for just over 5 per cent of the total acreage. Farms of 100 acres plus account for 4 per cent of the total number of farms and approximately 80 per cent of the acreage. About 53 per cent of the farmers who live off the estates occupy 3 per cent of the total land acreage, whereas farming families with farms of 500 acres plus occupy 75 per cent of the total acreage and comprise only 3 per cent of the total number of farming families. In the rice economy, one-half of the rice farms are under 5 acres in size and 90 per cent of the rice farms are outside the estates. Most of the farms on the estates are kitchen garden plots.

Significantly, only 15 per cent of the peasant households seem to depend on farming as a sole-source of income. Further, as many as 57 per cent of the households depend on sources other than farming for more than one-half of their income. It is clear from this that while farming contributes to the household subsistence, many farmers are forced to enter other lines of activity, which include plantation work, in order to make a livelihood. This highlights the strong links between the peasant economy and the plantation. The system of the part-time labourer rests on this foundation.

(b) The Peasant Sugar Sector

As far back as 1897 when a Royal Commission visited Guyana in response to a crisis in the sugar economy, it recommended that peasant cultivation of the sugar cane should replace the plantation system and that the plantation companies should confine their activities to processing. A system of land settlement schemes and government subsidies were recommended. Crop diversification was also advocated. Nothing along these lines occurred and by the end of World War II there was only individual scattered cultivation which in 1948 accounted for 3000 tons of sugar. Another Commission, when investigating the industry in 1948, also recommended an extension of acreage under peasant cultivation. This time the recommendation was not to replace the estates but to achieve peasant production of up to 10 per cent of the total crop. Without any apparent justification of it, this target seemed to have become acceptable to all groups in Guyana – including the political parties arguing for the 'decolonisation' of sugar.

While agreeing on this modest expansion, the question of farm organisation had to be settled. The individual fragmented holdings of the peasant seemed unfeasible because of water control problems and the logistical difficulties of getting the sugar cane to factories for immediate processing after it was cut. All proposals seemed to indicate some type of co-operative farming. What emerged out of this follows the now famous Belle Vue prototype, which came out of an experimental project at an estate of the same name. This prototype was based on strong centralised direction by the plantations. The plantations provided housing for the families, drainage and irrigation, specified the acreage to be planted, supervised pest control, directed the plant variety to be used, transported the crop, provided the ploughing equipment and fertiliser, charging a fee for all these services. In turn, the sugar cane crop was sold in its entirety to the plantations for processing in their factories. This pattern prevails on the farms attached to the plantations. As a pattern, it leaves little scope for initiative or technical innovation on the part of the farmer. It is not surprising that after the early 'heady' days of planting, many farmers just let the cane remain in the ground, often for more than a decade, without replanting and with no concern for proper ratooning. Alongside these developments, the system of sugar cane cultivation on the individual scattered holdings continued.

Several reasons have been advanced in favour of peasant cultivation of sugar. These have been as much social and political as economic. To the Colonial Office, first of all, and later to governments and the MNC, this appeared as one way of appeasing land hunger and the social discontent which this hunger generated. It was also seen as a means of committing the peasantry to support sugar as an economic activity and of placating the rest of the population, which by this time was hostile to the extreme mono-cultural agricultural specialisation in Guyana. At a more directly economic level, the MNCs were also concerned that mechanisation of their operations had contributed to the country's economic misfortunes by adding to the already large pool of surplus labour. To make matters worse, this and other factors were contributing to unionisation and labour militancy. One way of easing these difficulties was to divert from sugar cane cultivation and encourage, as in the rest of the Caribbean, peasant cultivation of the sugar cane and MNC processing of sugar. Lacking clear alternatives, even those who opposed MNC plantation domination could do little but urge the expansion of peasant cultivation of sugar.

By 1974–6, the peasant sector harvested 12 per cent of the acreage and produced 11 per cent of the crop. Compared with a decade or so ago (1961–3) this was a significant rate of increase, for at that time the peasant sector harvested and produced just about 3 per cent of the sugar cane crop. Over this period, the rate of expansion of the peasant sector exceeded that of the MNC. For the years 1961–6, farmers' yields averaged between 80 and 98 per cent of estate yields. The estates were responsible for the conversion of sugar cane to sugar. By 1976, there were 3952 peasant operations holding 17,440 acres – an average size of 4.4 acres. Of this total, about one-half were co-operatives (1886 farmers cultivating 8253 acres), and one-half individual operations (2066 farmers cultivating 9187 acres). Interestingly, the mean size per farmer was, in both instances, the same (4.4 acres). Detailed study of the data indicates a high level of stratification of the peasant sugar cane farmers.[7]

As was pointed out earlier, all of the sugar cane is processed in the plantation factories. To that extent indices of acreage and output overstate the importance of the peasant sugar sector. In addition, most of the transportation of the sugar cane and sugar, most of the drainage and irrigation, all of the scientific laboratories, management overhead, overseas marketing, etc. are in the control of the non-peasant sectors. Under the peasant marketing arrangements, the farmer is required to sell all his sugar cane to the estate to which he is attached. The estates in turn process and sell the sugar as part of their regular sales. At the end of the crop, the weighted average of the prices obtained for sugar sales to all markets is computed. From this the estates deduct insurance, selling charges, interest, transport, porterage, loading charges and government taxes. Out of the remainder, the farmer receives two-thirds and the estate keeps one-third. These payments are staggered with normally two interim payments and a final payment at the end of the crop. Interest is charged on the interim payments.

This arrangement severely disadvantages the farmer. (Interestingly, in Trinidad-Tobago and Barbados, the farmers' shares are 72 per cent and 80 per cent respectively. In Jamaica, the system operates on a cost plus basis.) Among the many difficulties are: (i) controversy over the manner of determining the sucrose content of farmers' cane; (ii) the farmers get none of the income produced by the further processing of their cane, i.e. alcohol; (iii) the estates direct the time of cutting the farmers' cane to suit their convenience;[8] (iv) the overseas sales were previously conducted through a subsidiary of the MNC and the mark-up was frequently disputed; (v) the estates refused to accept 'risk' for farmers' cane charging them for losses in manufacture, storage, shipment, etc., yet the farmers had no control over the cane once it was handed over to the estate. This also raises the question of the fairness of the peasants paying interest on interim payments. This practice presumes the cane is the farmers' until it is finally sold.

The above demonstrates that the marketing arrangements for sugar cane can result in a transfer of income out of this relatively impoverished peasant sector and into the relatively wealthy components of the sugar economy. (Significantly, since nationalisation these arrangements have not yet been modified.) It would be incorrect to place primary emphasis on the marketing arrangements for the sugar cane in determining the exploitation of the peasants who produce cane. The real source of this income transfer lies in the unequal access of the two sectors to the community's wealth and resources.

Insufficient land, marked inequalities within the peasant economy, an ineffective co-operative structure, inadequate drainage and irrigation, poor access facilities – all these and more testify to the reality that more than 130 years after the village movement had started, the peasant economy in sugar remains essentially marginal and dependent on the plantations. The historical and ecological factors indicated earlier explain the situation. The one that stands out above all is, given the effective scarcity of adequately drained land on a coastal strip hemmed in by sea and swamp and subject to the ravages of seasonal tropical rains, plantation dominance is incompatible with a strong peasantry.

If the nationalisation of the plantations simply represents a shift in property relations, while the dominant economic relations of the plantation remain intact, then the marginality and dependence of the peasantry on the estates will be transferred to the state. Thus the dualistic rural structure which has emerged is based on marked differences in land ownership and tenure, and the structure of the producing units. Additionally, in the plantation sector, agricultural surplus was extracted through the operations of the labour market and a foreign controlled technology and accumulation process. In the peasant economy, family labour is predominant, with agricultural surpluses precarious. In both instances, state intervention was important. But the state intervened much more decisively to support plantation interests. This was as the main plank of the economy, while the peasantry, despite its numbers, was viewed as essentially marginal.

6.6 Overcoming the plantation system: nationalisation

From the time of European conquest, Guyana's resources, people and their economic activities have been organised around the principle of producing for export and satisfying internal consumption and capital requirements through imports. The principal products exported have all been primary, i.e. agricultural staples (sugar and rice) and mineral resources (bauxite-alumina). Prior to nationalisation, the two largest industries (bauxite and sugar) were contributing 35–40 per cent of GDP and 70–80 per cent of exports. Both industries were foreign owned and controlled. Additionally, other important sectors of the economy: banking, insurance, manufacturing (assembly) plants producing for the local market, distribution (wholesale and retail), etc., were all managed and controlled by overseas interests. It was inevitable that this dependence of the economy combined with the specific forms of plantation domination of sugar and the rural economy indicated in this essay, would pose insuperable obstacles to the transition to a self-reliant, internally oriented, technologically independent and synergetic system of agriculture.

Taking the post-World War II period as a whole, there were striking rates of increase of MNC plantation acreage and sugar output, even though most of the increases were confined to the period from the mid-1950s to the early 1960s. This was accompanied by the construction of new factories, the modernisation of old ones and the elimination of some smaller inefficient factories. All the allied services – transportation, storage, etc. – were substantially expanded and modernised. Science and technology were rigorously applied in the industry from the selection of plant variety, control of pests, weeds, etc. right through to the construction of semi-automated factories. Although cutting and planting have not been mechanised, there was substantial reorganisation of planting methods. All these developments were financed by resources internal to the industry, thereby attesting to its profitability. But despite these achievements, the nationalisation of the estates continued to be seen as the only feasible strategy.

Many factors accounted for this. First, in the expansion of the industry great burdens were put on the work force. As we have seen, the mechanisation led to the redundancy of about one-third of the work force. This created important social problems, not least among which was stepped up rural-urban drift. Also, real wages and conditions of work did not keep pace with increases in worker productivity. Labour continued to be seasonal and temporary, with the estates avoiding responsibility for employment in the out-of-crop season and every year for the past twelve years a strike has been called during the major harvesting period. No wonder that the sugar industry alone accounted for just over two-thirds of all man days lost due to strikes in Guyana. Second, this expansion was at the expense of the peasantry. As the plantations expanded, so did their control of land and resources on the coastal belt deepen. In this situation, it was clear that diversification for the benefit of the peasantry could only take place at huge cost outside of the existing empoldered areas. Third, despite the growth and technical improvements, the latter did not last. As we indicated earlier, much of the technical gain was made by the parent company, as witnessed in its development of specialist services, and equipment production for the sugar industry. In addition, the threat of both nationalisation and relatively high prices in later years, led to some abandonment of technical standards. We have already noticed evidence of this in part, when observing the decline in yields and the rise in the conversion ratio of sugar cane to sugar. But the evidence suggests also that the high prices of sugar seemed to have encouraged the MNC to cut cane before its maturity to cash in on the good prices, to defer resting the land in crop, to encourage the extension of the crop in ratoon, and hence to reduce the rate of replanting.[9] Furthermore, as the level of industrial relations deteriorated, the standards of cutting, weeding, clearing, etc., also deteriorated, along with the standards of effective supervision and management. Two quotations from the first report of the state corporation which took over the industry highlight some of the points being made here. Thus during 1976 we find:

> non-sucrose in cane was so high that molasses production far exceeded estimates. . . . This was due to a combination of factors (including) poor cane quality.[10]

Or again:

> the problem of lack of cane stemmed from strikes and shortage of cane cutters on some estates, but our major concern was the number of hours lost for unscheduled maintenance. High absenteeism during the out-of-crop and weekend maintenance periods had tended to lead to low standards of workmanship with the situation being aggravated by poor inspection and supervision. Our ability to detect plant weaknesses before failure has been impaired by inexperienced staff at many levels.[11]

All these considerations were worsened by three further factors. The first factor was the social and working conditions on the estates. Cutting and planting sugar cane in water-filled terrain was a physically arduous task and housing conditions were extremely bad. But not only were the social services poor, there were also sharp contrasts of life styles between the plantation owners and managers and the rural communities living on and around these plantations. Second, the MNC structure of the plantations combined with inadequate commercial legislation gave rise to suspicions of super-profitability. The absence of any requirement for local incorporation plus the minimal disclosures inherent in a Companies Act based on the UK Company Act of 1929 gave grounds for suspecting that practices such as transfer pricing, over and under invoicing, profit disguising, etc. took place. Finally, and particularly in the earlier periods, the plantations were strong enough to influence the state to support their interests. This can be seen in state policies on subsidies, external trade, and public investments in agricultural infrastructure – all of which favoured the plantations.

As time progressed it became evident to the nationalist forces that only the removal of plantation dominance over agriculture could liberate resources and the rural community and thereby lay the basis for new conditions of rural life. In this policy, nationalisation emerged as the only valid option since the state was the only institution in Guyana which could match the strength of the plantation. Soon, therefore, the struggle became first to develop the nationalist movement in order to win independence and later, with control of the state, to undertake the nationalisation of the estates.[12] The shift in favour of the political influence of the masses brought about by universal adult suffrage in national elections (1953) and the rise of mass-organised political parties broke the political power of the planter class.

The nationalist forces felt that nationalisation was a necessary step in breaking the hold of the plantations. The MNCs also realised this and introduced strategies to prevent this development. These strategies can be categorised at three basic, but fundamentally interrelated levels. First, there was a political strategy. In general this meant supporting the colonial government until independence and, after independence, trying to secure representation of their interests in the various elected assemblies, in opposition to the rural based PPP party. Another allied aspect of this strategy was the refusal to recognise the majority union in the industry because of its political affiliation.[13]

The second approach was broadly economic and contained two main thrusts. The first was to pursue mechanisation in order to reduce dependence on a militant labour force.[14] The second thrust was to hedge investments by diversifying their activities internationally. In its Annual Report for 1953, the year in which the constitution was revoked, the company's policy was stated thus:

> Our policy must be, so long as politics do not make it economically unjustifiable and downright imprudent, to maintain our business in British Guiana . . . however this may be, your Board

> are keenly aware of the need that the group should continue to build up in the form of profitable interests elsewhere 'hedges' against catastrophe in British Guiana should the worst come to the worst . . .[15]

When nationalisation eventually came, the company had built on its foundations in sugar in Guyana operations in all continents over an extremely wide range of activities.

The third strategy can be broadly termed corporate and it had four basic elements to it. The first was to encourage peasant cultivation of the sugar cane in order to deepen commitment to sugar and dependence on the MNCs and minimise some of the problems of labour management. We have previously discussed the limitations of this approach. The second element was to rapidly promote Guyanese to the higher levels of the local managerial hierarchy. This policy was called Guyanisation. But as it was confined to a narrow social stratum who moved into the estates and accepted the segregated housing and other amenities, it was an inadequate response to the mass-based nationalist pressures. The third element was to issue shares on the local market in order to 'localise' the industry. This, too, failed to meet nationalist demands. But while the Guyanisation did succeed in a rapid replacement of expatriates by Guyanese on the estates, the policy of local share issues did not find a responsive local market. The final element was to promote local government in order to strengthen local participation and rebuild the corporate image. As it turned out, this too failed and the local government system has so deteriorated that since 1970, biannual elections to the local authorities have been postponed.

In 1975, Jessels Securities, which had previously taken over the smaller of the two sugar companies, was in the throes of a major financial crisis and their operations in Guyana were to be closed. Labour was being retrenched and in the midst of the industrial disputes which followed the government stepped in to nationalise the company. One year later, the assets of the major sugar company were also nationalised. This, too, was precipitated in part by crisis. Only this time, the cause was the boom in sugar prices in 1974. In that year, the government imposed a variable sugar levy to absorb excess profits. The company argued that by 1976 the levy had become punitive and excessive because of the fall in sugar prices. In the course of dispute over these claims, the government announced its intention to nationalise. While these immediate events no doubt precipitated the decisions to nationalise and determined their timing, their real causes lie deeper in the issues analysed in this paper.

Nationalisation was not confiscatory. It was, in fact, jointly announced at the end of negotiations by the two parties. The smaller company was purchased by the government for G$5 million in cash and a balance of G$15 million to be paid over 10 years at an interest rate of just under 7 per cent per annum.[16] The larger company was purchased for G$102.5 million. The method of payment was: (i) the immediate assumption by the Guyana government of G$32.5 million debt which was owed by the company to various creditors; (ii) G$8.5 million in cash; and (iii) the remainder of G$61.5 million to be paid over 20 years at 6 per cent per annum. This valuation of G$102.5 million represented about one-fifth of the declared book value of the company's assets world wide. It was agreed that all these payments were to be made out of profits of the nationalised companies. But, this represents no real risk to the previous owners since the nationalised industries continue to borrow in the same financial markets as they did before, and their continuance of this practice implicitly depends on their honouring of these obligations. In any event, the fact that the government has entered into marketing, technical services, and management contracts with the previous owner, Booker McConnell Ltd, is further indication that there is no serious risk attached to the 'payment out of profits' clause in the contracts. It is clear from all the available information that the terms of nationalisation are really those of a 'purchase and sale' agreement with a minimal confiscatory element attached to it. There was in the government's public stance little emphasis on the need to redress for previous exploitation of Guyana's resources. This has generated a great deal of criticism by opposition political parties and groups, who otherwise have welcomed the nationalisation of the estates, and indeed had advocated it for years before the government finally accepted its necessity.

6.7 Nationalisation and the agrarian impasse

Even though nationalisation of the plantation was a necessary step towards transforming the agrarian structure of Guyana, it is by no means sufficient for achieving this. Indeed, it is possible that if nationalisation remains unaccompanied by equally fundamental changes in other areas of state policy, it will degenerate into a simple juridical modification of the existing plantation structure. The reasons for this are obvious. The main burden of this essay has been to demonstrate that the plantation system embodies a complex set of economic, social, political, cultural and ideological relations. A shift from MNC property to state property cannot by itself transform these. While the state now owns the plantation assets, other non-legal structural relations of the countryside have not been altered. Of particular note is that there have been no other land reform measures nor policies for dealing with the disintegration of rural communities.

As we have seen previously, the local peasantry in Guyana is already highly stratified. In Guyana most of the land is held as 'Crown (state) Land'. This policy was developed by the British to exercise control over the use of lands outside the plantation empoldered areas. Indeed, at the time of nationalisation, just over 50 per cent of the land used by the larger of the two MNCs were rented from the state at pepper-corn rental. This policy has prevented the growth of an indigenous landlord class on the same scale as parts of Latin America. Nevertheless, through buying over peasant land or smaller plantations which were abandoned, a significant (although unmeasured) indigenous landlord class has come into existence. Usually these landlords combine ownership of land with the ownership of processing facilities in rice and coconuts; tractors and other equipment used in rice cultivation; groceries and other supply stores. At the other end of the rural scale are the landless and other poor smallholders who have come into existence under the pressure of scarce land resources. Although the man-to-land ratio in Guyana is low (approximately 10 persons per square mile), the necessity for water control systems in coastal agriculture, or heavy infrastructural investments to open up interior lands, has produced a great scarcity of cultivable land. Thus in the survey referred to earlier, the data collected indicate that while all plantation land is adequately drained and irrigated only 48 per cent of peasant lands had irrigation and 62 per cent drainage facilities.[17] Without suitably drained and irrigated land, the peasant sector is limited in its capacity to expand. The individual peasant cannot drain and irrigate his own land. This is so not only because of scale economies, but also because it would be practically impossible to exercise water control over one's own small holding without simultaneously doing the same for a wider catchment area.

The problem of the structure of land ownership is further complicated by the fact that as many as one-third of the peasant farmers did not have *bona fide* titles to their land.[18] Furthermore, those efforts at co-operative activity which have taken place have failed, and have resulted primarily in the creation of barely disguised agencies of state or estate control. Solving these problems requires a vigorous policy of land reform to accompany the nationalisation of the plantations.

The failure to address the wider problem of land reform has been accompanied by an equal neglect of the social problems of rural disintegration. It has long been recognised that the plantation system inhibited the growth of self-reliance and popular participation in rural communities. Indeed, the plantations themselves acknowledged this in that one of their strategies to contain the threat of nationalisation was the advocacy of local government reform. The two main structures of rural disintegration centre on the plantation-peasant duality, and the historically created pattern of African and Indian villages standing more in competitive than fraternal relation to each other. The latter in particular followed not only from the historical pattern of immigration, but also from the fact that these communities were subjected to opposing political influences which prevented the formation of a united peasantry.

The conversion of the MNC plantations into state property immediately raises the question of the political base of the government. The present government in Guyana draws its political support mainly from the African and urban communities. In the rural areas, the opposition forces prevail. The main rural support for the present government seems to come from the landlords and rich peasants, estate officials (and a few public officers) who may have objections to the ideology of the leading rural party and their vigorous espousal of land reform. Such a balance of political forces poses the threat of a reversal of the gains from nationalisation if it is not changed to give poor peasants and landless labourers greater participation in political processes.

Not only has there been a failure to tackle the issue of land reform and rural disintegration, but also developments since nationalisation point to the serious problems that still remain to be tackled before rural backwardness is overcome.[19] Since nationalisation, and despite the expiry of the 1972/73 Development Plan period, a new plan has yet to be completed and no policy has been announced on the role of sugar in the national economy. Prior to nationalisation, the MNCs had started preparatory work on expanding the industry's capacity by one-third, to about 500,000 tons of sugar annually. In the 1972–6 plan, there was the stated goal of expanding sugar acreage by 35,000 acres. But this deadline has already been passed and the expansion has not yet commenced. Public statements by officials in the industry and the government have casually endorsed the idea of expansion, but there is the absence of a policy on the optimal size of the industry and the balance of production and control which would exist among the three great rural interests, namely the state farms, the peasants and other rural interests.

There has also been no announced and definite programme of crop diversification on estate lands. Apart from statements expressing the desire to grow other crops (e.g. legumes) no programme has as yet been put forward. The seriousness of the failure to do so is worsened by the consideration that immediately prior to nationalisation (1975) an arbitration tribunal sitting on an industrial dispute had agreed to award the workers certain qualified rights to out-of-crop work.[20] The government has agreed to honour this obligation. The consequence is that efficient use of this labour in the out-of-crop season is now linked to the whole problem of crop diversification. To some extent plans for an optimal configuration of the industry have been affected by the current severe economic and financial crisis in Guyana. At the heart of the crisis is the fragility of a small, over-specialised, primary producing country. Consequently, solutions aimed at overcoming this distorted colonial production structure are the only truly feasible ones.

The long historical process of land and capital consolidation in sugar was finally achieved with nationalisation. This has laid the basis for the effective unification of the two separate management systems on the estates. It has also made more decisive the role of the state in marketing. Prior to nationalisation, state initiatives had led to some sales of sugar to such non-traditional markets as Algeria, China and the USSR. This policy has continued, but the UK-EEC market which Guyana participates in under the Lomé Convention, remains the primary overseas marketing outlet.

Despite a few gestures such as 'desegregation' of school buses and swimming pools on the estates, the system of industrial relations continues to be confrontationist. Evidence of this is seen in that within one year of nationalisation, the industry was plunged into its worse strike ever, lasting 135 days. During this period, a state of emergency was declared and the government introduced several thousand outside workers to harvest the cane. This heightened the conflicts on the estates and the fact that the strike ended without an agreed terms of resumption indicates that the situation is still grave. While the specific issue leading up to the strike was a dispute over profit-sharing, it was merged in the union's presentation of grievances into the more general issues of worker control, and a clear national policy for the sugar industry and the sugar workers.[21] A policy for this sector, however, cannot be formulated in the absence of a wider and more comprehensive set of policies for the entire society. Sugar is too important to be dealt with in isolation. And it is because of this interaction between sugar and society that the issues of agrarian transformation in Guyana go far beyond the mere creation of state property farms on the old estates.

Notes

1. This essay is based on a larger study entitled *Plantations, Peasants and State (Alternative Modes of Sugar Production in Guyana).* The statistical data cited in the text are derived from this study. The statistical tables presented are kept to the barest minimum in the interest of space. Full tables in the original text support the data cited here.
2. These forward contracts stipulated the number of hours of work each day, remuneration, place of employment, etc. This was permitted by an Ordinance of 1850.
3. See the study referred to in note 1 for fuller details.
4. The estimated capacity of the industry in 1976 was just over 400,000 tons. Only about 75 per cent of this output was achieved, and in years of labour dispute and strikes, e.g. 1973 and 1975, output was as low as 243 and 266 thousand respectively.
5. The sugar revenue figures for 1973–5 are $137.0m, and $211.0m respectively. For the years 1974 and 1975 this represented 43 and 44 per cent of government revenue, respectively.
6. The data cited here are also quoted from the study referred to in note 1. This, in turn, was based on the only recent survey available: *The National Food and Nutrition Survey of Guyana* (Washington: PAHO/World Health Organisation, Scientific Publication No. 323, 1976).
7. I have estimated that 19 individual farmers control 15 per cent of total acreage with a mean acreage of over 130 acres per farmer.
8. Thus the estates may leave the farmers' cane until late in the season when the cane is overgrown, or bring it forward, if, for example, rain may cause delays in processing. In other words, they operate on a schedule which puts most of the risk onto the farmer, as far as possible.
9. Detailed data supporting this argument are included in the original study referred to in note 1.

10. Quoted in study referred to in note 1. The original source is *Report and Accounts, 1976, Guyana Sugar Corporation Ltd*, p. 4.
11. Ibid., p. 11.
12. Space does not permit me to develop this line of argument. It is important to recall the dominance of radical ideology (socialism) in Guyana's politics. In 1953, this led to the election to national office (under the Colonial Authorities) of the first Marxist-led political government in the Western Hemisphere. After 133 days of rule, the British Government removed the government, revoked the Constitution and set up a period of interim rule until 1957. During this period, the nationalist movement split along racial and ideological lines. The original PPP remains under the control of Dr Cheddi Jagan. It is Marxist-led, mainly rural and Indian in its rank and file. The other faction became the PNC which forms the present government. It is led by Prime Minister Burnham. It is mainly African and urban, with a form of nationalist and socialist ideology. Currently 'co-operative socialism' is its main ideological plank.
13. The strength of this union was of such that when a poll was finally conducted in the industry on 31 December 1975, it won 98 per cent of the votes cast and the then recognised union won only 1.4 per cent.
14. In the long run, this policy aggravated the difficulties of the plantation rather than eased them – largely because of unemployment, the rural-urban drift, and inability to mechanise reaping and planting.
15. *Chairman's Statement, Annual Report of Booker McConnel Ltd*, 1953.
16. The dollars quoted are Guyana dollars. The current rate of exchange is $1US to $2.55 Guyana.
17. The Survey referred to is cited in note 6.
18. Ibid.
19. In the study referred to in note 1, state policy since nationalisation is treated more fully.
20. After decades of worker struggle over this issue, the Tribunal recommended an out-of-crop payment of $25 per week for men and $20 per week for women. To qualify for full payment, a worker had to work 90 per cent or more of the days on which work was available over the season of the three preceding crops. If the worker worked from 75 per cent up to 90 per cent of the available days, then men received 80 per cent of the payment and women 75 per cent. Note that strike days are counted as among the days available for work!
21. Because of the Sugar Levy Act of 1974, the huge profits which would normally have accrued to the sugar companies, and which the workers were entitled to share in because of a profit share agreement, were siphoned off as public revenue. The workers claimed to have been 'cheated' since they bore the burden of all the previous years of low prices. Underlining this dispute is the absence of a clear policy on priorities in the use of surpluses obtained from the rural sector.

7 Ujamaa and Villagisation in Tanzania

Dharam Ghai and Reginald Herbold Green

7.1 Introduction

The strategy for rural development pursued in Tanzania since 1967 has evoked widespread interest. The centrepiece of this strategy is the attempt to transform the rural sector through the creation of Ujamaa[1] villages. As originally conceived, the Ujamaa villages are to serve as the key institution of an agrarian system designed to achieve simultaneously the objectives of political democracy, economic growth and egalitarian distribution of income and wealth.[2]

It is now more than a decade since the initiation of the Tanzanian experience and its study may be of interest to other African countries such as Somalia, Mozambique, Angola and Ethiopia which in recent years have sought to restructure radically their rural institutions in pursuit of these objectives. Furthermore, this experience is of interest because it constitutes a rare, if not a unique, instance of an attempt to bring about far-reaching structural changes in a social context characterised by the relative absence of sharp inequalities in the rural sector and to achieve these changes rapidly without either violent revolution or a high level of coercion. At the time of the initiation of Ujamaa policies, the country was largely free from landlessness, landlordism, chronic indebtedness of the poor peasantry and other forms of oppression which characterise the rural societies of many countries in the Third World.

The main purpose of this study is to analyse the changes in organisation of agricultural production introduced by the Ujamaa and villagisation programmes since the Arusha Declaration in 1967. No attempt is made to discuss overall progress in the rural sector or to describe and evaluate all elements of the Tanzanian strategy for rural development. We first describe the main features of the pre-Ujamaa agricultural system and of the evolution of rural institutions and policies since 1967. This provides the overall context for a more detailed appraisal of Ujamaa and villagisation as alternative frameworks for agricultural production.

7.2 The agricultural system prior to Ujamaa

Tanzania is a predominantly rural country. Less than one-tenth of the population is resident in cities and towns of 20,000 or more inhabitants. In 1967 out of a total of two and a half million households about 350–400,000 were dependent on non-agricultural wage employment, 150–200,000 on non agricultural self-employment, 125–150,000 on agricultural wage employment and 1.8–2 million on agricultural self-employment. Almost a third of monetised, and over two-fifths of total, GDP in 1967 were derived from agriculture. Similarly, over four-fifths of exports consisted of either raw or processed agricultural commodities.

Tanzania is a relatively land abundant country. The population density was estimated to be 17 persons per square kilometre in 1976 and only five per cent of the total land area was under cultivation in that year. It has been estimated that irrigation and flood control could increase the present cultivated area by 40 per cent and thus a substantial land frontier exists.

The land-to-man ratio (total cultivated area divided by rural population) was estimated to be 0.31 hectares in 1972 and the average farm size was 1.76 hectares. The agricultural sector was, however, dualistic and the overall average conceals huge differences in farm size between the peasant and the estate sectors. In 1961 about 900,000 hectares (more than 20 per cent of the total cultivated area in that year) was in the estate sector. It would appear, therefore, that even after allowing for the decline in the size of this sector since 1961, the average farm size in peasant agriculture was somewhat lower than 1.76 hectares.

In the early 1960s Tanzanian agriculture consisted of about 2000 large farms and plantations, and nearly 2 million peasant holdings. The former comprised sisal, tea and coffee grown on large farms or plantations, small sugar estates and mixed farms largely centred on grain, pyrethrum and cattle. In terms of ownership patterns, about 20 per cent were company owned, 50–60 per cent were European individual holdings, 10–15 per cent Asian family ventures and a handful belonged to emergent African capitalist businesses. In the years immediately preceding independence, the large farm sector was estimated to account for 35 per cent of marketed output and 45 per cent of agricultural exports. In addition it provided about half of the total African wage employment.[3]

The dual system did not give rise to large numbers of landless peasants or sub-marginal peasants whose survival depended on working for large landholders. Inequality of landholding among Tanzanians was relatively low. In 1967, the typical small peasant family probably cultivated about 0.5–2 hectares, the typical middle peasant 2–7 hectares and perhaps the 50,000 *proto kulaks* (defined in terms of at least one-year round employee) 5–25 hectares. However, over 1961–7 there were signs of growing inequality. The *proto kulak* group had become stronger largely through its control over co-operatives and links with government officials, and was perceived by many party leaders as poised to secure adoption of a rural capitalist strategy similar to that practised in Kenya. While freehold land tenure had been abolished in 1964, the communal allocation system as operated via district councils had in practice allowed some concentration of control over better quality land by emergent capitalist farmers – perhaps most notably at Ismani in the Iringa Region.

While lack of access to some land was not a problem in Tanzania, the level of agricultural productivity and incomes was very low. Less than three per cent of the total cultivated area was irrigated, rainfall was erratic and insufficient in many parts of the country and soil fertility was generally low. Peasant agriculture was characterised by extensive and shifting cultivation with almost exclusive reliance on manual labour working with simple tools. Moreover, apart from regions where cash-crop production has become well established, peasant production was predominantly subsistence level and used no modern inputs.

The nature of agricultural production and the pattern of rainfall and soil fertility were among the factors which had given rise to a highly dispersed distribution of population.

It is estimated that by 1967 only some 15 per cent of the rural population lived in villages or small towns. The majority lived in scattered homesteads or hamlets of two to five families. Another 15 per cent were pastoralists with seasonal residential rotation patterns. This pattern of settlement meant that, purely from the standpoint of the conditions of agricultural production, the various villagisation programmes might be regarded as attempts to change from an extensive to an intensive, non-shifting mode of cultivation.

Hours worked by peasants and pastoralists probably amounted to an annual average of about 1000 for men and 1250 for women (including water and firewood collection and housebuilding but excluding other non-agricultural economic activity) except in the minority of the country with two rainy seasons and two crops. This low level of labour utilisation (125 to 160 eight-hour days) was caused by a number of factors: (i) the absence of irrigation which prevented the growing of a second crop; (ii) the constraints on acreage from peak labour requirements in the land preparation and harvesting phases; (iii) the limits on joint infrastructure or directly productive investment imposed by isolation, limited peasant organisation and low governmental support; and (iv) the related limited opportunities for non-agricultural income generating activities.

7.3 The evolution of agricultural organisation and policy: 1967–77

The agricultural system as it had evolved during the colonial era and in the early years of independence was generally recognised to suffer from a number of weaknesses. As indicated in the previous section, these included the existence of dualism, growing economic differentiation among smallholders and regions, low productivity and a low level of labour utilisation in the smallholder sector and the pattern of rural settlement with a preponderance of isolated homesteads spread over a vast country. It was mainly to overcome these weaknesses that the concept of an agrarian system based on Ujamaa villages was first put forward by President Nyerere.

Although experiments in communal farming and village settlement had been pursued in the years preceding and following independence, it was the Arusha Declaration, issued in February 1967, which marked a decisive turning point in the search for institutions and policies aimed at building rural socialism. Its central themes were self-reliance, rural development, public control over the economy and social equality. While these have now become familiar themes in the literature on development policy, in 1967 they represented a sharp break from the established priorities in most post-independence African societies. The Declaration itself was followed by a wide-ranging programme of nationalisation of large enterprises in banking, insurance, external trade, manufacturing and agriculture. Equally critical was the enforced separation of party office and middle or higher level civil service and parastatal employment from private business involving hired labour or rental of property. But the Declaration contained no overall framework for the development of rural socialism. This was provided by the presidential paper *Socialism and Rural Development*, published in September 1967. Since this paper constitutes the most authoritative and systematic Tanzanian statement on the ultimate goals and forms of rural socialism and has continued to exercise major influence on rural policy, it is important to highlight its key points.

In this paper President Nyerere argues that rural socialism in Tanzania must be built on the principles embodied in the African extended family system. These were respect for the individual, sharing of 'all the basic goods', reasonable equality in ownership of land and cattle and 'an obligation to work'. However, the traditional system had been seriously eroded by colonialism and was both too small and technically too weak to be suitable for simple revival. The modern version of the traditional family should be co-operative communities: villages with common ownership of basic productive assets and sharing of proceeds 'according to the work done and to the needs of the members'. Organisation of collective agricultural production at the village level would also make it possible to benefit from modern techniques of production. The creation of such Ujamaa villages must be guided by three principles: first, moves towards Ujamaa must be made voluntarily by the members and through their own efforts – 'socialist communities cannot be established by compulsion'; second, there must be a step-by-step approach, starting with co-operation in a few activities and processes and gradually extending it to other areas; third, 'village democracy must operate from the beginning; there is no alternative if this system is to succeed'.

Nyerere envisaged three distinct stages. In the first stage the scattered smallholders would move into villages to facilitate the distribution of services and to organise production more efficiently. In the second stage each village would initiate communal activities in an embryonic form. In the third and final stage the villagers would be persuaded that communal activity was the superior form of organisation and, as a result, would convert themselves into predominantly communal farmers with some provision for private plots. To date the last stage has not been reached; individual smallholdings remain the overwhelmingly dominant form of organisation.

This idealistic vision of self-reliant, democratic, egalitarian and co-operative Ujamaa villages employing progressively more modern techniques of production inspired earlier efforts in communal farming and villagisation and continues to be the ultimate objective of rural policies. But in the implementation there have been shifts of emphasis in relation both to communal production and to the timing and mode of establishment of villages.

Alongside these proposals for the socialist transformation of the peasant sector measures were also taken to nationalise the estate sector. We shall first describe the changes in the estate sector before going on to analyse the villagisation programmes.

(a) The estate sector

The estate and large farm sector was largely converted into parastatal enterprise units over 1967–77. In 1967/68 and 1971–3 sisal estates were nationalised, and in 1974/75 nationalisation was extended to the mixed farms and coffee estates. A principal aim of nationalisation was to end the anomaly of large, private European farms in the midst of small African peasant holdings designed eventually to lead to communal farming. Very few large units were broken up or turned into worker co-operatives, as opposed to parastatal farms. Thus the parallel systems remain although with different patterns of ownership, income structure and surplus control. A few private plantations remain in sisal, tea, sugar and tobacco but the sector is now small and constricted and is not perceived as a threat to the implementation of rural strategy.

Tanzania has put little in the way of resources into state farms beyond parastatal ventures in sugar, sisal, tea, cattle and wheat. State farms have occasionally enjoyed brief vogues but never a significant share of resources. Large-scale parastatal or joint venture units in grain have been limited to wheat (where technical problems limit, or are believed to limit, smallholder production) and seed multiplication. Discussions of large-scale joint venture farms with foreign firms have occurred on and off for a decade and a half, but no actual deals have been concluded. While the division of labour is not clear, it is roughly the case that the parastatal ventures should limit themselves to acquired plantations, products posing special problems for smallholder output (namely, sisal, sugar, wheat and seed multiplication) and activities to complement (e.g. cattle fattening) or to stimulate (e.g. export horticultural core farms) peasant production. The large-scale rural sector will not be considered further in this study.

(b) Patterns of villagisation

Before proceeding to describe the progress in villagisation it will be useful to clarify a few key concepts. The term 'villagisation' will be used to describe the process of resettling hitherto scattered or pastoral population in permanent settlements. 'Ujamaa' involves villagisation but has broader political and organisational objectives including eventual collectivisation of farming; it is thus a specific type of villagisation. Until 1973, the two terms could be used almost interchangeably since virtually all villagisation was inspired by the Ujamaa model. Since then, however, there has been a massive drive towards villagisation *per se* with the major emphasis on grouping people in permanent settlements. Collective cultivation as a short-term goal has been abandoned in these development villages.

Table 7.1 shows the growth in the number, average and total population of villages over the period 1967–76. After a relatively slow start in 1967–8, following a call by President Nyerere to 'go Ujamaa', there was a big spurt in the growth of villages over the

Table 7.1 Growth in Number, Average and Total Population of Villages

Year (March)	*Number*	*Average Population*	*Total Population ('000)*	*Percentage of National Population*
1967	48	104	5	0.04
1968	180	322	58	0.5
1969	809	371	300	2.5
1970	1956	272	531	4.3
1971	4484	345	1545	12.3
1972	5556	357	1981	15.3
1973	5628	360	2028	15.3
1974	5008	511	2560	17.6
1975	6994	1260	9140	61.6
1976	7684	1703	13061	85.0

Source United Republic of Tanzania, *Maendeleo ya Ujamaa,* Prime Minister's Office (various years).

period 1969–71, raising the total village population from 58,000 in 1968 to over 1.5 million in 1971. The lull from 1972 to early 1974 was again followed by massive regional and national 'operations' under which vast numbers were moved to villages raising the proportion of village population from less than 18 per cent in March 1974 to over 60 per cent in March 1975. Nearly 90 per cent of the rural population was reported to be living in villages at the end of 1976.[4] Thus, in little over two years, nearly 10 to 11 million people were incorporated into villages.[5]

There is a great deal of variation in the size of the village population, but the average has tended to rise, especially since 1970. There were particularly sharp rises in 1975 and 1976. The average population of a village in 1976 was nearly five times greater than in 1971.

Typically, the establishment of an Ujamaa village involved settlement on new lands. In the early years, and indeed right up to 1973, the greatest progress in villagisation had been made in areas of low potential with sparse population depending largely on subsistence farming. Conversely, densely populated districts, in which cash crop production was most developed and where individual rights in land were most entrenched, were least affected by the Ujamaa movement.[6]

Impatience with the slow progress in villagisation led to the adoption of a 'frontal' approach in 1969. Instead of relying solely on exhortation, an element of inducement was added through provision by the government of transportation, building materials, machinery, food, funds, etc. The years of consolidation – 1972/73 and the early part of 1974 – were followed in 1974–6 by large scale 'operations' under which vast numbers of peasants were moved to villages. This marked a turning point in the approach towards villagisation. The government and the party came to play the dominant role in planning and implementing village settlements. The voluntary nature of the movement gave way to compulsion and instances of the use of force to move peasants were not unknown. At the same time, the emphasis shifted away from the creation of Ujamaa villages with communal production to mass villagisation based on block farming (in which peasants cultivate individual holdings within an overall block of land allocated to the village). While in the earlier phase of villagisation, the Ujamaa villages received favoured treatment in the allocation of funds, machinery, agricultural staff, fertilisers, etc., this ceased to be the case from 1973 onwards.

The Villages and Ujamaa Villages Act of 1975 provides the legal institutional framework of the villages. It provides for registration of villages by a Registrar, thereby conferring legal status and certain rights and obligations. Under the Act, the village is deemed to be a multi-purpose co-operative society. There is provision for the constitution of a village assembly, consisting of all persons aged 18 years or over, and for the election of a village council by the village assembly. The village council is empowered to do 'all such acts and things as are necessary or expedient for the economic and social development of the village'. It can plan and co-ordinate the work of villagers engaged in agriculture or other activities. It is also empowered to set up committees dealing with such subjects as finance and planning, production and marketing; education, culture and social welfare; works and transport; and security and defence. A village may be designated as an Ujamaa Village by the Registrar when it is established that a substantial portion of its economic activities are being undertaken on a communal basis.

(c) Parallel policy changes

These changes in policy relating to organisation of agricultural production were accompanied by equally far-reaching changes in other aspects of rural policy. In 1972/73, there was a major attempt at decentralisation of rural development programmes. The regional and district administrations were strengthened and invested with additional powers, responsibilities and resources. The role of the parent ministries was reduced to preparing broad policies and guidelines. In 1976 the government abolished all primary marketing co-operatives and co-operative unions. Their purchasing and marketing functions were to be shared between the parastatal crop authorities and the newly-established village multi-purpose co-operative societies.

There were also significant shifts in agricultural pricing policy. After a period of relatively constant producer prices for most of the export and domestic crops for nearly a decade ending in 1973/74, the prices were raised sharply in the subsequent years (Table 7.2). The price of maize was raised more than three-fold between 1972/73 and 1976/77; that of wheat more than doubled between 1973/74 and 1976/77 and that of pyrethrum and fire-cured tobacco nearly doubled over the same period. The terms of trade for

Table 7.2 Producer Prices for Selected Agricultural Crops (shs. per kilo)

	1969/70	*1970/71*	*1971/72*	*1972/73*	*1973/74*	*1974/75*	*1975/76*	*1976/77*	*1977/78*
Maize	0.28	0.26	0.24	0.26	0.33	0.50	0.75	0.80	0.85
Wheat	0.57	0.57	0.57	0.57	0.57	0.77	1.00	1.20	1.25
Rice (paddy)	0.52	0.58	0.52	0.56	0.57	0.53	1.00	1.00	1.20
Cotton BR Gr.		1.10	1.10	1.13	1.19	1.50	2.00	2.00	2.30
Cotton BR Gr.		0.55	0.55	0.60	0.60	0.65	1.00	1.00	1.15
Groundnuts (shs. per ton)				1028	1150	1500	2000	2500	
Cashewnuts	0.95	0.95	0.95	0.95	0.95	1.05	1.05	1.10	1.10
Tobacco (flue cured)			5.80	5.85	5.85	5.85	7.00	7.40	7.40
Tea (Bukoba)			0.60	0.60	0.60	0.60	0.60	0.65	0.65
Cassava (Grade I)				0.25	0.31	0.36	0.40	0.50	

Source Ministry of Agriculture, *Price Policy Recommendations for Agricultural Price Review* (relevant years).

agriculture declined significantly between 1961 and 1967 and again from 1970–3. There was, however, a sharp improvement in 1973–5, followed by a deterioration again in 1976/77.[7]

7.4 Impact on distribution and production

It is virtually impossible to make a quantitative estimate of the impact of Ujamaa and development villages on overall agricultural output and income distribution in the rural sector. This is so primarily for three reasons. Firstly, the data on rural income distribution are grossly inadequate. Secondly, the changes in settlement and production patterns have been phased over a number of years, culminating in mass villagisation only over 1974–6. Thus the period since the completion of this programme is too recent to permit any worthwhile quantitative analysis. Thirdly, the period since 1967 has been marked by other major policy changes which have an independent and significant effect on agricultural production and income distribution. There have been changes, for instance, in the terms of trade faced by the agricultural sector, in the role of co-operatives and in the government administration of rural development programmes. Thus, while a quantitative analysis is ruled out, it is still possible to assess in qualitative terms the impact of the institutional changes on some aspects of agricultural production and income distribution. This assessment is based on the characteristics of the new institutional arrangements and on information contained in some micro-level studies.

In an Ujamaa village, a certain amount of land is set aside for communal farming. Members of the village are also allocated personal plots for construction of dwellings and for family cultivation. They are required to work for a specified time on communal farms and other joint activities, the balance of the time being available for work on personal plots. Income from communal farms is distributed among members either on a time- or piece-rate basis. In a 'development' village, on the other hand, members cultivate personal plots which form part of a block, thus facilitating, if necessary, mechanised land clearing and cultivation. There may be little or no communal farming or other co-operative activities.

Among the potential economic benefits of the Ujamaa system are a relatively egalitarian distribution of wealth and income, and greater efficiency in production and economies in the provision of public economic and social services. In this section, we investigate the extent to which these benefits have been realised in practice under Ujamaa and villagisation.

(a) Impact on land and income distribution

In the rural peasant economy of Tanzania, the main sources of wealth are land and, to a much lesser extent, livestock. Disparities in landownership arose primarily from the existence of a few plantations and large mixed farms. The majority of these were nationalised in the wake of the Arusha Declaration and became known as state farms. The distribution of land in the smallholding sector which accounts for over 90 per cent of the rural population was relatively egalitarian. The pattern of land distribution in the country is illustrated in Table 7.3.

Table 7.3 Size Distribution of Land, 1972

Farm size (ha)	*Number of holdings*	*% of Total Holdings*
<0.5	771954	31.5
0.5–1.0	651386	26.6
1.0–2.0	605291	24.7
2.0–3.0	218375	8.9
3.0–4.0	88696	3.6
4.0–5.0	49985	2.0
5.0–10.0	53252	2.2
>10.0	11625	0.5

Source 1972 Agricultural Survey (preliminary results)

It will be noticed that well over 90 per cent of the holdings are less than three hectares, and slightly less than 60 per cent are below one hectare. Only 0.5 per cent of the holdings exceed 10 hectares.[8] While in most parts of the country arable lands are equitably distributed, there are high potential, densely populated areas such as in Arusha, Kilimanjaro, West Lake, Iringa and Mwanza regions where disparities in land distribution are more pronounced and land is becoming increasingly scarce. However, in relation to the situation in many developing countries, land distribution even in these areas would appear to be relatively even as shown in Table 7.4 for the West Lake Region.

The establishment of Ujamaa villages should in principle contribute to a more egalitarian pattern of landholdings. This is because

Table 7.4 Land Distribution in West Lake Region's Three Districts

Category of Holdings	*Bukoba*		*Karagwe*		*Ngara*	
	Mean Size (ha)	*% Land Owned*	*Mean Size (ha)*	*% Land Owned*	*Mean Size (ha)*	*% Land Owned*
Largest 10%	3.7	23	7.9	19	9.9	27
Next 30%	2.0	39	5.3	42	4.8	38
Smallest 60%	1.0	38	2.6	39	2.2	36

Source Jannik Boesen et al: *Ujamaa – Socialism From Above* (Uppsala: Scandinavian Institute of African Studies, 1977, p. 28).

part of the land is collectively cultivated while the remainder is distributed equally as personal plots. In actual practice, it would seem that because of the small area of land under communal cultivation, and the geographical pattern of the location of Ujamaa villages, the impact of the Ujamaa system on land distribution has been relatively marginal. There are no countrywide figures on the share of the communal sector in total cultivation and production but one estimate put collective production at less than five per cent of total agricultural production in 1973.[9] The share of communal cultivation may have increased marginally in the intervening period, but as indicated earlier, the big leap in villagisation in 1974–6 did not involve an appreciable increase in communal farming. On the other hand, there is scattered evidence that some original communal farms have been distributed as personal plots.[10]

A high proportion of Ujamaa villages were set up in new settlements with no previous cultivation. Furthermore, there was a tendency for such villages to concentrate in areas with low population density, a preponderance of subsistence activities and low-potential land. Ujamaa villages tended to attract poorer peasants and landless persons and to that extent, they undoubtedly made a contribution to better land distribution. But this contribution consisted more in opening up and settlement of new areas than in redistribution of the already cultivated land especially in high potential, land scarce areas.[11]

The mass villagisation of 1974–6 has been quantitatively more important in promoting better land distribution. Although in these villages, there is insignificant collective farming, each family is allocated an equal amount of personal plot whose size, however, may vary from one village to another, depending upon soil and climatic factors.[12] In some of the more densely populated parts of the country with high value and perennial crops grown on individual holdings, however, villagisation amounted to little more than demarcation of boundaries leaving the pattern of land distribution essentially unchanged.

To summarise this section, the new institutional arrangements associated with Ujamaa and development villages have contributed to a more egalitarian land distribution with the latter playing a more important role.

Regardless of the minimal contribution of communal ownership, it could still be argued that the equal distribution of privately-owned land, carried out on all new settlements, represented a significant improvement over the pre-existing distribution of land among the previously non-villagised population. Perhaps of even greater significance is the framework they have created for an equitable growth pattern in the future. Ceilings were imposed on the size of personal plots in Ujamaa and development villages and restrictions were placed on the hiring of workers on individual farms. These measures were designed to prevent the growth of inequality in these villages.

Apart from land, the other main form of rural asset is livestock. There is considerable disparity in livestock ownership in areas where livestock accounts for a major proportion of economic activity. But while it is quite common for an Ujamaa village to have a small communal herd, there were very few instances of private herds being pooled together to create communal herds. Likewise the villagisation of the past few years seems to have had practically no impact on the structure of livestock ownership.

Hand tools – hoes, axes, shovels – are the most widespread aids to human labour. These continue to be personal property under all institutional forms – the traditional smallholdings, Ujamaa and development villages. As we shall see later, at one stage an effort was made to promote the use of more advanced technology – ox-ploughs and tractors – in Ujamaa villages under communal ownership. But tractors and ox-ploughs continue to be individually owned in development villages.

Several factors in the Ujamaa and development village system could be identified as being conducive to egalitarian growth. The ceiling on the size of personal plots in Ujamaa and development villages has been effective but restrictions on the hiring of permanent workers on individual farms have not always been enforced. The methods of remuneration for work on communal farms reinforce the egalitarian thrust of the Ujamaa system. In the majority of cases, payment is based on number of days worked and except for a very few jobs requiring higher skills, workers are paid the same per unit of time worked. Furthermore, the gradual expansion of the communal sector in non-farming activities is likely to limit economic differentiation. Typically, in an Ujamaa village, enterprises such as village shops, blacksmithing, milling, butchery, carpentry etc. tend to be communally owned. While thus there is a strong presumption that Ujamaa tends to generate a more egalitarian pattern of income distribution than alternative production systems, there are unfortunately few empirical studies which investigate these issues in a systematic and rigorous manner.[13]

(b) Organisation and efficiency of production

One of the reasons for the move towards Ujamaa was its alleged superiority in efficiency over smallholder agriculture. It was argued that collective cultivation would lead to improved yields and higher productivity through specialisation and division of labour, adoption of more advanced technology, fuller utilisation of labour and more effective and lower cost provision of services such as extension advice, marketing, credit, and supply of inputs like fertilisers, seeds, etc. We examine here the Ujamaa experience with respect to production technologies, employment promotion and yields.

Although experience with the capitalised settlement schemes of the early 1960s had been largely negative and Ujamaa villages were expected to be self-reliant and move gradually to greater mechanisation, in actual practice the Government attempted in many of the schemes initiated in 1968/69 to bring about swift transformation in production methods through the introduction of heavy machinery and modern inputs such as fertilisers and insecticides. These were either provided freely or heavily subsidised. For instance, the Ismani maize credit scheme initiated in 1972/73 represented an attempt to transform maize production through the use of tractors, hybrid seeds, fertilisers and insecticides on communal farms.[14] Similarly, in the Ujamaa villages started around 1968 in Karagwe and Butoba Districts in West Lake Region, 'new and improved seeds . . . were introduced and some agricultural mechanisation was begun. Tractors with ploughs and planting machines replaced human labour input in cultivation and planting'.[15]

On the whole, the attempt to transform agricultural production on communal farms through heavy machinery did not prove economical. The increase in yields, except in tobacco growing areas, were not adequate enough to justify mechanisation. Production often suffered because of machinery breakdown and delays in repairs or in receipt of spare parts. Gradually there was a shift of emphasis in favour of intermediate types of technology such as ox-ploughs.[16] The bulk of the Ujamaa villages established in subsequent years did not benefit from such provision of machinery and inputs from the government and the techniques of production on such farms differed little from those on private holdings.[17] Ujamaa villages might be expected to lead to a higher level of labour utilisation because of greater possibilities of crop diversification, easing of seasonal labour bottlenecks through selective mechanisation, compulsory work on communal farms and non-farming communal activities. There is little hard empirical data on comparative labour utilisation in Ujamaa villages and on traditional smallholdings.[18] The potential for generating higher levels of employment through crop diversification and selective mechanisation seems to have been exploited only exceptionally. To the extent that Ujamaa villages have been able to enforce regulations on compulsory work on communal farms, this is likely to have increased the overall supply of labour.

Labour mobilisation on non-farming projects has been far more important in generating higher levels of labour utilisation. The establishment of new villages is associated with a massive amount of construction work involving houses, communal buildings, roads, etc. In addition, there is evidence of labour mobilisation on a continuing basis in the construction of new classrooms, health centres, workshops, storage facilities, day-care centres, irrigation works, water supplies, etc.[19] Villagisation has also made it profitable to establish a wide range of manufacturing, processing and servicing activities such as flour-milling, blacksmithing, furniture-making, brick-making, manufacture and repair of farm implements and handicrafts. For instance, in Dodoma district, by February 1975, 51 grain mills were being operated by villages and 7 villages had blacksmiths' or carpenters' shops.[20]

The stimulation of non-farming activities has been made possible by villagisation as such and not necessarily by Ujamaa, although mobilisation of labour for communal infrastructural projects is clearly facilitated in situations where co-operative methods of work are understood and well-established and where there is an assurance of equitable distribution of the benefits generated by projects of this nature.

Indicators of efficiency in cultivation are provided by such measures as overall factor productivity, output per acre and per man. A number of attempts have been made to measure the efficiency of Ujamaa production and to compare it with cultivation on individual smallholdings. But most of them suffer from a number of drawbacks which limit the validity and usefulness of their results. In the first place, the Ujamaa experience is characterised by such diversity of conditions and results that generalisations based on one or a small group of villages are apt to be misleading. Secondly, the data used in many of these studies are recognised even by the authors to be deficient in several respects. Thirdly, many of the studies were carried out in the early phases of the establishment of Ujamaa villages and therefore reflect the inevitable difficulties associated with a major change in systems of production. It is significant that there have been practically no studies of production at the village level over the past three to four years. Fourthly, in comparing Ujamaa and smallholder production, few if any studies have attempted to isolate the impact of such relevant variables as the quality of land, background of farmers and access to other inputs. Finally, hardly any of these studies has compared production performance in a fully-fledged Ujamaa village with smallholder production.[21] They have generally sought to compare production on communal farm and individual holdings in the same village. Since communal farming in such situations is regarded as additive to family holdings, it is not surprising that it shows lower yields. Such limitations of data and methodology must severely limit the value of their findings on the efficiency of Ujamaa production.

There are instances of some Ujamaa villages where there is significant collective production and output growth, and yields and labour productivity compare favourably with the individual smallholdings in the neighbourhood.[22] But it would appear that such villages constitute a relatively small proportion of total Ujamaa activity. Certainly, the majority of the studies on communal farming indicate a variety of production difficulties and show lower yields than on individual smallholdings.

Boesen et al found relatively poor yields in the ten Ujamaa villages in Ngare, five in Karagwe and six in Bakoba Districts.[23] Sumra found the average yield for all crops in the Handeni Ujamaa villages that he studied lower than on smallholdings.[24] Likewise, in a survey of 14 Ujamaa villages in Ismani, the average yield of maize for 1971/72 was 595 kilograms per hectare, while the individual farm averaged 869 kilograms despite the fact that Ujamaa farms applied fertilisers while individual farms did not.[25] In a study of Ujamaa villages in Iswiga, De Vries and Fortmann found that returns on communal farms averaged five shillings per man-day as compared with seven shillings on individual smallholdings.[26]

These results should be interpreted in the light of the remarks made earlier on the data and methodological limitations of many of these studies. Nevertheless, there appears to be adequate evidence from studies carried out in different parts of the country to warrant the conclusion that barring some cases, communal farming in Ujamaa villages has encountered serious production problems and has experienced lower yields and lower productivity than on comparable individual smallholdings. Perhaps the most important reason for lower yields and productivity on Ujamaa farms is to be found in the priority given by members to work on personal plots. This in turn is both a consequence and a cause of lower returns to labour in the communal sector. Under Ujamaa farming system, each member is allocated a personal plot in addition to being required to work on the communal farm. For the great majority of Ujamaa villages, the size of personal plots available to a household is several times its share in a communal farm. The peasants are expected to meet the household food requirements from personal plots while the bulk of the output from the communal farms is intended for sale. Each Ujamaa village specifies the required days of work on communal projects.[27] But it proves difficult in practice to enforce these rules. In particular, communal farms find it difficult to secure adequate labour during periods of peak labour demand.[28] Faced with a choice between allocating his scarce labour on a personal plot to grow foodcrops and the communal farm with uncertain and distant returns it is understandable if the peasants opt for the former.[29] While the returns from work on personal plots accrue entirely to the family, this is not the case with the communal farm part of whose produce must be allocated for future investment, or to meet other charges. Many Ujamaa villages were also dogged by poor leadership, financial mismanagement, corruption and use of communal property and means of production for accumulation of personal wealth.[30]

It is important, however, to place the relatively poor performance of communal farming in proper perspective. Communal farming had, in many instances, to overcome two difficult obstacles simultaneously. In Ujamaa villages based on new settlements, the first obstacle was the technical one of a transition in cultivation methods from an extensive to an intensive type. At the same time, however, large-scale collective farming, a new form of agricultural organisation for the peasants as well as for government officials, had to be introduced. This raised new problems of work organisation, distribution of income and devising appropriate incentives. In the absence of qualified and experienced persons, satisfactory solutions to these problems could only be found over a period of time. Some of these problems were compounded by the fact that often Ujamaa villages were set up in haste in disregard of such crucial factors as soil suitability or nearness to water. An obstacle suffered by most Ujamaa villages, and especially the larger ones, in comparison with smallholdings, was the time taken in travelling to and from the communal farm.[31] Furthermore, as noted earlier, in the early years

Ujamaa villages tended to attract landless or poorer peasants with less knowledge and experience of improved cultivation practices than the better-off farmers who seldom showed much enthusiasm for communal farming. All these factors have thus to be borne in mind when comparisons are made between the performance of communal and smallholder farming. In particular, in comparing communal farming on new settlements with smallholder agriculture in long-established settlements, due allowance would have to be made for the dual transition involved in establishing communal farming.

(c) Provision of social services

One of the major arguments in favour of villagisation has been the possibility of provision of social and economic services to the rural areas in a more effective and economical manner. This argument has been used, for example, with respect to health services, including family planning; education, especially adult literacy and universal primary education; water supply; transportation; irrigation and electricity. Indeed one of the main criteria used in the siting of villages was the existence or proximity of certain core services such as schools, medical facilities and a transport network. There has been a rapid expansion of many of these services in the past few years; the achievements in adult literacy and primary education, health and water supply, as shown below, have been especially noteworthy. The overall progress has been accompanied in most cases by sharp reductions in regional disparities as shown below.

		Population Per Rural Health Centre	*Access to Rural Water Supply*	*Illiteracy*	*Enrolment Rate, Primary Education*
National level:	1970	187901	13%	67.7%	34.7%
	1975	98650	25%	38.7%	54.4%
Coefficient of regional variation:	1970	0.58	0.53	—	0.22
	1975	0.30	0.25	0.26	0.23

Source R. van der Hoeven, *Meeting Basic Needs in a Socialist Framework: The Case of Tanzania* (Geneva: World Employment Programme Working Paper, ILO, 1978).

It has been argued that the villagisation programme made an important contribution to the rapid expansion of such services, especially of adult literacy.[32] With the bulk of the rural population living in villages, the potential for accelerated expansion of these services to meet the ambitious national targets has greatly increased. There should thus be major gains in this area in the coming years.

7.5 Conclusion

In the decade since the Arusha Declaration, the pattern of rural settlement has been transformed from isolated homesteads to communal living in villages. The villages have been endowed with institutions to facilitate self-government and economic and social planning. The mode of production exhibits wide diversity. There are a small number of villages where communal cultivation is significant, but in the overwhelming majority, smallholdings continue to be the dominant mode of agricultural production. In the cases of the newly settled villages in which smallholding predominates, the difference from the earlier pattern is that instead of practising shifting cultivation in isolated holdings, each family engages in settled agriculture on relatively equal size plots forming part of a block farm. In some of the more densely populated areas such as parts of the Arusha and Kilimanjaro regions, the existing settlements were designated as villages with minimal changes in patterns of land distribution. While collective agricultural production forms a negligible proportion of total output, communal initiatives have been far more important in rural public works and non-farming productive enterprises. This is so even in those villages which may not practise any collective agriculture.

It is not possible yet to evaluate the impact of these institutional changes on the overall production, accumulation and income distribution in the rural areas. On *a priori* grounds, and on the basis of some fragmentary empirical evidence, it would appear that at the micro-level Ujamaa and development villages have had a positive impact on land and income distribution. They have also facilitated a rapid expansion and equitable distribution of social consumption. Further, there has been a significant rise in labour utilisation and in rural capital formation associated with an increase in the volume of rural works.

The communal farms have, however, encountered production difficulties with resultant low yields. These difficulties derive mainly from the fact that communal farming has been regarded as additive to basic household plots and has thus suffered from irregular, and a low level of, labour participation. Not only are personal plots in general several times the household share of communal farms, but they are expected to provide the family with its food requirements. Furthermore, the returns from communal farming are subject to various deductions. It is, therefore, not surprising that in their decisions on labour allocation the peasants have tended to give preference to personal plots over communal farms. Other contributory factors to low yields have been lack of experience with work organisation and remuneration methods in collective farming. For collective farming to advance significantly it would be necessary to resolve the problem of balancing incentives for private and communal work.

The new system provides the institutional framework for an egalitarian pattern of growth, especially in the newly settled villages. Its essential features are distribution of land by villages on a relatively equal basis; restrictions on permanent wage employment on family farms; existence of collective activities of varying importance in cultivation, infrastructure development, commerce, processing, manufacturing and transport and public organisation of marketing and distribution of inputs. In the short term, the critical questions are whether direct labour investment, commerce, processing and manufacturing can be consolidated and expanded on a largely communal basis. Peasant production on relatively equal holdings within a public sector marketing framework have established the pre-conditions for rapid growth of communally organised physical investment and non-agricultural productive activity. Indeed, the experiments with village implement services and work teams in some block farm development villages suggest that if the other communal production is successful and provided appropriate incentives are given, a gradual shift towards more communalisation of agricultural production is not unlikely.

On balance, the new institutional framework should have a positive effect on growth of agricultural output. Greater use of joint extension demonstrations, grouping peasant farms in block farms, and village planning and provision of supporting services such as storage, marketing and transport are likely to raise output. Agricultural output should also benefit from improvements in social and economic infrastructure made possible by villagisation. Balancing these factors are some problems associated with the new pattern of settlement. These include the greater distance between farms and dwellings, danger of soil deterioration from settled agriculture and over-grazing by cattle and animal damage to crops. The government policy on agricultural prices and, more generally, on the provision of appropriate individual and collective incentives will be a critical factor in determing the prospects for growth. Furthermore, given the heterogeneity of the country and the low level of technology, agricultural growth will depend heavily on success in devising chemical and biological innovations suited to the ecological and climatic diversity of the country.

Notes

1. Ujamaa is loosely translated as socialism, but its literal meaning of 'familyhood' is perhaps nearer to the usage in Tanzania connoting communal action and decision-taking through direct participation.
2. Julius K. Nyerere, *Socialism and Rural Development* (1967), reprinted in *Freedom and Socialism* (Dar es Salaam: Oxford University Press, 1968).
3. H. Ruthenberg, *Agricultural Development in Tanganyika* (Springer-Verlag, 1964).
4. Julius K. Nyerere, *The Arusha Declaration: Ten Years After* (Dar es Salaam: Government Printer, 1977). It is likely that the 1975/76 figures represent a slight overestimate of the actual number living in villages.
5. This did not always represent physical movement of people and their belongings. In many instances, especially in the densely populated regions of Arusha and Kilimanjaro, the process of villagisation consisted simply of demarcation and designation of a certain area as a village.
6. As of March 1973 about one-third of the total number of villages were located in only two regions while five regions, which together contributed less than 25 per cent of 1969 GDP, accounted for over 60 per cent of villages with nearly 70 per cent of total village population. Conversely the six most affluent regions, which contributed over 50 per cent of 1969 GDP, accounted for only about 15 per cent of the number of villages and less than 10 per cent of the Ujamaa population.
7. *Towards Self-Reliance: Development, Employment and Equity Problems in Tanzania*, Report of the Employment Advisory Mission, 1977, Appendix II: Agriculture's Terms of Trade (Addis Ababa: Jobs and Skills Programme for Africa, ILO, 1977).
8. It should be noted that especially with respect to small plots, multiple holdings are common and some rural wage earners and pastoralists also have small plots to supplement their income. In addition because of tremendous ecological and climatic diversity in the country, there are major differences in land fertility within and across regions. The data in Table 7.3 therefore systematically exaggerate disparities in land distribution by households.
9. This estimate by the Minister of Agriculture is quoted in J. De Vries and H. Hänsel, 'Ujamaa Villages as a Strategy for Rural Development in

Tanzania', in *Agricultural Extension in Ujamaa Village Development*, edited by H. Hänsel, J. De Vries and P. C. Ndedya (Dar es Salaam, 1975).

10. Colin Barnes, *A Study of the Peasant Fishing Economy in Tanzania* (Economic Research Bureau, University of Dar es Salaam, hereafter referred to as ERB, 1976).
11. For substantiation of this argument at the high point of the Ujamaa movement, see Lionel Cliffe, 'The Policy of Ujamaa Vijijini and the Class Struggle in Tanzania', in *Socialism in Tanzania*, L. Cliffe and J. Saul (eds.), (Dar es Salaam: East African Publishing House, 1974).
12. The Kigoma villagisation pattern is typical in this respect. Each family was given an acre of land around the homestead and a plot in the block farm. The size of the plot varied from less than one-quarter hectare per household in Mahembe to two hectares per household in Mugere. See R. B. Mabele and M. Schultheis, *Evaluation of the Kigoma Rural Development Project* (ERB, 1977).
13. Ndissi, Barnes and Boesen et al, all op cit, quote some data on income distribution in some Ujamaa villages and under other modes of production. But their data are not sufficiently reliable or detailed for any firm conclusions to be drawn, though Ndissi tentatively concludes that income distribution is more even on the Ujamaa tobacco farm than on the neighbouring smallholdings.
14. A. T. Mohele, *The Ismani Maize Credit Programme* (ERB, 1975).
15. Boesen et al, op cit, p. 105.
16. John Markie, 'Ujamaa Villages in Tanzania: A Possible Solution to the Problems of the Rural Poor', in *Land Reform, Land Settlement and Co-operatives* (Rome: FAO, 1976, No. I).
17. Suleman Sumra, *Problems of Agricultural Production in Ujamaa Villages in Handeni District* (ERB, 1975).
18. One exception is C. Ndissi, *Ujamaa Villages as a Collective Development Strategy in Tanzania's Economic Development.* He found a higher average of workdays on an Ujamaa tobacco growing village than on smallholdings.
19. Visits to Ujamaa villages in 1977, as well as discussions with a wide range of government field staff, confirmed this point.
20. Markie, op. cit.
21. The one exception is Ndissi, op cit.
22. Ndissi, op cit, shows that total factor productivity, net yield and labour productivity, and incomes were all higher on the communal tobacco farm than in the neighbouring smallholders' scheme. Frank Kürschner, 'Ujamaa Villages in Tanzania: the Example of Mahiliti' in *Afrika-Spectrum* (No. 1, 1973), documents another instance of a highly successful Ujamaa village. There are, no doubt, many other examples of successful, efficiently organised Ujamaa farms.
23. Boesen et al, op cit, pp. 69, 103.
24. Sumra, op cit.
25. Mohele, op cit.
26. J. de Vries, and L. Fortmann, 'A Study of Ujamaa Villages in Iringa Region' (Morogoro, 1974).

27. There is a good deal of variation in the amount of compulsory work for communal activities. In a study of Ujamaa villages in Mwanza, Mapolu found the following distribution of compulsory communal work:

No. of Days Communal Work Per Week	*Percentage of Ujamaa Villages*
2	25
3	42
4	21
5	8
6	4

H. Mapolu, *Social and Economic Organisation of Ujamaa Villages* (Dar es Salaam: M.A. Thesis, 1973).

28. Fluctuating, irregular and inadequate supply of labour by members on communal farms has been documented in several studies. In this connection, see Boesen et al, op cit, Zinnat K. Bader, *Women, Private Property and Production in Bukoba District* (Dar es Salaam, M.A. Thesis, 1975); Sumra, op cit, Reuben R. Matango, *Peasants and Socialism in Tarima District* (Dar es Salaam, M.A. Thesis, 1976); Mohala, op cit; R. D. Sharma, *Manpower Utilisation in Ujamaa Villages* (Dar es Salaam, undated, mimeo).

29. Extremely low or delayed returns from communal farms in the early years of their establishment have been mentioned in several studies as a major disincentive to collective agriculture on the part of peasants. A. M. Sendaro, *The Alienation of Peasants from the Process of Development: A Bottleneck of Development* (Dar es Salaam, M.A. Thesis, 1975) notes that for more than three years there was no income distribution from the Ujamaa production surplus in Nkulati village. Sumra op cit lists exceedingly low incomes from communal farming in Handeni Ujamaa villages: in 1972, the high was sh. 78 p.a. per member and the low sh. 1, with the great majority of villages distributing less than sh. 20 p.a. Matango and Bader also mention low, zero or delayed returns from communal labour as a major constraint to expansion of collective farming. Boesen et al, p. 68. 'In none of the Ngara Ujamaa Villages had the members by mid-1971 (two to four years after establishment) earned any income from the communal sector.'

30. For a documentation of such malpractices, see a series of case studies in J. H. Proctor, *Building Ujamaa Villages in Tanzania* (Dar es Salaam: Tanzania Publishing House, 1975); Matango, op cit, Kemal Mustefa, *The Development of Ujamaa in Musoma: A Case Study of Butiama Ujamaa Villages* (Dar es Salaam, M.A. Thesis, 1975); Zinnat K. Bader, op cit, Matango, op cit.

31. In their base-line survey of ten villages in Kigoma Region, Mabele and Schultheis found 'Some of the sample villages already have an average distance from the household to the block farms of approximately 5 miles', Mabele and Schultheis, op cit. This situation is not typical of Ujamaa villages.

32. Julius K. Nyerere, *The Arusha Declaration – Ten Years After.*

8 Achievements and Incentives in Communal Agriculture: The Case of China

Azizur Rahman Khan and Ng Gek-boo

8.1 A summary of achievements

The achievements of China's agricultural sector have been very substantial by way of both increasing output and promoting an egalitarian distribution of income. In terms of each of these objectives, however, China's agriculture is still far short of the goals set by itself. The inadequacy of the rate of output growth is particularly emphasised by the national leadership. They also recognise that further improvements in the distribution of income have to be brought about.

Table 8.1 shows the estimated trend rates of growth in the output of all grains (which, in the Chinese estimates, include potatoes and beans duly converted into grain-equivalents), rice and wheat. These rates of growth have been shown for three time-periods (of which the last encompasses the other two). The first column of the table refers to the period that ranges from 1950–7. We shall designate this as the pre-commune period. The next column refers to the years from 1969–76, a period which might be called the post-Cultural Revolution period. The third column refers to the entire post revolutionary period. No trends have been shown for the 1958–68 period – the period of the Great Leap Forward and Cultural Revolution (GLF-CR). Output estimates for much of this period are either unavailable or unreliable. The comparison of the three columns in the table, however, sheds some light on the growth performance in this period.

During the pre-commune period agricultural production increased at a very high rate and the rates of growth during the post-Cultural Revolution period have also been reasonably high.

Table 8.1 Trend Rates of Growth in Grain Production (per cent per year)

	1950–7	*1969–76*	*1950–76*
All grain	5.16	3.03	2.53
Rice	6.07	2.90	2.54
Wheat	7.39	5.09	2.98

Note For 1950–7 the production data are from the official Chinese source (published in *Ten Great Years*). For 1969–76 the production data are from the US Department of Agriculture estimates. Observations are not available for 1959–64 and 1966–8. An estimate for 1958 is available from *Ten Great Years* but is believed to be an exaggeration and hence not used. The 1965 estimate, used for the calculation of 1950–76 trend, is from USDA. The trend for 1950–76 is based on an incomplete set of observations, those between 1950 and 1957, the one for 1965 and those between 1969 and 1976. The trends have been derived by fitting log $X=a+gt$ where X=output, t=time (=1 for 1950, 2 for 1951 etc.) and g=trend rate of growth.

The trend rates of growth during the entire post-revolutionary period are lower than the corresponding rates in either of the above time-periods. This implies that the rates of growth during the GLF-CR period have been lower than in any of the other two periods. It is well known that China suffered natural calamities of massive proportions during some of these years. But one should look for additional reasons to explain the unfavourable growth performance in these years. During this period the incentive structure in agriculture was in a bad state. For a period the commune – the highest of the three-tiered organisational structure – was made the basic accounting unit and the value of a work-point was the same for the members of the highly productive and unproductive teams and brigades within the same commune. Also, communes used team labour without compensation and free food and services were provided in the 'public dining rooms'. Meanwhile, individual economic operations were mostly prohibited. These policies were inconsistent with the level of consciousness of the rural population. As a consequence there were inadequate incentives to attain higher productivity.[1]

It is sometimes argued that China's growth performance has not been significantly higher than that of, say, India over the comparable period. While the comparison of growth rates over a relatively long period would tend to confirm this claim, it is necessary to recognise a few points to understand the full significance of China's sustained growth rate of 2.5 per cent over more than a quarter of a century. First, both in terms of per capita consumption and in terms of per hectare output, China was initially (i.e. around 1950) considerably above the major Asian developing countries like India. It was obviously a greater achievement to attain a sustained rate of growth of over 2.5 per cent from this higher level than that of those who achieved a similar rate of growth from a lower starting point. Secondly, the rate of population growth in China has been lower than that in the major developing Asian countries and the difference has probably been a half of a percentage point per year or more.[2] Thus the trend rate of growth sustained over more than a quarter century provided a steady gain (of over half a per cent per year) in per capita consumption in China while the same rate of growth elsewhere was more or less eaten up by population growth. Thirdly, an important achievement of China's agriculture, especially over the last decade, was to reduce the variability in production in the face of natural disasters. It seems probable that investment in agriculture has been substantially directed to achieve a reduction in variability and to that extent investment in immediate output growth was probably less than what it might have been.[3]

The egalitarianism of the distribution of income in rural China is documented in Table 8.2 (including the note). Major gains were made by land reforms which were completed in the early 1950s. Since then, despite some probable fluctuations, steady improvement has been maintained. The shares of quintile groups shown in Table 8.2 refer to the early to mid-1960s and the distribution today, though perhaps somewhat better, is unlikely to be qualitatively

Table 8.2 An Approximate Distribution of Collective Income among Quintile Groups in Chinese Communes

Quintile Groups in Ascending Order	*Share of Income*	*Cumulative Share of Income*
Lowest	9	9
Second	13	22
Third	17	39
Fourth	22	61
Top	39	100

Note The above table is from ILO, *Poverty and Landless in Rural Asia*, Geneva, 1977, Chapter 12 which discusses the method of estimation and shows that the distribution would be more equal once allowance is made for incomes from private plots and collective consumption. A crude estimate of the allowance due to such factors would raise the *share of the lowest quintile group to 11 per cent and reduce that of the top quintile group to 34.6 per cent.*

different.[4] The share of the bottom quintile in China is about twice as much as that of the corresponding group in a typical Asian rural society. Once collective consumption is taken into account the per capita income of the top quintile is just over three times that of the bottom quintile. In a typical rural society of a developing Asian country the corresponding ratio ranges from 6 to 14 or more. The achievement of China's communal agriculture in terms of promoting an egalitarian distribution of income would appear to have been truly remarkable, more so than the performance in terms of growth.

One of the basic principles of China's rural development strategy has been to encourage self-reliance and decentralisation. The emphasis has been on getting the individual communes to save at a high rate to finance the major part of their investment requirement. Table 8.3 shows the proportion of gross collective revenue that is saved by a number of communes, brigades and teams for which the writers have been able to assemble information for recent years. Although there is considerable variability among the collective units the picture is completely different from the one that is observed among the debt-ridden and aid-dependent co-operatives in the contemporary Third World. Even after allowing for the relatively small amount of welfare fund that is included in some of the figures the unweighted average of savings for the 17 collective units in the table would probably be about 15 per cent of gross income. We have no information on the average proportion of current inputs to gross revenue, but even on the conservative assumption that it is 25 per cent, savings would be 20 per cent of net income (value added). This is a very high rate of accumulation by any standard.

It should also be recognised that most communes engage in a substantial amount of direct capital construction through the allocation of work-points to individual members who supply necessary labour. This is not reflected in the collective savings shown in Table 8.3. Admittedly, the ratios of Table 8.3 refer only to collective saving as a proportion of gross collective income. We have no corresponding information about personal savings, but it is known that households save a significant proportion of their earnings (which consist of both collective earnings and earnings from private sources) through financial institutions. Thus the above estimates of saving rates are, if anything, too low an indicator of the aggregate accumulation generated in China's communal agriculture in recent years.

Table 8.3 Saving as a Percentage of Gross Collective Income in Selected Communes, Brigades and Teams

Collectives and Their Province of Location	*Saving as a Percentage of Gross Collective Income*	*Year of Reference*
Communes		
Liuchi, Hupeh	11.1	1975
Huatung, Guandong	6.8	1972
Chachiao, Kiangsu	20.4	1976
Huangto Kuang, Peking	9.5	1971
Lokgang, Guandong	14.0[1]	1973
July 1st, Shanghai	13.0[1]	1973
Production Brigades		
Sipu, Hopeh	19.3	1973
Tachai, Shansi	25.6[1]	1975
Huasi, Kiangsu	26.2[1]	1975
Huangzhong, Hunan	30.0	1977
Tonghsin, Hunan	15.9	1977
Chiaoli, Chekiang	14.9[1]	1974
Rainbow, Shanghai	11.0	1974
Nancheng, Shansi	18.8	1974
Hungshan, Shansi	10.0	1974
Production Teams		
Chenkuan No. 8, Kiangsu	8.0[1]	1973
Kuaiti, Hopeh	24.0[1]	1971
Unweighted average	16.4	

Sources *China Reconstructs*, No. 9, 1972, and Nos. 3–4, 1978; *People's Daily*, 15 March 1972; Harold Champeau 'Five Communes in the People's Republic of China', *Foreign Agriculture*, 18 August 1973, p. 16; Wu Chou, *Report from Tungting*, Foreign Language Press, Peking, 1975; *People's Communes are Leaping Forward: The New Experience of People's Communes in Suburban Shanghai*, People's Publishing House, Shanghai; Party Committee of Liuchi Commune, The Poor and Lower-Middle Peasants in Liuchi Commune, and Students of the Department of Economics at Wuhan University (1972), *The Agricultural Mechanisation of Liuchi Commune*, People's Publishing House, Peking; Maxwell N. 'Learning from Tachai', *World Development*, Vol. 3, Nos. 7 and 8 (July–August, 1975). *History of July 1st Commune in Shanghai County*, People's Publishing House, Shanghai, and A. Z. M. Obaidullah Khan 'The Three Great Struggles in the Chinese Countryside: A Case Study Part II', in *The Bangladesh Development Studies*, No. 3, Vol. IV.

Note 1. Including Welfare Fund.

8.2 The need for further progress and an incentive policy

While China has made substantial, even remarkable, achievements in terms of growth, distribution and accumulation, the need to make further progress is obvious. In terms of output per hectare China is far behind her neighbours like Japan and Korea (see Table 8.4). Taking rice as an example, yields per hectare in Japan, North Korea and South Korea in 1976 were respectively 67, 56 and 81 per cent above that in China. The average difference for all cereals is even greater.[5] We do not have comparative figures for yield per worker, but it is well known that a disproportionately high amount of rural China's labour force is engaged in the production of food. Labour productivity is low and it must go up quickly if China is to redeploy labour in a way that is more consistent with a high level of material prosperity. The need to achieve higher productivity and output are of paramount importance. The current agricultural plans emphasise this objective and the consequent need to mechanise and increase the absorption of modern inputs. To achieve these objectives the rate of accumulation at the communes must continue to be high.

On the distribution of rural income China has achieved a degree of egalitarianism matched by few contemporary societies. But given the stated objectives of Chinese political leadership, the existing inequality, however limited in comparison with that elsewhere, must be considered unacceptably high. This, notwithstanding the recent fight against the 'Gang of Four's' unrealistic strategy of too rapid a reduction in inequality, continues to be stressed as a development objective by the Chinese leadership.

The problem is that the objective of high growth and accumulation can conflict with the objective of greater equality. The precondition for achieving high growth with efficiency is a proper system of material incentives. Such an incentive system must reward those who are able to produce more with greater efficiency in

Table 8.4 Yield per Hectare in 1976 (in metric tonnes)

Crop	*China*	*Japan*	*North Korea*	*South Korea*
All cereals	1.94	5.28	3.02	4.50
Rice	3.29	5.50	5.13	5.97
Wheat	1.39	2.50	2.01	2.22

Source FAO, *Production Yearbook*, 1976, Vol. 30 (Rome 1977). Japan and Korea are not major producers of wheat.

relation to those who are less productive and less efficient. But the price of depending on such production incentives is the need to tolerate the greater inequalities that they produce. Moreover, these effects are exacerbated by the incentives to accumulate. The collective units with higher incomes are known to devote a higher share of earnings to accumulation, thus leading to further polarisation.[6]

The question of incentives is, therefore, of crucial importance for the evolution of communal agriculture in China. There must be adequate incentives in the form of material rewards for differential production performance to generate a high enough rate of growth. On the other hand, differential rewards should be kept within the limits of tolerable income inequality.

In a system of collectivised agriculture it is very difficult to devise a system of incentives that would induce individuals and collective units to organise their efforts in a way that would maximise output and economise on scarce resources. The inability to work out and implement such a system is perhaps the most important explanation of the failure of some programmes of collectivisation to fulfil the objective of rapid agricultural growth. Historically, in the less successful cases of collectivisation a large part of the problems has been due to the 'primitive socialist accumulation' through the imposition of very high rates of direct or concealed taxation on agriculture. But this has not been the only problem of incentive in a typical system of collective agriculture. It has been difficult to establish an appropriate system of remuneration for individual workers within a collective unit and to work out a system of marketing, prices and taxation that would induce collective units to undertake the socially necessary effort.

The recognition of the difficult problems of organising incentives in collective agriculture has frequently led to extreme kinds of deviations from socialist objectives. At one extreme, such a deviation, illustrated by the example of some East European countries, has amounted to the non-adoption of collectivisation and the partial or complete organisation of agriculture on the basis of private peasant farming. Quite apart from its inconsistency with socialist objectives, such a system may not be feasible as a permanent solution in many land-scarce countries. A highly egalitarian system of peasant agriculture in many parts of South and South-east Asia would in the long run condemn the peasantry to extremely small subsistence farms.

The other type of extreme deviation has consisted of attempts to do away with incentives altogether, the argument being that the need for incentives arises out of the persistence of basic human instincts nurtured by a class society over thousands of years. What one should aim at is to create 'socialist men and women' by changing human beings. Something of this nature was attempted in Cuba for a short period in the 1960s and in China during the Great Leap Forward. The result was largely unfavourable. That 'socialist men and women' must ultimately be created if the final objective of a completely equal society is ever to be achieved is hardly a disputable statement. But all non-utopian builders of a socialist society must recognise that the process of such a transformation is necessarily very slow. As yet there is no example of a large community which has successfully experienced such a transition on a durable basis. As Karl Marx himself recognised, in the meantime

> what we have to deal with here is a communist society, not as it has *developed* on its own foundations, but, on the contrary, just as it *emerges* from capitalist society; which is thus in every respect, economically, morally and intellectually, still stamped with the birth marks of the old society from whose womb it emerges. Accordingly, the individual producer receives back from society – after the deductions have been made – exactly what he gives to it.[7]

In the process of transforming agriculture along a collective path, the planners must, therefore, face up to the need to devise an appropriate system of incentives. This is an essential pre-condition for the working of a collectivised agriculture in the process of transition from private to socialised farming. This does not mean that the task of creating 'socialist men and women' has to be abandoned or the ultimate objective of creating an equal society in which all payments are proportional to needs has to be given up. The process of transition must entail conscious effort to achieve these objectives gradually. But, in order to avoid the creation of disorder which will endanger the transition to collective organisation of agriculture, it must be remembered that 'right can never be higher than the economic structure of society and its cultural development conditioned thereby'.[8]

The development of collective agriculture in China over the last twenty years has evolved through the interaction of these two types of conflicting policies, viz., the concession to the backward (though continuously developing) socialist consciousness of the population in the form of material incentives and the gradual effort to raise the level of consciousness of the people in order to induce them to give up their rights to unequal incomes. At times one or the other of the objectives received greater emphasis than would appear to be appropriate. Two decades after the beginning of the communes and a decade after the beginning of the campaign of 'learning from Tachai' the need to have a system of material incentives continues to be strongly recognised by the official policy towards agriculture.

One could conveniently divide incentives in a system of collective agriculture into three broad categories. Firstly, the individual members of a basic accounting unit must be induced to perform with greater productive efficiency through the provision of *individual incentives.* These consist of differential rewards to individuals of different productive capacity. Secondly, different *collective* units – teams, brigades and communes – must be induced to make greater efforts at attaining productive efficiency by making the return to such effort at the margin as high as possible. Finally, there must be adequate incentives for the *sector* as a whole so that the results of its overall performance are translated into higher rewards and not neutralised by macroeconomic measures such as open or concealed taxation.

For the system to operate successfully, the three types of incentives must be integrated on a consistent basis. Indeed, the classification is artificial insofar as a change in one type of incentive would affect the same at other, especially lower, levels. Thus, the provision of individual incentives (through, for example, measures such as a system of graduated distribution of work-points, a carefully devised and implemented method of piece-rate working, etc.) would be inadequate to motivate the individuals to greater effort if returns at the margin to collective efforts are low. This is because the returns to the collective determine individual earnings from collective labour. Such earnings at the margin would be low if collective incentives are low and the result will be an inadequate effort by individuals, even though the individual incentives themselves are well conceived. Similarly, high collective incentives may produce little effect if individual incentives are such that rewards to individual members have no correspondence to their relative efforts and abilities. Thus, while each of the three above types of incentives is a necessary condition for productive efficiency, a sufficient condition for the promotion of such efficiency is to have a *well integrated* system of all three. The adjective in italics deserves to be noted. It is possible to design each of the three types of incentives in a way that independently, each is geared to the promotion of productive efficiency and yet the whole system may contain elements of inconsistency.[9]

8.3 Individual incentives

Individual incentives are designed by varying the rewards to members with different productive capacities in order to induce those with greater ability to contribute more to production. Given the three-tier organisational structure of the people's communes, the individual incentive policy is implemented by the production team, the basic accounting unit in most communes. An average production team consists of about 30 households. Since the level of agricultural mechanisation is still relatively low, this is a small enough unit to permit effective monitoring of work performance and the determination of differential rewards on the basis of a reasonable evaluation of individual performance.

The work-point system is essentially based on the socialist principle of distribution: to each according to his work. It is operated in two stages. The first is to allocate jobs to members and then to reward them with a certain number of work-points based on the assessment of their work performance. The second is to determine the value of a work-point and to distribute net collective income to member households according to their accumulated number of work-points or work-days at the end of the year. Work assessment, which is the most difficult task of team management, is undertaken through the application of either a piece-rate and/or a time-rate payment system.

The Chinese term for piece-rate system is literally translated as 'norm management'. It is operated by classifying all farm tasks into different categories, with agreed norms and number of work-points for each. The final allocation of work-points is thus based on the evaluation of work done on the basis of these agreed norms.

This piece-rate system has been very popular in the communes since 1962, the year in which the 'Sixty Regulations' were introduced to solve the problems of motivation caused by the collapse of the incentive system during the 'three bad years' (1959–61). A great deal of evidence suggests that the piece-rate system still prevails in a majority of communes. Thus, in most communes there are up to 200 categories of tasks for which norms and the number of work-points are separately specified. These tasks are allocated to small groups each consisting of a few individuals. The work-points awarded to each group are distributed among the members of the group according to each individual's respective contribution to the fulfilment of that particular task. Tasks which are more technical, heavy and dirty are generally awarded higher numbers of work-points.

The piece-rate system needs periodical readjustment to take into account changes in production patterns and farming techniques. Failure to provide due revisions of norms and rates could easily give rise to biases in payment and difficulties in the allocation of tasks among groups of members. Thus, the introduction of the piece-rate system involves arduous tasks of classifying and readjusting hundreds of types of farm work and finding the appropriate number of work-points for each under conditions of technical change.

The strict application of the piece-rate system can aggravate income inequality between the strong and skilful members on the one hand, and the weak and unskilled members on the other. At the present the Chinese communes seem to prefer a piece-rate to a time-rate system because the defined norms and rewards are more effective in raising the rate of attendance and productive efficiency.

The time-rate system was first introduced through the method of 'basic work-points with flexible assessment', under which members of a production team were classified into various categories according to their skill and physical capacity. Each category of worker had a specified number of work-points assigned for a day's work. These standard number of work-points were given adjusted through 'flexible assessments' designed to take into account the quantity and quality of work done by each individual.

There was no general rule for the number of categories and the number of work-points that should be attached to each. In 1963, for example, 59 per cent of male workers of Hu-li brigade of Fukien province were in category I with eight work-points per day, 24 per cent in category II with seven work-points, and 17 per cent in category III with six work-points. The number of daily work-points for various categories of workers in other communes was known to have ranged from five to twelve.

A major problem of the time-rate system was that when it was applied in politically backward collectives, it was slow and inefficient in motivating individuals to undertake collective work to the best of their capacities. Members in some classified categories tended to refuse to undertake heavier and dirtier tasks. Flexible assessment could also lead to disputes over the quality of performance.

On the basis of 'basic work-points and flexible assessment', the famous Tachai brigade of Shansi province developed the method that came to be known as 'standard work-points, self-reporting and public assessment'. The general practice under this system is that each member's working time is duly recorded at the end of the day, and the overall assessment of the work performance of each member takes place periodically. The calculation of work-points for each person is made by comparing his performance with that of the 'standard worker', who is elected and allocated 'standard work-points' by the membership in a general meeting on the basis of his attitude, rate of attendance and quality and quantity of work. Each member evaluates himself in a public meeting with reference to the work performance and work-points of the standard worker. This self-evaluation is subject to approval and modification by the majority.

The Tachai method of allocating work-points is based on work done over a period of time. Since the public assessment takes place on a monthly, quarterly or yearly basis, time-saving is an undeniable advantage of this method. Moreover, under this system the attitude of members is an important criterion, so that the physically weak members with collective enthusiasm are likely to benefit. It is relatively egalitarian, but the prerequisite for successful application of this method is a high level of consciousness among members. Few communes have attained the required level of such collective consciousness.

8.4 Collective incentives

In discussing collective incentives, we shall try to see how the major instruments of public policy affect the incentives of the various collective units – the teams, brigades and communes – to attain higher output. On many aspects of the operation of the instruments of public policy our information is sketchy. Thus, the discussion, at this stage, is partly speculative and hypothetical. As and when we are able to obtain reliable information on many of the specific questions discussed in this paper, we shall be able to see the effect of the working of the incentive system more clearly.

Table 8.5 Notations(1)

T	=	Total tax on agriculture, direct and concealed.
T_1	=	Total direct tax on agriculture.
T_2	=	Total tax concealed in compulsory procurement.
T_3	=	Total tax concealed in voluntary procurement.
The lower case t refers to the corresponding *rates* of taxes (e.g., t_1 = direct tax *rate* on agriculture)		
X	=	Actual farm output.
X_b	=	Benchmark or normal farm output.
X_p	=	Basic (compulsory) procurement quota.
X_e	=	Amount sold voluntarily to the state over and above X_p.
P_a	=	Price on compulsory procurement.
P_e	=	Price on voluntary procurement.
P^*	=	Possible alternative price.
r_1	=	Direct tax as proportion of benchmark or normal output.
k	=	Compulsory procurement as proportion of benchmark output.
q	=	Proportion of excess output over benchmark level that is voluntarily sold to the state.

Note (1) These notations will be used throughout.

(a) Agricultural taxation

In China there is 'agricultural tax on units and individuals engaged in farm production with income derived from it' but there is no tax 'on the pay of people's commune members based on work-points and distributed collectively and their income from small plots for their personal needs and household side line production'.[10] Agricultural tax consists of a fixed amount in kind which is determined on the basis of the benchmark output in a normal year. 'Once fixed, this norm remains unchanged for a certain period of time', usually five years.[11] Thus, tax revenue in kind, T_1, would be some fixed proportion, r_1, of the benchmark output, X_b:

$$T_1 = r_1 X_b.$$

It appears that r_1 itself has been allowed to go down fairly sharply. 'The nation's total agricultural tax accounted for 12 per cent of actual farm output in 1952 and it now [1975] has dropped to only 5 per cent.'[12] If, as it appears from available writings to have been the case, both 1952 and 1975 were the benchmark years in which taxes were reassessed then 'normal' output would not be different from actual output for those years so that the above rates would be reasonable approximations of r_1 prevailing in those years.

In any given year the actual rate of agricultural tax, t_1, would, of course be different from r_1 depending on the difference between actual output, X, and normal output:

$$t_1 = r_1 \quad \frac{X_b}{X}.$$

A collective unit has a complete 'tax holiday' on the increment in output over the normal level until the date of the next assessment of normal output, usually five years later. Even at the time of reassessment a good deal of advantage is preserved by those units which have higher than average performance. This is because X_b is estimated on the basis of average yield in a relatively large homogeneous area and any production unit which has succeeded in attaining the levels of yield higher than the average in the locality will pay a lower rate of tax than the average.[13] Similarly a collective unit that lags behind the average performance in the locality pays a higher rate of tax than the average.

(b) Compulsory and voluntary procurement

Direct agricultural tax is not the only contribution that agriculture makes to state revenue. It also makes a contribution through sales to the state if prices are below what it might otherwise have obtained. From available information from various sources it clearly appears that there are two types of procurement of agricultural goods.[14] A lower price is paid on a basic quota which must be fulfilled by all collective units. Over and above the compulsory quota, all collective units are free to sell additional quantities to the state. On such voluntary sales they receive a higher price. The rate of this price premium has been different in different regions and has been varied over time. In the early years it used to be about 10 to 20 per cent of the basic procurement price. Recent reports suggest that it currently ranges between 20 and 50 per cent in different localities. It seems to us that an estimate of 30 per cent is probably a reasonable average. This is the figure that we shall use in the subsequent calculations.[15]

It can be argued that this state monopoly in the purchase of the marketable surplus constitutes a concealed tax if the procurement price is below what a collective enterprise could otherwise obtain from its sales.[16] Estimating the extent of such a concealed tax is difficult because of the problem of defining the principles of quantifying the possible alternative price that the collective enterprises could get. Later in the paper we shall try to indicate a plausible range of values for such an alternative price in relation to the two sets of procurement prices.

The amount of concealed tax (T_2) imposed through the operation of the compulsory procurement quota (X_p) can be found by comparing the price paid for compulsory procurement (P_a) with the alternative price (P^*) that the collective units could have obtained:

$$T_2 = (P^* - P_a)\, X_p.$$

It seems that the compulsory procurement quota is an approximately fixed proportion (k) of normal or benchmark output. The rate (t_2) of this concealed tax is given by the ratio of T_2 to the value of actual output. For the valuation of output for this purpose the appropriate price to use would appear to be the alternative price. Thus:

$$t_2 = \frac{T_2}{P^* X} = K\left(1 - \frac{P_a}{P^*}\right)\frac{X_b}{X}.$$

Two things are clear: (i) the smaller the gap between the possible alternative price and the compulsory procurement price, the lower would be the rate of concealed tax. In the limiting case of the two prices being the same the tax vanishes altogether. (ii) The greater the increment in output over the normal or benchmark level, the smaller the rate of tax.

Next we shall consider the voluntary procurement over and above the compulsory quota. The concealed tax (T_3) through such procurement would amount to the following:

$$T_3 = (P^* - P_e)\, X_e,$$

where P_e = price paid for voluntary procurement and X_e = amount voluntarily sold over and above the basic quota. Let us assume, further, that the quantity sold voluntarily is a fixed proportion (q) of the additional output over and above normal output.[17] Thus, the rate of concealed tax (t_3) implicit in the working of voluntary procurement would be as follows:

$$t_3 = \frac{(P^* - P_e)\, q(X - X_b)}{P^* X} = q\left(1 - \frac{P_e}{P^*}\right)\left(1 - \frac{X_b}{X}\right).$$

Once again, the greater the amount by which the possible alternative price is higher than the price for voluntary procurement, the higher would be the concealed tax. But it is not so clear that the alternative price is higher than the price on voluntary procurement. As we shall see later, it is possible that in China the two-tier procurement price is a device to have a fairly high average rate of tax on agriculture while rendering the marginal rate of tax very low, even negative. If the marginal rate of concealed tax through voluntary procurement is indeed negative (i.e., if $P^* < P_e$), then the greater the excess of actual output over normal, the smaller (greater) is the (negative) tax rate.

(c) The possible alternative price

To carry the discussion any further it is necessary to try to guess where the possible alternative price that the collectives would get in the absence of state trading and distribution of farm products would lie. The procurement price that would equilibrate supply and demand for foodgrains at the current levels of income in urban and rural areas can be taken as the alternative price at which farm output should be valued. The sale price to consumers is probably about the same as the compulsory procurement price.[18] At such a price, there is some excess demand for food, but, although rationing prevails, it is widely believed that the amount of excess demand, at prevailing prices, is modest. Furthermore, the levels of consumption of foodgrains are quite high for the average incomes of the consumers. Thus, it is unlikely that it would take a large increase in the sale price to consumers to equilibrate supply and demand. Hence the alternative procurement price, P^*, will be higher than the compulsory procurement price – but not by very much. We do not know definitely whether it is as high as, or higher than, the voluntary procurement price also. On our present trend of argument, it is unlikely to be higher than the voluntary procurement price. In fact, it may well be lower than the latter price.

Let us summarise our very tentative hypotheses. It seems quite plausible that the possible alternative price would lie somewhere between the two procurement prices. The possibility that it is lower than the voluntary procurement price is quite real and must be taken seriously. It seems unlikely that the possible alternative price would be significantly higher than the voluntary procurement price.[19]

The above discussion has been carried out in terms of the

procurement of grain. But non-grain agricultural products are also subject to similar procurement rules. It is not known to what extent the above generalisations about grain would apply to the cash crops. We shall assume that the overall effects are similar to that for grains.

(d) Average and marginal rates of taxes on agriculture

To make some illustrative estimates it is necessary to find the estimates of values of the various parameters. As we argued earlier, the value of r in the benchmark year should be approximately the same as the proportion of actual grain output collected in the form of direct taxation. Thus, if taxes were re-assessed during the mid-1970s, then 0.05 (which was the actual rate of direct agricultural tax collected in kind at that time) would be a reasonable approximation of the value of r_1 around that date.

Despite a reasonable rate of growth of foodgrain production shown in Table 8.6, the marketed proportion does not seem to have gone up: it is reported to have declined from an average of 28 per cent during 1953–7 to only 20 to 23 per cent in the early 1970s.[20] Table 8.6 shows that the proportion in 22 counties in the early and mid-1970s was 32 per cent. The national average was almost certainly lower because most of the counties in the table belong to the category of the so-called Tachai-type counties which are more advanced in terms of productivity and procurement. In the following calculations we shall use a value of 0.20 for k and, with less confidence, a value of 0.5 for q.

Following the discussion in Section 8.4(*b*) we shall put the value of P_e at 1.3 times P_a. We shall use three different values of P^*, the possible alternative price. The first is to assume that $P^* = P_e$. The second is based on the assumption that P^* is 5 per cent below $P_e (P^* = 0.95\, P_e)$ and the third is based on the assumption that P^* is 5 per cent above $P_e (P^* = 1.05\, P_e)$.

The total tax rate, direct and concealed through procurement, would be:[21]

$$t = t_1 + t_2 + t_3$$

$$= r_1 \frac{X_b}{X} + k\left(1 - \frac{P_a}{P^*}\right) \frac{X_b}{X} + q\left(1 - \frac{P_e}{P^*}\right)\left(1 - \frac{X_b}{X}\right)$$

$$= (r_1 + r_2 - r_3) \quad \frac{X_b}{X} + r_3,$$

where:

Table 8.6 Marketed Proportion of Foodgrain Output in Twenty-two Counties in Fifteen Provinces

County, Province	*The Proportion*	*Year of Reference*
Payen, Heilungkiang	54.5	1974
Aihui, Heilungkiang	46.3	1971
Wusu, Sinkiang	41.7	1975
Liaozhong, Liaoning	41.5	1974
Yitong, Kirin	40.0	1974
Yingkou, Liaoning	37.0	1974
Hsiyang, Shansi	36.4	1977
Shougang, Shansi	35.7	1977
Anhsiang, Hunan	34.5	1974
Hsinhua, Kiangsu	32.8	1974
Linhsien, Honan	29.2	1973
Wumei, Kansu	28.8	1974
Wuming, Guangsi	27.9	1973
Huanghsien, Shantung	26.6	1976
Qonggya, Tibet	24.4	1975
Hsiahsien, Anhwei	24.1	1975
Pingkou, Peking	21.2	1974
Pintin, Shansi	20.0	1976
Tsunhua, Hopeh	19.1	1975
Sougian, Kiangsu	18.8	1974
Hsinhua, Hopeh	15.0	1971
Tsinhai, Hopeh	11.5	1971
Weighted average	31.6[1]	

Sources *People's Daily*, various issues; and booklets published by the Agricultural Publishing House on the case studies of Tachai-type Counties during 1976–7.

Note 1. The total foodgrain output of these 22 Counties was 5,076,000 tons, and the total marketed foodgrain was 1,602,725 tons.

$$r_2 = k\left(1 - \frac{P_a}{P^*}\right)$$

$$= 0.046,$$

if P^* is assumed to be equal to P_e,

$$= 0.038,$$

if P^* is assumed to be 0.95 P_e,

$$= 0.054,$$

if P^* is assumed to be 1.05 P_e,

and

$$r_3 = q\left(1 - \frac{P_e}{P^*}\right)$$

$$= 0,$$

if P^* is assumed to be equal to P_e,

$$= 0.026,$$

if P^* is assumed to be 0.95 P_e,

$$= 0.024,$$

if P^* is assumed to be 1.05 P_e.

In a benchmark year during the mid-1970s the average tax rate was between 8.8 per cent (on the assumption that $P^* = 0.95\ P_e$) and 10.4 per cent (on the assumption that $P^* = 1.05\ P_e$). In the early 1950s direct agricultural taxation alone was 12 per cent of output. Thus, clearly, the average tax burden on agriculture today is much less than it was before.

However, once account is taken of the fact that the agricultural collectives directly finance a large part of rural investment out of their own savings, it must be recognised that the above tax rates represent a fairly high rate of contribution to the nation's capital accumulation. As discussed in Section 8.1, the rate of internal saving is probably around 20 per cent for most collectives. Thus, the total rate of surplus generation is probably close to 30 per cent of output. In few countries of the developing world does agriculture make a comparable contribution to capital formation.

The next important point to note is that the marginal rate of tax is much less than the average rate. It may actually be negative. Marginal tax being defined as the additional tax, directly levied or concealed in the procurement practices, that would have to be paid in any given year if output exceeds that in the recent benchmark year. Total tax revenue is:

$$T = T_1 + T_2 + T_3$$

$$= r_1P^*X_b + k(P^* - P_a)X_b + q(P^* - P_e)(X - X_b)$$

$$= r_1P^*X_b + k(P^* - P_a)X_b - q(P^* - P_e)X_b$$

$$+ \frac{q(P^* - P_e)}{P^*}V,$$

where $V = P^*X$ = 'Value' of actual output.[22] The marginal rate of tax (t') is:

$$t' = \frac{dT}{dV} = \frac{q(P^* - P_e)}{P^*} = q\left(1 - \frac{P_e}{P^*}\right) = r_3.^{23}$$

As we have shown above, under the assumption of a plausible range of value for P^*, the maximum value of r_3 would be 2.4 per cent – about a quarter of the average rate of tax. Under the plausible assumption that $P^* < P_e$, the marginal tax rate would be negative.

Thus, during the five-year period between any two dates of assessment a collective unit could pay a considerably lower average rate of tax by increasing its output. If a collective unit is able to increase its output by 25 per cent over that in the benchmark year, then it would pay an average rate of tax between 6.5 and 8.8 per cent as compared to the range of between 8.8 and 10.4 per cent in the benchmark year.

An intriguing question is whether a collective unit's current performance is affected by its expectation that a few years hence its taxes would go up if it performs better today. We do not have enough detailed information about the working of the system to answer such a question. All we shall note is that at the time of re-assessment a collective would preserve the advantage of exceeding the norm to the extent its performance is above that of the average for the area for which the norm is the same.

(e) The collectives' perception of the incentive system

The above analysis is a valid description of the incentive system in Chinese agriculture only if its major assumptions coincide with the perception of the collectives and the rural population. Specifically, we should consider whether their own expectation of the possible alternative price is roughly the same as the range indicated by us or not. If they expect that P^* would be much above P_e, then they would still be subjectively under a heavy rate of marginal tax.

We have stated above that the average retail distribution price of grain is

> virtually the same as the state's purchase price paid to the producers. All costs of storage and transportation and the losses in the course of handling by commercial departments are covered by a state subsidy which amounts to about 2.5 million *yuan* for every 50 million kilogrammes (50,000 metric tonnes) of grain.[24]

Similar claims have been made by numerous official and non-official sources in recent years. Indeed, some have claimed that in certain areas the retail price is below the procurement price.[25] Although these statements do not clearly say so, it appears that the reference is to compulsory procurement price (i.e., the above statements should be interpreted to mean that retail sale price is the same as compulsory procurement price).[26] Thus, the price on voluntary procurement would be 30 per cent higher than the retail distribution price. From available information it appears that the unsatisfied demand at the ruling retail price is modest. There is no evidence that the black market in foodgrains or other agricultural products is widespread. Under these circumstances it appears unlikely that the collectives would expect that the possible alternative price would be much higher than the voluntary procurement price. Indeed, it is quite likely that they would recognise an element of subsidy in the latter.

The argument will still hold, though less strongly, if the above statements are interpreted to mean that retail price is equal to the weighted average of voluntary and compulsory procurement prices. Since the weight of the amount compulsorily procured is very high (perhaps about 0.8), this price would be closer to that on voluntary procurement. Thus, the price on voluntary procurement would still be 23 per cent higher than the retail distribution price.

In the above discussion of collective incentives we have abstracted from many of the detailed features of the system. There are many additional incentives to promote collective effort in specific directions. It may be useful to give a few examples. There are tax exemptions to promote the development of new land. Agricultural tax is exempted in legally reclaimed areas for up to three years. The tax on waste land reclaimed by immigrant workers is exempted for three to five years (seven years in tea estates and forestry developed in mountainous areas).[27] Another source of collective incentives is that in addition to price premium the collectives which overfulfil the basic procurement target are awarded special ration coupons to purchase fertilisers and other essential goods.

8.5 Sectoral incentive

That the marginal tax rate is lower than the average tax rate is a positive factor in providing incentive for higher output. But this would hardly be enough if the collectives feel that they are under a crippling burden of average tax. We should, therefore, consider if the average tax burden on the sector as a whole was excessive or not.

The average tax rate on agriculture, both direct and concealed, is not excessively high today. As argued above, together they probably amount to no more than 10 per cent of the value of collective output. What is more important from the standpoint of sectoral incentive is that the tax burden on agriculture has been falling over time.

The evidence on the falling trend of direct tax has already been cited. In 1952 it was 12 per cent of farm output. By the mid-1970s it had declined to only 5 per cent.

To get some idea of the change in the concealed taxation of agriculture one might look at the evidence of changes in the prices of agricultural goods sold to the state relative to that in the prices paid by agriculture for its purchases from the rest of the economy. In Tables 8.7, 8.8 and 8.9 changes in some of these prices over more or less the last two decades have been shown. Generally, the prices of goods sold by agriculture have gone up while those of the agricultural inputs have declined over the periods for which information is available.

Table 8.10 shows general indices of the state purchase price of farm products, the agricultural sector's purchase price for manufactured goods and the agricultural sector's terms of trade. According to this evidence the state purchase price of farm products has been increasing rapidly over the last quarter century. Over the same period, especially since 1951, the prices of industrial goods sold to the agricultural sector remained virtually unchanged.[28] Thus, the extent of concealed taxation must also have come down very sharply.

The Chinese strategy in state trading and distribution of farm products seems to have been as follows: agricultural purchase prices have been allowed to rise sharply while the retail distribution prices of farm goods have been held constant. As a result, state trading in agricultural goods has gradually stopped producing a surplus and, in recent years, has resulted in a fairly large deficit. This deficit has been financed by a portion of the sharply rising profit in the non-agricultural sector where the rise in wages has been kept way below the rise in output per worker.

Table 8.7 Procurement Prices of Selected Farm and Side-line Products, Hanshou County, Hunan Province 1957–75, yuan/kg

Year	*Wheat*	*Ginned Cotton*	*Jute*	*Pork*	*Fish*	*Eggs*
1957	0.146	1.57	0.74	0.764	0.44	0.80
1965	0.228	1.77	0.84	0.920	0.58	1.24
1975	0.260	2.10	0.84	0.920	0.78	1.30

Source Hsiang Jung and Chin Chi-chu, 'Socialist Commerce: A Vast Rural Market', *Peking Review* Nos. 32–33, 1976, p. 26. It is not clear whether these refer to compulsory, average or premium procurement prices.

Table 8.8 Increases in Compulsory Procurement Prices of Selected Crops

Crops	*Reference Period*	*Price increase (%)*
Foodgrain[1]	1950–76	69.0
Rice	1956–72	19.5
Soyabean	1958–64	20.2
Sorghum	1958–71	145.3

Sources 'Special Report to FAO Conference', *Hsinhua News*, 30 October 1977. S. J. Burki, *A Study of Chinese Communes, 1965*, Harvard East Asian Monograph No. 29, 1969, p. 74: Etienne, op cit, p. 26 and S. Ishikawa, 'China's food and agriculture: A turning point', *Food Policy*, May 1977, p. 101.

Note 1. Refers to average procurement price.

Table 8.9 Retail Prices of Selected Industrial Goods for Farm Use in Hanshou County, Hunan Province 1957–75, yuan/kg

Year	*Chemical Fertilisers*	*Pesticides*	*Diesel Oil*
1957	0.42	1.35	0.264
1965	0.38	1.15	0.164
1974	0.29	1.04	0.164

Source Same as that of Table 8.7.

Table 8.10 Agricultural Sector's Terms of Trade (1952=100)

	(1) *State Purchase Price for Farm Products*	*(2)* *Prices of Manufactured Goods Sold in the Agricultural Sector*	*(3)* *=100 [(1)÷(2)] Agriculture's Terms of Trade*
1952	100.0	100.0	100.0
1953	110.1	98.5	111.8
1954	113.8	100.2	113.6
1956	116.6	100.4	116.1
1957	122.4	101.6	120.5
1958	125.1	101.0	123.9
1963	154.7	114.3	135.3
1971	156.2	104.0	150.2
1974	164.4	100.3	163.9

Source The calculations were made by Nicholas Lardy, 'Economic Planning and Income Distribution in China', *Current Scene*, November 1976. For 1952–8 he uses the data in *Ten Great Years* (Peking: Foreign Languages Press, 1960). For 1963, 1971 and 1974 his calculations are based on fragmentary references to percentage changes in state purchase price etc. between these and earlier years made in *Peking Review*, No. 47, 1964, No. 40, 1972 and No. 41, 1975. These must be regarded as approximate. It appears that in China a lot of attention is paid to the calculation of what is called 'the scissors' difference between the prices of industrial and farm products'. In *Peking Review* No. 41, 1975 (p. 9) it is claimed that such difference in 1974 'has been narrowed 45 per cent as compared with 1950'. Although references to 'scissors' difference are frequent, it is not possible to get a description of the method of how it is calculated.

8.6 Incentives and inequality

Without adequate information it is not possible to quantify the income distribution effects of the policy of incentives. It is, however, clear that the system of incentives tends to aggravate income differences between collective units of different productive capacity. In Table 8.11 some crude calculations have been made to show the addition to per capita collective income due to the premium on voluntary sales to the state in the 1970s in certain counties and in a prefecture and a commune. Such addition ranges from a negligible 2 *yuans* in a Shantung prefecture to 45 *yuans* in Wumei county. Clearly, the overall effect is to aggravate income inequality between collective units and regions.

Table 8.11 Additions to Per Capita Collective Income Derived from Price Premium for Excess Delivery

Locality	*Additional Collective Income* ('*000* yuan)	*Additional Per Capita Income* (yuan)	*Year of Reference*
Hsiyang County, Shansi Province	1950	10.33	1977
Huanghsien County, Shantung Province	2730	6.11	1976
Wumei County, Kansu Province	2940	4.55	1974
Pintin County, Shansi Province	997	3.66	1976
Huatung Commune, Guandong Province	164	3.11	1972
Chanwei Prefecture, Shantung Province	18300	2.03	1976

Source See Table 8.6.

Note The amount of additional collective income is obtained by multiplying the quantity of voluntary procurement by the difference between basic and voluntary procurement prices. This is then divided by the estimated size of agricultural population of the locality.

It is reasonable to wonder whether the preoccupation of the Chinese policy with the provision of incentives would not lead to polarisation and inequality. It may appear that from the standpoint of income distribution a more appropriate policy would be to keep the marginal tax rate higher than average and use the tax revenue to help the poorer collectives. But such has not been the orientation of Chinese policy.[29] The stated policy in the matter is to curtail the 'bourgeois right' in the realm of ownership by transferring the ownership of land and other assets gradually from the lower levels of collectives to the higher levels and, ultimately, to the whole people.[30] This is a slow political process which will take a long time, if ever, to be implemented. The ultimate success of this policy requires that the whole question of incentives be rendered irrelevant by the creation of 'new men and women' who are inspired by motives other than personal well-being. The advance towards this final goal will not be steady through time and one cannot rule out the possibility of an irreversible deviation away from this path at some future date.

In the meantime, however, the Chinese policy-makers seem to have a set of 'short-term' policies to reduce inequalities between collective units. There are specific types of state assistance for the poorer collectives. Financial assistance to the poorer units is being provided in the form of credit. The recruitment policy of the industrial sector has been aimed partly at relieving the extremely unfavourable land to man ratio in the poorer areas. Advanced units lend experts and model workers to poorer units.[31]

The emphasis on self-reliance is, however, overwhelming. The incentive system is designed to help such attempts at self-reliant growth. To the extent that unequal performance is due to unequal resources (e.g., better and more land) the average tax rates are different (since 'normal' output estimates take these differences into account). But the collective units retain the results of unequal performance due to harder work, greater ingenuity and thrift.

Notes

1. For a discussion of the incentive problems during this period, see ILO, *Poverty and Landlessness in Rural Asia* (Geneva, 1977), Chapter 12 and references therein, and Ng Gek-boo, *Operation and Control of Individual Economic Activities in Collective Agriculture: The Case of China* (Geneva: ILO, WEP Working Paper, 1978).
2. Estimates of Chinese population growth are difficult to arrive at with confidence. The present statement is based on Joint Economic Committee of the Congress, *China: A Reassessment of the Economy* (Washington, July 1975) which shows two different sets of population figures – one representing a high rate of growth (2.21 per cent per year compounded between 1954 and 1976) and the other representing a lower rate (1.76 per cent). The average of these two rates is 1.99 per cent.
3. Although the discussion above has been carried out with reference to grain, it should be noted that on the whole production increases in non-grain crops have been at least as high and, almost certainly, higher.
4. For detailed accounts of income distribution in rural China, see A. R. Khan, *The Distribution of Income in Rural China* (reprinted in ILO, op cit, in a revised form) and Ng Gek-boo, *Rural Inequalities and the Commune System in China*, both ILO WEP Working Papers (Geneva, 1976).
5. The comparison of yield per hectare of foodgrain between China and other countries should take into account the fact that the importance of foodgrain production in China in terms of its share in total arable land is fairly uniform among most collectives. Geographical specialisation is relatively less important in Chinese agricultural planning as compared with the other countries. The very poor performance of localities with

comparative disadvantages in foodgrain production have pulled the average yield level of the country as a whole down below a level where it would have been if, like other countries, China had abandoned grain cultivation in relatively uneconomic areas.

6. See A. R. Khan, op cit, for the evidence that accumulation and income were positively related among the 18 communes for which information was available for the mid-1960s.
7. Karl Marx, *Critique of the Gotha Programme* (Peking: Foreign Languages Press, 1972), p. 15. The deductions referred to by Marx are for replacement and accumulation of capital and public consumption.
8. Ibid.
9. Thus, consider the source of bias due to the treatment given the private plots in China. Incomes from private plots are not taxed. Nor are they subjected to deductions for collective accumulation and collective consumption. Thus, in the allocation of an individual household's labour between private and collective work there is a built-in bias in favour of the former. This would be so even if individual and collective incentives are each well formulated unless a method is found to subject the earnings from private plots to equivalent deductions (which is practically impossible to do). For details, see Ng Gek-boo, *Operation and Control of Individual Economic Activities in Collective Agriculture: The Case of China* (ILO, June 1978), WEP Working Paper.
10. These quotations, from Wei Min, 'China's Tax Policy' in *Peking Review*, No. 37 (1975), are slightly confusing: it is claimed that individuals engaged in farm production are taxed on their production, but at the same time it is clearly stated that personal incomes, including those from private plots and side lines, are not taxed. We shall assume that the reference to 'individuals engaged in farm production' is to the exceptional cases of people wholly or mainly engaged in private production outside the commune system.
11. Ibid. The period of tax re-assessment was extended to five years in 1958. See *Agricultural Taxation* (in Chinese), (Peking: People's Publishing House, 1958).
12. Ibid. The insertion within parentheses has been added by the present writers.
13. Wei Min, op cit, says that normal output is estimated 'on the basis of yields in a normal year assessed according to the natural conditions of the land and local farming conditions'. It is not known to us how the norm is decided at different levels, e.g., county, commune, brigade and team. It is probable that for each level the higher authority sets the norm which is then divided by the authority of the given level into norms for the collective units at the next lower level.
14. The following are some of the Chinese and Western sources which have recorded this phenomenon: Rewi Alley and Wilfred Burchett, *China: The Quality of Life* (Pelican Books, 1976), p. 26; Gilbert Etienne, *La Voie Chinoise* (Paris: Presses Universitaires de France, 1974), Annexe II; Kwangtung Provincial Branch of the People's Bank of China, 'Accounting in the Production Team of the Rural People's Commune (1974)', in *Chinese Economic Studies* (White Plains, New York,

Fall–Winter 1976–7). The voluntary procurement system was first introduced in November 1959 for foodgrain procurement.

15. See FAO, *Information Gathered on a Private Visit to China* (Rome: W/K6874); G. Etienne, *China's Agricultural Development* (Geneva: Asian Documentation and Research Centre, 1977), and Henle, *Report on China's Agriculture* (Rome: FAO, 1975).
16. Audrey Donnithorne, in her *China's Economic System* (London: Allen and Unwin, 1967), pp. 361–2, quotes the former Chairman of China's National Price Commission, Hsüeh Mu-chiao as having made the following statement in 1963 (in *Economic Research* [in Chinese], May 1963):

> in practice the state sometimes accumulates funds through the operation of price relationships, that is to say the state slightly lowers the procurement prices of agricultural products, and the margin between these prices and the 'values' of the products represents state accumulation.

On the question of the desirability of such concealed taxation Hsüeh says: 'these questions are still considered from different points of view and must be subject of further full discussion.'

17. This is a simplifying assumption which is almost certainly inaccurate. Since the alternative to sale to the state is retention for consumption by the members of the collective, a more realistic assumption would be that the marketed proportion increases with the balance of output over normal level. Note that voluntary procurement cannot be negative. Unless this restriction is employed the following equation will give a negative rate of tax if $X < X_b$. Thus, $q=0$ if $X < X_b$.
18. For example, consider the following statement: 'The average retail prices for grain staples in China today are . . . virtually the same as the State's purchase price paid to the producers.' See Peng Kuang-hsi, *Why China Has No Inflation* (Peking: Foreign Languages Press, 1976), p. 6. Peng Kuang-hsi's statement could be interpreted on the basis of the following arithmetic: the state buys 100 kilogrammes of paddy rice at the compulsory procurement price of 0.196 yuan per kilogramme so that the total cost is 19.6 yuan. After processing, 67 kilogrammes of cleaned rice is obtained. The retail price per kilogramme of cleaned rice is 0.30 yuan, so the total retail sale proceeds are 20.1 yuan which is almost the same as total procurement cost at compulsory procurement price. Thus, a subsidy is needed to cover the cost of processing, storage, transportation and marketing. Note that the argument that follows would remain valid even if we were to interpret Peng's statement as meaning that the retail sale price were as high as the weighted average of the voluntary and compulsory procurement prices (which would be closer to the latter because of its much greater weight in total procurement).
19. It may be asked why some kind of 'shadow' prices – for example, the world prices – should not be used to measure the alternative price P^*. As we shall argue below, the rationale behind the alternative price is that it should be perceived by the collective units. It is highly doubtful if the Chinese collectives have any perception of the so-called world prices (or,

more accurately, the *relative* world prices). Secondly, the world prices are supposed to measure the marginal import costs or the marginal export revenues. This would be the case if the country in question is marginal to world trade. China is not such a marginal case. Our final comment on the point is that the use of world prices (or of marginal import costs and export revenues calculated by applying the elasticities in world market to the world prices) as an approximation of P^* would almost certainly leave our qualitative conclusions unchanged.

20. FAO, op cit.
21. The term for t_1 has been shown as the ratio of real quantities in this equation. Unlike t_2 and t_3 it is independent of prices. The 'value' measure of t_1, obtained by multiplying both the denominator and the numerator by P^*, is exactly the same as the 'quantity' measure.
22. Unlike the equation in Section 8.4 above, T_1 is expressed here in terms of 'value' to make it comparable with T_2 and T_3. Hence the previous expression for T_1 is multiplied by P^*.
23. Since T_1 and T_2 are fixed, the marginal tax rate is simply

$$\frac{dT_3}{dV} = r_3.$$

24. Peng Kuang-hsi, op cit, p. 6.
25. Jan Prybyla cites the case that the procurement price of rice of a given kind was 0.156 yuan per catty as compared to the retail price of 0.14 yuan for the same kind. See Jan Prybyla, 'Note on Incomes and Prices in China', *Asian Survey* (March 1975), p. 272.
26. See note 18 above.
27. *Agricultural Taxation*, op cit.
28. To reconcile this with the information in Table 8.9 one must assume that the prices of goods other than those shown in Table 8.9 – mainly manufactured consumption goods purchased by the agricultural sector – must have increased during 1957–74 period.
29. Progressive taxation was used extensively in the 1950s until the setting up of the communes. As the rich peasant economy vanished after collectivisation, progressive taxation became a penalty against the efficient and hard-working collectives. One of the present writers argues elsewhere that progressive taxation and fiscal redistribution

   > would be in conflict with the ideological basis of the policy of restricting and finally abolishing the 'bourgeois right' in property ownership by lower level collectives. Progressive taxation is based on the concession of such rights and is aimed at reducing some of their extremely unfavourable effects.

   See A. R. Khan, *The Distribution of Income in Rural China* (Geneva, ILO, July 1976) WEP Working Paper.
30. In the wake of the current struggle against the 'Gang of Four', it is being clearly implied that the leadership considers the objective of curtailing the 'bourgeois right' a relatively distant one.
31. A. R. Khan, op cit and Ng Gek-boo (1976) op cit.

9 Collective Agriculture in Soviet Central Asia

Azizur Rahman Khan and Dharam Ghai

9.1 Introduction

The territory that now forms the Soviet Central Asian economic region was annexed by Tsarist Russia in the second half of the nineteenth century. The present national boundaries were drawn in the decade following the Soviet Revolution. Although the criteria of delimitation were linguistic and cultural each of the republics is in reality today a multinational entity. The region consists of four republics:[1] Uzbekistan (14.47 million people in 1977), Tajikistan (3.59 million), Kirghizia (3.44 million) and Turkmenistan (2.65 million). Together the four republics have a territory of 1.28 million square kilometres and a population (in 1977) of 24.2 million. The average population density of 19 per square kilometre is misleading. Vast parts of the region consist of desert or mountains where very few people live. But in the fertile river basin the density of population frequently exceeds 200 per square kilometre.

A few decades ago Central Asia was an overwhelmingly rural and extremely poor society. In recent decades it has undergone a remarkable transformation and the rate of progress has been remarkably high. Social institutions have been transformed in unprecedented proportions; the rural society of the region was catapulted from the middle ages into modern collective farms and state farms at one go. In contrast to many other parts of the world, the development of the rural economy in Central Asia was an integral part of the material advancement. As a result, the region continues to be predominantly rural. Thus it is interesting to analyse the Central Asian experience in agricultural and rural development in order to determine the factors that led to such a different performance in the region as compared to that in the immediate neighbouring areas.

The two main forms of agricultural organisation in the USSR are *Kolkhoz* (the collective farm) and *Sovkhoz* (the state farm). In a collective farm the assets are owned or held collectively by the members who provide labour and share among themselves the net earnings. In a state farm the assets are owned by the state and the labour force is paid wages just as in any industrial enterprise. Over time the proportion of the agricultural labour force in the *Sovkhozy* has been increasing and that in the *Kolkhozy* has been falling.[2] But this process has been slower in Central Asia than in the rest of the USSR. In 1976 Sovkhoz workers constituted 37 per cent of the total *Kolkhoz* and *Sovkhoz* workers in Uzbekistan as compared to 43 per cent in the USSR. Besides *Kolkhoz* and *Sovkhoz* there are certain other types (e.g., inter-*Kolkhoz* and/or *Sovkhoz* enterprises) of agricultural organisation, but they employ only a small proportion of the agricultural labour force.

The main focus of the study is on the collective farms – the *Kolkhoz* – to determine their achievements from the standpoint of productive efficiency, egalitarianism of income distribution, generation of surplus and promotion of employment and participation. The information on which the study has been based comes from two main sources: published official data and the limited amount of data generated by the authors during a field trip in October 1977. The field trip was made to the republics of Uzbekistan and Tajikistan, the two most populous republics of the region.[3] A considerable amount of detailed data have been collected for five collective farms that the authors visited in these republics. This is why in the study these two republics receive much greater attention than their other two Central Asian neighbours and frequent references are made to the five collective farms that the authors visited.

Gaps in statistical information will be only too obvious to the reader. On many interesting questions we have been able to do little more than raise questions or suggest hypotheses. To arrive at less tentative results, one will have to await the availability of information on a whole range of issues related to this experience.

In 1913 the region constituted only 4.6 per cent of the total population of the area that now forms the USSR. By 1977 the share had more than doubled to 9.4 per cent. About the time of the Revolution the rate of population growth in Central Asia was probably no higher than that in the USSR as a whole. Even by 1940 the Central Asian population growth rate was only a little higher than that for the rest of the USSR. But since then the demographic trends in Central Asia have diverged sharply from those in the rest of the Union. While the latter experienced a demographic transition through sharply reduced birth rates, the rate of population growth in the Central Asian republics continued to accelerate.

The main factor in this expansion has been the sharp reduction in death rate. At the time of the Revolution the death rate was probably similar to, or higher than, that in the present-day developing Asian countries. Even as recently as in 1940 the weighted average of the death rates in the four Central Asian republics was as high as 14.5 per thousand. By 1970 it had fallen dramatically to 6 per thousand, one of the lowest rates ever recorded in any country of the world. By 1977 the death rate in the region was slightly higher, 7.5 per thousand, but still significantly below the average for the USSR, Europe and North America. As regards the birth rate, the region is showing no signs of a demographic transition. The wighted average of birth rates in the four republics is 35.1 per thousand, about twice as high as in the USSR as a whole.[4]

The resulting natural rate of growth of population – 2.8 per cent per year in 1976 for the region as a whole – is one of the highest in the contemporary world and completely out of line as compared to nations with a similar living standard. The actual rate of growth in population has been even greater due to net immigration.

The high birth rate and population growth have important economic and social consequences. The age distribution is skewed relatively sharply in favour of children below working age. Consequently, the proportion of the population in the potential labour force (which is defined somewhat arbitrarily to include all persons between the ages 15 and 65 years) is low. Even when compared with the developing countries of South Asia, including India, the Central Asian republics appear to be at a disadvantage from the standpoint of the age-distribution of the population. The difference in the actual labour force tends to be even greater due to the limitations imposed on the participation of women in the labour force by high birth rates and large families.

The total effect of the above factors is to provide the Central Asian region with a much lower ratio of labour force to population as compared to the USSR (or, for that matter, any country that is comparable to Central Asia in terms of economic development). For every 100 persons the USSR has 45 per cent (56 per cent) more workers than does Uzbekistan (Tajikistan). As a consequence the Central Asian republics have a much higher dependency ratio than does the rest of the USSR. This is a very significant factor in understanding the economic circumstances in the Central Asian republics. Gains made in terms of income and productivity per worker in comparison with her past and with the rest of the USSR are partly offset by the unfavourable dependency ratio that obtains in this region.

On the eve of the Revolution the level of urbanisation in Central Asia was no lower than for the area that later became the USSR. Indeed the territory that now forms Uzbekistan was more urbanised than the then territories of any other republic except Georgia. During the period of Soviet power, Central Asia experienced a much lower rate of urbanisation than did the USSR as a whole. Rural–urban migration within the region was remarkably small.

By any ordinary standard the growth rates of industrial production in Central Asia have been impressive. But they have generally been lower than the industrial growth rates in the USSR as a whole. Table 9.1 compares national income, its sectoral composition and per capita income in Uzbekistan, the overwhelmingly largest Central Asian republic, with the corresponding measurements for the USSR during the decade since 1965. In both, the share of agriculture in national income declined rapidly,[5] but the rate of decline was significantly greater for the USSR than for Uzbekistan. By 1976 the percentage share of agriculture in national income in Uzbekistan was three-quarters higher than that in the USSR. Correspondingly, the share of industry in national income in Uzbekistan is lower than that in the USSR. A comparison between the rest of Central Asia and the USSR will give results which are qualitatively similar.

9.2 Trends in sectoral incomes

Table 9.1 shows that the per capita income in Uzbekistan in recent years has varied between 62 and 65 per cent of the per capita income for the USSR.[6] The disparity in terms of per capita consumption has been slightly lower. Tables 9.2 and 9.3 contain some additional comparative information about sectoral incomes.

Table 9.2 compares payment per man-day in *Kolkhoz* in Uzbekistan and Tajikistan with that in the USSR as a whole. The most striking feature is that by 1965 payment in Central Asia was more than a fifth higher than that in the USSR as a whole. Indeed, the data available from an alternate source[7] indicate that in 1958 payment per man-day in the Central Asian *Kolkhoz* was 70 per cent above that for the USSR as a whole.

Thus we have the important fact that around the mid-1950s the

Table 9.1 National Income, Its Composition and Per Capita Income and Consumption

Uzbekistan

	1965	*1970*	*1975*	*1976*
National income (current price million rubles)	5495.9	8702.5	12483.1	13191.6
% share of industry	37	34	38	40
% share of agriculture	38	37	32	30
% share of transport, construction trade and services	25	29	30	30
Per capita national income (current prices)	544	738	912	937
Per capita consumption	412	543	675	698

USSR

	1965	*1970*	*1975*	*1976*
National income (thousand million rubles)	193.5	289.9	363.3	382.0
% share of industry	52	51	53	53
% share of agriculture	23	22	17	17
% share of transport construction and services	25	27	30	30
Per capita national income	843	1199	1434	1495
Per capita consumption	611	833	1052	1095
Uzbek per capita national income as % of USSR's	64.5	61.6	63.6	62.7
Uzbek per capita consumption as % of USSR's	67.4	65.2	64.2	63.7

Source *Narodnoe Khozyaistvo Uzbekskoi SSR Za 60 Let* (hereinafter referred to as *Nk Uz. 60*) and *Narodnoe Khozyaistvo SSSR Za 60 Let* (hereinafter referred to as *NK SU 60*).

Kolkhoz workers' collective earnings in Central Asia had reached a very high level in relation to that for the USSR. Unfortunately, we do not have the comparable data for the earlier years, but according to all available evidence the Central Asian agriculture in the early 1930s was less prosperous than the agriculture in most parts of the

Table 9.2 Payment per Man-day in *Kolkhoz*

	All USSR (rubles)	*Uzbekistan (rubles)*	*Tajikistan (rubles)*	*Uzbekistan as % of USSR*	*Tajikistan as % of USSR*
1965	2.68	3.29	3.21	123	120
1970	3.90	4.24	4.17	109	107
1972	—	—	4.10	—	—
1973	—	—	4.37	—	—
1975	4.54	4.60	—	101	—
1976	4.77	4.96	—	104	—

Source *NK SU 60*; *NK Uz. 60*; and *Sovietski Tajikistan Za 50 Let* (hereinafter referred to as *ST 50*).

USSR. It would be reasonable to assume, therefore, that the disparity in *Kolkhoz* payments between Central Asia and the rest of the USSR started to move in favour of the former after these early years.

Since the middle of the 1950s, the disparity between Central Asia and the USSR in terms of the earnings of the *Kolkhoz* workers has been narrowing. By the mid-1970s the disparity virtually disappeared.

In Table 9.3 the earnings disparity in state agriculture (predominantly *Sovkhoz*) is shown. Once again, in the early years (e.g., in

Table 9.3 Monthly Payment in Industry and State Agriculture

Monthly Payment to Employees in Uzbekistan as % of that in USSR			*Monthly Payment to Employees in Tajikistan as % of that in USSR*		
	Industry	*State Agriculture*		*Industry*	*State Agriculture*
1940	81	104	1965	89	100
1965	90	92	1970	90	88
1970	93	97			
1975	94	96			
1976	92	94			

Source These calculations are based on wage data shown in: *NK SU 60*, (p. 472) for the USSR, *NK Uz. 60* (p. 217) for Uzbekistan and *ST 50* (p. 190) for Tajikistan.

Note 'State agriculture' refers to *Sovkhoz* and other (numerically less important) state enterprises in the agricultural sector.

1940) the monthly payment in Central Asia was higher. In more recent years the disparity has been reversed in favour of the rest of the USSR. One important fact to note is that the regional disparity in payment in state agriculture is smaller than that in *Kolkhoz* payment.

When one looks at the regional disparity in industrial earnings, one finds a very different trend. In the earlier years earnings in the Central Asian industries were relatively low as compared to the rest of the USSR. Over the years this disparity has narrowed, but even in the 1970s earnings in the Central Asian industries remain significantly lower.

In summary, by 1970 the collective earnings per worker from *Kolkhoz* in Central Asia were still somewhat higher than in the rest of the USSR, although the disparity had come down from about 70 per cent in 1958 to less than 10 per cent. The earnings in state agriculture had by 1970 already become higher in the rest of the USSR than in Central Asia. Since earnings in Sovkhoz are higher than those in *Kolkhoz* and since the USSR has a higher share of *Sovkhoz* in total agriculture as compared to Central Asia the weighted average earning per agricultural worker in Central Asia was probably only a shade higher than that in the rest of the USSR.[8] But the dependency ratio in rural Central Asia was much above that in the USSR as a whole. As a consequence per capita agricultural income in Central Asia was much lower than that in the USSR.[9] Earnings per industrial worker were lower in Central Asia and this disadvantage was further aggravated by the unfavourable dependency ratio so that the per capita industrial income in Central Asia was much lower than that in the USSR. To this one should add the fact that the USSR has a higher share of industry – the high income sector – in total economic activity. The total effect of all these factors is to reduce the per capita national income and per capita consumption in Central Asia to about two-thirds of the corresponding quantities in the USSR as a whole.[10]

9.3 An exception to the strategy of primitive socialist accumulation

In development literature the Soviet experience is frequently characterised as the strategy of 'primitive socialist accumulation' – extracting large surplus from the agricultural sector by imposing on it unfavourable terms of trade through the use of the monopsonistic power of the state as the buyer of agricultural goods and the monopolistic power of the state as the seller of non-agricultural goods. Indeed, the use of monopsonistic power was supplemented by the imposition of compulsory delivery quotas on rural producers. The main justification for the strategy was the lack of an alternative source of capital to finance socialist industrialisation in a pre-capitalist country in which socialist revolution had taken place.[11]

It appears that the main Central Asian cash crop, cotton, was no exception to this strategy in the very early years. The price offered to the growers was very low and, as a consequence, output fell. It was not too difficult, especially once collectivisation got under way, to increase the area under cotton and thereby raise production to the pre-revolution peak and beyond. But the disincentive of low price was too strong an impediment to promote the necessary effort to restore the yield per hectare to anything like the pre-revolution level. Yields continued to decline and by 1932 output per hectare was 35 per cent below the level of 1913.

For a period attempts were made to compensate the cotton growers by offering them grain, tea, sugar, seeds and fertiliser at fixed prices, but in January 1935 these deliveries at fixed prices were discontinued and a major shift in policy was initiated by nearly quadrupling the procurement price of cotton. The result, in terms of production and yield, was dramatic: by 1937, in Uzbekistan, the production of raw cotton was 1522 thousand tons – nearly three times the pre-revolution peak – and yield per hectare was at an all-time high of 1.6 tons. The result in the rest of Central Asia was similar.

For the next two decades cotton retained this extraordinarily favourable position in comparison with the other major agricultural products, which continued to be subjected to 'non-equivalent exchange' through the imposition of compulsory delivery quotas at extremely low prices. For the 1930s and 1940s actual procurement prices for agricultural goods are not available to us. Information about such prices is available from 1952 onwards. In 1952 relative procurement prices of the major agricultural goods still retained the same general pattern as they did over the two preceding decades. Thus, an examination of the relative procurement prices in 1952 should give us a general impression of the relative advantage enjoyed by cotton during the period under review.

In 1952 the procurement price per ton of cotton was nearly 37 times that for grains. No estimates of relative costs of production in those days are available, but in recent years the average cost of production per ton of cotton was about seven times that for grain.[12] It is possible that, depending on whether techniques and relative input prices were different in the earlier period, the ratio of costs was different in those years, but such a difference could not have been very great. Thus net return per ton was many times higher for cotton than for grain. Indeed, the net return on grain was negative and that on cotton highly positive.

Another way to look at relative prices is to compare them with the prevailing international prices. Without implying that the prices in international trade were in some sense optimal one could argue that such a comparison shows the extent to which the state monopoly of external trade and the system of procurement have entailed asymmetrical degrees of income transfer from the producers of different agricultural products.

Such comparisons are, of course, extremely difficult. Since our main purpose is to look at the *relative* ratios of procurement prices to international prices we need not worry too much about the appropriateness of the official rate of exchange which has been used throughout. But the problem of the comparability of quality is a serious one. Another point is that the procurement prices should properly be compared with the growers' prices elsewhere. Instead we have compared them with the wholesale or *cif* export prices in major producing or trading centres.

And yet the results are so strikingly different for different goods that one is left in no doubt about the direction of bias in official policy. For grains and meat the average procurement price paid to the producers in the USSR in 1952 was less than one-seventh of the international price. For cotton the procurement price paid was nearly a third above the international price.[13] Thus relative to the production of other major agricultural goods (namely, grains and meat) the production of cotton was being subsidised (or taxed far less heavily).[14]

The first important effect of the exceptionally favourable terms of trade for cotton was the sharp shift of sown area away from grain into cotton. In the pre-revolution days grain was the main crop in Central Asia both in terms of area and physical volume of production. In 1913 about 75 per cent of the sown land in Central Asia consisted of grain and only 15 per cent of cotton. By 1965 the share of grain had fallen to 41 per cent and that of cotton has risen to 36 per cent. What is even more important to note is that most of the irrigated land came to be diverted to cotton while grain was gradually shifted to marginal and dry land. In terms of physical weight grain fell from 2.82 times of cotton in 1913 to only 0.39 times in 1961–5. Between these time periods cotton production increased sevenfold while grain production declined in absolute terms.

This remarkable shift in cropping pattern was the result not of coercion or administrative fiat but of a systematic use of price incentives.[15] This constitutes an outstanding exception to the strategy of squeezing out surplus from agriculture that is usually attributed to the USSR.

The labour requirement per hectare is on the average six times higher for cotton than for grain in *Kolkhoz* in Uzbekistan. For *Sovkhoz* (which have generally a higher degree of mechanisation and a lower degree of labour-intensity as compared to *Kolkhoz*) the difference in labour requirement between cotton and grain is even greater. In the rest of Central Asia the orders of magnitude are roughly the same.

The labour requirement for grain cultivation has fallen quite dramatically over the last eleven years: in Uzbekistan it fell by more than 50 per cent while in the rest of the USSR the decline was as much as 70 per cent. In comparison, there has been a very slow reduction in the labour intensity of cotton cultivation. Over the corresponding period it fell by about 12 per cent in Uzbekistan and by about 13 per cent in the rest of Central Asia. Thus the gap in the labour requirement between cotton and the other crops has been increasing over time.[16]

It is clear, therefore, that the increased specialisation in cotton has led to a much greater demand for labour in agriculture in Central Asia than would be the case otherwise. This is probably one of the most important factors contributing to the lower rate of urbanisation for the Central Asian republics over recent decades. Cotton cultivation continues to absorb such a high proportion of the labour force that despite its rapid growth the proportion of the labour force engaged in the non-agricultural sectors increased only slowly.

In order to make the increased demand for labour effective, it was necessary to offer workers a high enough income to induce them to stay on the farm. It is here that the favourable terms of trade for cotton played a crucial role. After the mid-1930s agricultural incomes grew very fast and the income payments, especially to the *Kolkhoz* workers, were high enough to render the pull of wages offered by the urban industries relatively weak.

We do not have information about the differences in sectoral wages in the earlier years but Table 9.4 summarises some information for 1965 and 1975. As has already been noted, the difference between Central Asia and the rest of the USSR in terms of agricultural income had already narrowed substantially by 1965. And yet it is revealing that the difference in earnings between urban industries and rural *Kolkhoz* was so much lower in Central Asia than in the rest of the USSR, even at as recent a date as 1965.

Table 9.4 A Comparison among Incomes in Industry, *Kolkhoz* and State Agriculture: 1965 and 1975

	Industry Incomes Divided by Kolkhoz Income		*State Agriculture Incomes Divided by Kolkhoz Income*		*Industry Incomes Divided by State Agriculture Income*	
	1965	*1975*	*1965*	*1975*	*1965*	*1975*
USSR	1.56	1.43	1.12	1.12	1.39	1.28
Uzbekistan	1.14	1.32	0.84	1.05	1.36	1.26
Tajikistan	1.16	1.19	0.93	0.87	1.25	1.36

Note For industry and state agriculture monthly wages have been used. For *Kolkhoz* payment per man-day has been multiplied by 25. In reality a *Kolkhoz* worker works less than 25 days a month but this number has been used due to the fact that in industry a worker works for about 25 days a month. For state agriculture (mainly *Sovkhoz*) the monthly payment is presumably also standardised for a full month of approximately 25 days. The sources are the same as for the data in Table 9.2.

9.4 Changes since 1953

Beginning in 1953 the policy of extremely low procurement prices for agricultural goods began to change. Over the next decade the prices of grains, meat and milk were increased steadily and rapidly and the rise in the prices of livestock products continued at a rapid rate to date. By 1962 grain prices had risen more than eight-fold over the 1952 level. The procurement prices for meat and milk increased, respectively, by more than 15 and four times over the same period. During this period of spectacular rises in the prices of other goods the procurement price of cotton remained virtually unchanged.

By 1962 the procurement price per ton was only 4.76 times higher for cotton than for grains. The cost of production per ton of cotton is 7.42 times that of grains for the USSR as a whole. Thus the price/cost ratio had altered sharply in favour of grains.[17]

As a result of the sharply lowered relative incentive for cotton, its production and yield ceased to grow as rapidly as in the earlier decades. In many areas production and yield fell between the early 1950s and early 1960s. The effect on peasant income was severe. During these years the *Sovkhoz* specialising in cotton were making substantial losses. It has also been claimed that incomes on the Uzbek collective farms declined by 17 per cent between 1957 and 1960.

After unsuccessfully trying out various administrative methods of persuading the cotton growers to produce more, the procurement price was increased substantially in 1963 after a long period of stagnation. Thereafter prices were increased in a number of further steps and such increases have generally been at a higher rate than that for grains, the main competing crop. By 1976 some of the relative price advantage lost over the 1953–62 period was restored. Especially in the Central Asian republics the relative procurement prices of cotton and grain were carefully balanced to keep the price to cost ratio higher for cotton.[18] Thus, the ratios of procurement prices of cotton and grains in 1976 were 3.73 and 4.13 respectively for Uzbekistan and Tajikistan. The ratios of costs of production at *Kolkhoz* were respectively 3.53 and 3.62.[19]

A comparison between procurement prices and the prices in the international market in 1976 at the official rate of exchange shows that while in 1952 grains were paid only 13 per cent of the international price, in 1976 grains received a little more than the price in the world market. The position of cotton remained the same as in 1952. Thus cotton still received higher protection (or was subjected to a lower rate of concealed taxation depending on how different the appropriate rate of exchange was from the official one) than grains.

But, clearly, cotton had lost the extraordinarily privileged position it had had until 1953. This is reflected in the comparatively slower rate of expansion in cotton production in recent decades. In both Uzbekistan and Tajikistan the rates of growth of cotton production have been declining steadily over time. In Uzbekistan the growth rate in the output of grain has been substantial over the last decade – nearly twice as high as for cotton – as compared to the negative or negligible rates of growth in the preceding four decades.

Several aspects of the system of procurement pricing deserve attention. First, it is an instrument for providing collective incentives. There is a basic price for the given quota that the collectives are expected to fulfil and a 50 per cent premium on the quantity that a collective voluntarily sells to the state in excess of the quota.[20] This system provides an incentive to exceed quota sales by obtaining an appropriate expansion in output. Secondly, procurement pricing is an instrument for the reduction of inter-regional inequality. The procurement price of any product is not uniform all over the USSR, or even within each republic. For grain, for example, there are 17 procurement zones in the USSR. The lowest price is offered to the Krasnodar zone where natural conditions are best and the cost of production lowest. The highest price is paid to the producers in the Far Eastern zone where natural conditions are worst and cost highest. For cotton there are three procurement zones within the Uzbek Republic alone. Price is lowest for the Ferghana-Tashkent zone where conditions are best and highest and in the semi-arid Karakalpak-Bokhara zone where conditions are worst.[21]

As is well known, there is no land rent in the USSR. From what is known about the principle of differentiating procurement prices between regions it appears that this practice is a substitute for land rent in preventing polarisation in regional income inequality arising out of vast regional differences in the quality of land. Had procurement prices been uniform, the differential rent of land in a fertile region would have augmented the income of the workers of the region. This is a benefit from which the workers in a less fertile region would be excluded. Differential procurement pricing is a mechanism designed to siphon off differential rents. In principle, it could serve as a generalised tool of regional income equalisation to offset the extremes of inequality arising out of differences in resource endowment. It should be noted that translating into practice what is possible in principle is by no means an easy task. To set procurement prices for many products in many regions and yet not distort relative incentives away from the socially desirable pattern must be a very difficult task.

9.5 Productive efficiency

(a) The success story of cotton

Soviet agricultural performance is frequently indicted by critics for low factor productivity. This critique does not apply to cotton, the overwhelmingly dominant product of Central Asian agriculture, so long as output per hectare is used as the indicator.[22] By all standards the achievement in terms of output per hectare has been remarkable. In comparison with a pre-revolutionary level of 1.22 tons (and only about 0.8 ton at the beginning of the 1930s) the output of raw cotton per hectare reached the level of three tons in 1976. Table 9.5 shows that the Central Asian republics have the highest yield per hectare of all the significant producers of cotton in the world. Even within the USSR these republics, which supply the bulk of the cotton, are far above the rest in terms of yield per unit of land.

Table 9.5 Major Cotton Producers of the World (1976)

Country	*Total Production (thousand tons of lint equivalent)*	*Yield per hectare (tons of raw cotton)*
USSR	2800	2.82
China	2400	1.47
USA	2298	1.35
(Uzbekistan)	(1800)	(3.00)
India	1146	0.46
Pakistan	515	0.84
Turkey	470	2.10
Brazil	390	0.66
Egypt	386	1.95
(Tajikistan)	(288)	(3.00)
(Turkmenistan)	(356)	(2.13)
(Kirghizia)	(71)	(2.88)
Total World	12695	1.16

Note Uzbek, Tajik and USSR data have been quoted from *NK SU 60*. Data for the other countries have been compiled from FAO, *Production Yearbook*. Vol. 30 (Rome 1977).

For the world as a whole the ratio of lint to raw cotton is about 0.35. For the USSR the ratio is 0.34. These ratios have been used to convert raw cotton into lint (and *vice versa*) in deriving the figures shown in the table. In 1976 only three countries had a higher yield per hectare than the USSR. These were Guatemala (3.35 tons), Israel (3.21 tons) and Sri Lanka (3.00 tons). Each of them is a small producer. Their outputs, in lint equivalent, were respectively 100, 50 and 2000 tons.

In the five collective farms that the authors visited the yield per hectare was, on the average, higher than that for the republics. But these farms were not far out of line with the average for the republics.[23]

(b) Relative performance of Kolkhoz and Sovkhoz in growing cotton

Table 9.6 contains some information about the relative performance of *Kolkhoz* and *Sovkhoz* in the production of cotton. The performance of the collective farms is respectively 29 and 26 per cent better in Uzbekistan and Tajikistan in terms of output per unit of land as compared with the state farms. In terms of the officially reported unit cost of production, a concept whose content is not clearly defined, the collective farms have a 5 per cent advantage in Uzbekistan and a 4 per cent disadvantage in Tajikistan in comparison with the state farms. Direct labour cost is higher in the collective farms, as compared with the state farms, by a quarter or more.

Table 9.6 Yield and Cost of Cotton in *Kolkhoz* and *Sovkhoz* in Uzbekistan and Tajikistan

	Uzbekistan		*Tajikistan*	
	Kolkhoz	*Sovkhoz*	*Kolkhoz*	*Sovkhoz*
Output per hectare (tons of raw cotton)	3.30	2.56	3.08	2.45
Cost of production (rubles per ton)	417	441	467	450
Direct labour cost (Man hours per ton)	330	260	400	320

Note Figures showing yield per hectare in Tajikistan are for 1973. All the other figures are for 1976. *Sovkhoz* figures include a small amount produced by 'other state enterprises'. Cost of production and direct labour cost are quoted from *NK SU 60.* The definition of these concepts, beyond what is implied by the terminology employed, is not spelled out in any detail. Outputs per hectare have been estimated from the data shown in *NK Uz. 60* and *ST 50.*

Table 9.7 contains some facts about the relative levels of mechanisation of the collective and state farms. In terms of tractorisation (i.e., tractors per unit of land) the collective farms are significantly more mechanised than the state farms in the two republics. In both the republics the level of tractorisation is considerably above that for the USSR as a whole, where an average collective farm had 11

Table 9.7 Mechanisation of *Kolkhoz* and *Sovkhoz* in Uzbekistan and Tajikistan

	Uzbekistan (1976)		*Tajikistan (1973)*	
	Kolkhoz	*Sovkhoz*	*Kolkhoz*	*Sovkhoz*
No. of tractors per thousand hectares of sown land	47	30	34	30
Cotton harvesters per thousand hectares of cotton land	15	20	10	13
No. of workers per tractor	13	10	16	11

Source The estimates have been based on the data in ***NK Uz. 60*** and ***ST 50.***

tractors per thousand hectares of sown land in 1976. What is even more surprising is that the overall rate of tractorisation in Central Asian agriculture would appear to have been greater than that in the highly capital-intensive agriculture in the USA, since in 1972 in the USA there were 33 tractors per thousand hectares of cropland.[24] Whether this comparison means an effectively higher rate of tractorisation of agriculture (especially of the collective farms) or merely a less efficient rate of use of these machines in Central Asia, as compared to the USA, is something that can be determined only after a good deal of additional information becomes available.

Although the *Sovkhozy*, as compared with the *Kolkhozy*, have fewer tractors per unit of land they have more tractors per worker. Also, in terms of tractors per worker, an average *Kolkhoz* in the USSR, with 14 farm workers per tractor, is no less capital-intensive than their Central Asian counterparts. Agriculture as a whole in the USSR, with 10 workers per tractor, is somewhat more capital-intensive than that in Central Asia by this criterion. In the agriculture of the USA in 1972 there was one tractor per worker.[25]

The crucial advantage that the *Sovkhozy* have over the *Kolkhozy* in terms of the mechanisation of cotton growing is their access to about a third more of harvesters per hectare. In terms of harvesters *per worker* the *Sovkhozy* have an even greater advantage over the *Kolkhozy*. This must be the major explanation for the much lower direct labour cost of producing cotton in the *Sovkhozy*. Manual cotton picking is still widespread at the *Kolkhozy* and this is a highly labour-intensive process.[26]

One can summarise the relative performance of the *Kolkhozy* and the *Sovkhozy* in growing cotton in Central Asia in the following way: (i) Output per unit of land is unambiguously higher on the *Kolkhozy* by more than a quarter on average. (ii) Output per unit of labour is about 25 per cent higher in the *Sovkhoz*. (iii) The evidence about cost of production is less unambiguous, but the weighted average cost would appear to be somewhat lower for the *Kolkhozy*. (iv) The greater labour-productivity in the *Sovkhozy* is probably explained largely by their greater capital-intensity (measured as the amount of capital equipment per worker) and the more balanced composition of capital equipment (as reflected in the relative availability of tractors and harvesters). (v) The higher output per unit of land at the *Kolkhozy* has been achieved at the cost of employing considerably more labour per hectare. One may tentatively conclude that the lower output per worker at the *Kolkhozy* is a reflection of diminishing returns arising out of the more intensive application of labour and a somewhat unbalanced composition of capital equipment (in the form of too few harvesters and, perhaps, too many tractors).

What can one say about the relative efficiency of the two forms of organisation in growing cotton? Not much in the absence of more information. If prices were accurate reflections of social scarcity and if costs were comprehensively accounted for then one could argue that the lower cost of production at the *Kolkhozy* is testimony to their greater efficiency. It is, unfortunately, difficult to argue that either of these conditions is fulfilled.

Again, if one knew that land was relatively scarce and labour relatively abundant, then the performance of the *Kolkhozy* would appear to be superior. There are powerful considerations to suggest that such may indeed be the case. In spite of its vast geographical expanse Central Asia is in short supply of cultivable land. In 1976, in Uzbekistan, there were only 1.6 hectares of cultivated land per *Kolkhoz* worker. This was less than a quarter of the corresponding figure for the USSR. Indeed, this is less than the amount of land available per agricultural worker in Pakistan. In Tajikistan the amount of cultivated land per *Kolkhoz* worker is 1.9 hectares – only a little higher than in Uzbekistan. Although actual figures are not available, it appears that an augmentation in the supply of cultivated land is very costly in terms of resources. Expensive irrigation projects have to be undertaken to bring new land under cultivation.

On the other hand, one could argue that the supply of labour probably is less of a constraint at present. The rate of population growth has been accelerating and, so far, there is no sign of deceleration in growth. Moreover, the region has a positive rate of net immigration. Thus, in years to come, the rate of increase in the labour supply is likely to continue to accelerate.

If the above characterisation of relative factor supplies were approximately correct, one could justifiably claim that the *Kolkhozy* are more efficient than the *Sovkhozy*. It is, however, uncertain how accurate the above characterisation is. A possible alternative hypothesis is that the existing allocation of labour is not dictated by conditions of supply in relation to production possibilities but by the lack of adequate supply of farm equipment. Agriculture in Central Asia generally uses more labour than agriculture in the USSR as a whole. Compared to the leading agriculture in the world in terms of labour-productivity, that of the United States, labour input in Central Asian agriculture is exceptionally high. But the available evidence shows that the Central Asian collective farms have a *relatively* higher labour productivity in cotton than in grain or livestock when their performance is expressed as indices of those achieved by the Soviet or US agriculture. Thus, compared with the average in the US during the mid-1960s, Uzbek collective farms a decade later needed 19 times as much labour per unit of grain and 10 times per unit of cattle but only 6 times per unit of cotton. The ratios are broadly similar for Tajikistan.

It is only natural that Central Asian agriculture, having a resource endowment that is very different from the average for the USSR and the USA, would use much more of labour per unit of land and thereby having a lower output per unit of labour, but to ensure continued progress in the living standard of the rural population it will be necessary to improve output per worker by reducing labour requirements per unit of cotton and of other agricultural produce.

(c) Grain crops

So far the discussion of productive efficiency has been centred around cotton. This is natural in view of the overwhelming importance of this crop in Central Asia. It is, however, of some interest to have a quick look at the performance in grain crops. These are, admittedly, grown in many areas on relatively inferior land.

Table 9.8 shows their yield per hectare. In every republic of Central Asia there has been a very large increase in output of grain per hectare in recent decades. The rate of increase has been phenomenal in recent years. Much of this growth must have been the result of the policy of improving incentives for grain cultivation. During the 1960s yields in the Central Asian republics, with the exception of Kirghizia, ranged between 60 and 70 per cent of those in the USSR as a whole. By 1976, the average yield in Central Asia was above that for the USSR, and Tajikistan – a minor producer contributing less than eight per cent of the region's grain output – was the only Central Asian republic that lagged significantly behind the all Union average. This was a remarkable achievement.

Table 9.8 Grain Yield per Hectare (in Metric Tons)

	1940	*Average 1961–5*	*Average 1966–70*	*Average 1971–5*	*1976*
USSR	0.86	1.02	1.37	1.47	1.75
Central Asia Average	—	—	—	—	1.82
Uzbekistan	0.41	0.67	0.74	0.95	1.69
Tajikistan	0.57	0.61	0.66	0.83	1.14
Kirghizia	0.76	1.02	1.56	1.87	2.28
Turkmenistan	0.57	0.72	0.86	1.40	2.10
All Asia	—	—	—	—	1.69
India	—	—	—	—	1.22
Iran	—	—	—	—	1.27
Iraq	—	—	—	—	0.97
Pakistan	—	—	—	—	1.44
Turkey	—	—	—	—	1.89
Afghanistan	—	—	—	—	1.31

Source Soviet data quoted from: *NK SU 60*. The rest quoted from FAO, *Production Yearbook*, Vol. 30, 1977.

Productivity per hectare in Central Asia still is well below that achieved by Japan (5.28 tons per hectare), Korea (3 in North and 4.5 in South), Europe (3.1) and North-Central America (3.1). But compared to its more immediate Asian neighbours the performance must be rated as highly satisfactory.

We do not have separate data about productivity of grain per hectare in the *Kolkhozy* and the *Sovkhozy*. Table 9.9 shows unit costs of production and direct labour inputs into grain in these two forms of organisation. The general pattern is the same as in the case of cotton: cost of production is significantly lower (Turkmenistan is

Table 9.9 Relative Performance of *Kolkhoz* and *Sovkhoz* in Grain Production

	Cost of Production: (rubles per ton)		*Direct Labour Input: (man-hours per ton)*	
	Kolkhoz	Sovkhoz	Kolkhoz	Sovkhoz
Uzbekistan	118	149	74	43
Tajikistan	129	139	88	91
Kirghizia	62	72	21	25
Turkmenistan	120	103	58	27

Source *NK SU 60.* The figures are for 1976.

the only exception) while direct labour input is higher (Tajikistan and Kirghizia are the exceptions, but each by a relatively small margin) on the *Kolkhozy*. It would appear that the lower cost of production on the *Kolkhozy*, despite a much higher (in the case of Tajikistan and Kirghizia marginally lower) labour input, is probably the result of higher output per hectare. Thus one is tempted to conclude that all that was said about the relative efficiency of the *Kolkhozy* and the *Sovkhozy* in connection with cotton would apply also to their relative performance in grain production although, in the absence of direct measurements of output per hectare in the two systems, one cannot be certain about this.

9.6 Collective agriculture and personal plots

The issue of personal plots has been a controversial one in Soviet agricultural development. The performance of this sector has been the basis of criticism levelled against the Soviet system both from the right and the left. Western economists have long argued that personal plots have a much higher output per hectare than collective agriculture. This, according to such economists, is undisputed testimony to the greater efficiency of private, as compared to collective, agriculture. The critique from the left has been based on the argument that the continuation and prosperity of personal plots is inconsistent with the avowed objective of a transition to communism – a society in which distribution should be based on need and not on unequal entitlements based on unequal capacities to produce.

The Soviet answer to these criticisms has been to claim that the produce of personal plots, as a proportion of total marketed farm output, has been declining and that by now this sector is not of much quantitative significance. Unfortunately, the scanty time-series data published by the official sources refer to the USSR as a whole.[27] Our discussion of Central Asia must be based on the small amount of data that we generated for the five *Kolkhozy* and some comparative figures of an aggregative nature for the republics. These data are presented in the appendix table.

Personal plots in the Central Asian *Kolkhoz* are on the average much smaller than the ones in the rest of the USSR. The average sizes in Uzbekistan and Tajikistan are respectively 0.12 and 0.13 hectare as compared to 0.33 in the USSR. This is probably due to the relatively adverse land to man ratio in Central Asian agriculture – a fact to which reference has already been made. The average size of a personal plot in the five *Kolkhoz* we examined is 0.14 hectares, which is marginally higher than the corresponding averages for the two republics.

If the data from the five Kolkhozy are representative, the contribution of personal plots to family income certainly is not small. Income from a personal plot, including self-consumption, varies rather widely, namely, from 18–40 per cent of the total household income from collective *Kolkhoz* labour, earnings from employment outside and income from personal plot (excluding items such as imputed rental value of owner-occupied house and collective consumption). There is a perfect negative rank correlation between this ratio and family income from collective *Kolkhoz* labour. Although the 'sample' is very small, one is tempted to note a strong negative association between high collective prosperity and low income from personal plots.

The value of output per hectare in a private plot is, on the average, more than four times as much as the value of output on collectively farmed land. There is a lot of variation even for our very small 'sample', however. This is a ratio that needs to be explained and understood clearly.

The produce of collective farms is largely sold to the state at fixed prices. The state in turn sells these goods at controlled prices to urban and rural buyers through state retail outlets. The produce of private plots does not have to follow the same route but can be taken by the *Kolkhozniki* to the '*Kolkhoz Markets*' in nearby towns. At these markets prices are freely determined by the interaction between buyers and sellers. It is well known that prices at these *Kolkhoz markets* are higher than the prices ruling at the state shops, often located on the same premises, frequently by as much as one to two hundred per cent. Thus, even if output per hectare were the same on collective and personal, the *value* of output, defined in this case as the actual sale proceeds, would be greater for the personal plots.

How can one explain the difference in prices? One possibility is a shortage and consequent rationing at the state retail shops. This probably is responsible for a large part of the price difference between the state shops and *Kolkhoz markets*. The difference in revenue due to this factor must be seen as the result of a concealed tax on collective production by the state procurement machinery – a tax that is passed on to the consumers as a subsidy through the state retail shops.

But part of the difference in price is due to the inherent difficulty encountered by a centralised procurement and marketing agency in dealing with certain goods of a perishable nature. Thus, one is struck by the remarkable difference in quality for vegetables, fruits and similar items between the state shops and *Kolkhoz markets*. The price difference is highest for these products.[28] Sometimes the high revenue is due to the non-competitive nature of the goods, e.g., home processed food, sold by the *Kolkhozniki*.[29]

To the extent that the difference in revenue is due to such factors one must treat the higher revenue from personal plots as a return to the *Kolkhozniki* labour as traders and manufacturers. Indeed, there seems to be a tendency on the part of the personal plots to specialise in the goods to which such activities add considerably to retail prices. Thus, in the case of the five collective farms, there seemed to exist a pattern of specialisation in items such as vegetables and fruits.

It must be recognised that even if the factors noted above were absent there would be an incentive for the *Kolkhozniki* to apply greater effort to personal plots. This is because the value added by collective labour is subject to deductions for taxation, collective accumulation and provision for social services. The weighted average of the ratios of labour payment to value added in the five collective farms was 0.69. Thus, even if an individual *Kolkhozniki* has no additional uncertainty about the process of collective production, he would allocate his effort between collective and personal plots in such a way that the product of the marginal unit of labour was lower in the latter. As a result labour input and volume of output would tend to be greater per unit of personal plot than per unit of collective land.[30]

9.7 The distribution of income

(a) The degree of inequality among collective farms

Even given the limited variation in the natural conditions of the five collective farms that the authors visited, a good deal of inequality in average incomes and earnings can be observed. The basic information has been summarised in Table 9.10. Payment per man day from work in the collective farm in the richest *Kolkhoz* is about three-quarters above that in the poorest of the five. The difference in earnings per worker is even greater due to the greater average intensity of employment in the richer farms. The difference further widens in terms of per capita collective income because the richer farms are able to employ or attract a higher proportion of the available family labour. The difference in terms of per capita income from all sources is smaller because earnings from non-*Kolkhoz* employment and personal plots exert a strong equalising influence.

For the Central Asian region as a whole the difference in average earnings and incomes is likely to be greater among the collective farms. In Uzbekistan payment per man day in an average *Kolkhoz* in 1976 was 4.96 rubles. The lowest such payment in any *Kolkhoz* was reported to be 3.50 rubles while the highest known rate was 11.00 rubles, the ratio of the highest to the lowest being 3.14.[31]

Another way of indirectly measuring inequality among collective farms is to look at the differential in productivity. Table 9.11 shows the distribution of the Uzbek Kolkhoz according to the yield of cotton, frequently the major determinant of collective income. Unfortunately, the size distribution of *Kolkhoz* is not shown and the highest class interval (3.5 tons per hectare and above) is too large to permit a comparison between, say, the top and bottom deciles. But it is quite clear that differences are substantial.

Not enough quantitative information is available to analyse the impact of state policy on the *inter-Kolkhoz* income differentials. But, on the whole, such policies may have reduced such inequality. There are at least three instruments which have substantially promoted this objective. The first of these is the procurement price which has been regionally varied to mop up the 'differential rent'

Table 9.10 Earnings and Incomes in Five Collective Farms (rubles)

Name of the Kolkhoz	*Average Payment Per Man-Day*	*Average Annual Earning from Collective Labour*	*Per Capita Annual Income from Collective Labour*	*Per Capita Annual Income from Personal Plot*	*Per Capita Annual Earning from Outside Employment*	*Per Capita Total Annual Income*
Leninism	7.00	1287	386	102	81	569
Kholkabad	6.15	1187	321	103	118	542
Karl Marx	5.49	783	237	137	89	463
XXII Party Congress	4.07	721	170	138	87	395
Rossiya	4.02	681	69	155	168	392

Note These data have been reproduced or calculated from Appendix Table. The last column includes income from *Kolkhoz* labour, outside employment and personal plots.

Table 9.11 Distribution of Uzbek *Kolkhoz* According to the Yield of Cotton (tons per hectare), 1976

Yield	*Percentage of* Kolkhoz
Below 2.0	0.9
2.0–2.5	7.3
2.5–3.0	21.8
3.0–3.5	34.0
3.5 and above	36.0

Source *NKUz. 60.*

(and perhaps a part of the cost advantage due to the favourable endowment of resources other than land) in the regions with more favourable conditions and thereby to compensate for the cost disadvantage of the regions with unfavourable conditions.

The details of the operation of the system of taxation are not available but it is known that the collective farms with low profits (below 15 per cent) are exempted from taxation. Thus of the five collective farms for which detailed information is available *Kolkhoz* XXII Party Congress paid no tax in 1976 because of low profits.[32]

The final instrument in equalising income among collectives is the contribution of the state to the collective consumption and pensions of the *Kolkhozniki.* These contributions are distributed far more equally than are the incomes of the collectives.

(b) Intra-Kolkhoz inequality

Each *Kolkhoz* has a guaranteed minimum pay. For 25 days of work such pay in most *Kolkhoz* is 70 rubles per month, although in some it is only 60 rubles. Since not all *Kolkhozniki* work a minimum of 25 days on collective labour the *actual* minimum payment per month may be less than this guaranteed minimum. But those who work fewer than 25 days a month on the farm are usually able to find work for the remaining days, frequently at a wage higher than the minimum guaranteed by the *Kolkhoz.*[33] Thus, in effect, no *Kolkhoznik* appears to receive a money income which is less than the guaranteed minimum. Such an income probably is adequate in most cases to secure the basic human needs given that the unfavourable dependency ratio (one worker for 2.5 to 3 members) is partly compensated by the payment of family allowances (described later), that housing is provided at low cost, that personal plots add considerably to the income of the family and that adequate provision for health and education are made by the *Kolkhoz* and the state.

It is difficult to generalise about differences in income among various categories of employees of a *Kolkhoz*. Even among the five collective farms studied, the pattern seems to vary quite a bit.

One may distinguish three different categories of *Kolkhoz* employees. At the top are the Chairman, members of the *Kolkhoz* Board and specialists such as Chief Agronomist, Chief Economist and Chief Book-keeper (some of these specialists are members of the *Kolkhoz* Board). These employees are on a fixed monthly salary. In the middle category are the technically skilled workers such as tractor drivers. In terms of income, middle level officials such as ordinary economists and agronomists and the brigadiers appear to be at the same level, although these officials are paid fixed monthly incomes. At the bottom are the 'field workers', the production workers with little or no specialised skill. These workers are all paid on the basis of piece-rate work.

Little is known about the dispersion of earnings among field workers. It is known, however, that there are six grades of such workers and that there are significant differences between the grades. Some idea of such differences is conveyed by the fact that the *average* earning of a field worker is about twice as high as the *minimum* earning (see Appendix Table).

In the three Tajik collective farms the earnings of one belonging to the middle category – a tractor driver or a brigadier – are one and three-quarters to twice as high as that of a fully employed field worker. The ratio appears to be much lower in the two Uzbek collective farms.

The Chief Specialists are relatively highly paid. The Chairman, frequently himself a specialist, is the only official who receives a salary higher than that of the Chief Specialists. The Chief Specialists are paid more than twice as much as a fully employed field worker. The Chairman's salary is three to four times that of a fully employed field worker. But the true income of a chairman is not fully reflected in the salary. Frequently he or she is provided with a rent-free house of generous proportions and, in the case of the prosperous collectives, personal cars. In some of the collectives the authors studied such privileges were also given to the Chief Specialists.

On the whole one is left with the impression that there is considerable, but not excessive, inequality in the distribution of earnings. Indeed, it is frequently admitted in discussions that differences in remuneration are kept sufficiently large to promote incentives.

(c) Personal plots and income distribution

Incomes from personal plots in a *Kolkhoz* must be fairly equitably distributed. This is because the size of the personal plots is based on the criterion of allocating roughly the same amount per *person*. Thus the system may be compared with one of peasant farming with an absolutely egalitarian per capita distribution of land, a ban on the hiring of labour and a compulsory quota on each household for collective labour. It is not surprising that the resulting distribution of income is highly egalitarian, although complete equality is ruled out by the inter-family differences in quality and quantity of labour available to work on personal plots.

Between collective farms, as we have already observed on the basis of our small 'sample' of five, the rank correlation of per capita earnings from personal plots is almost perfectly negatively correlated with per capita collective income. If this is representative of the region as a whole, then the contribution of the personal plots to *inter-Kolkhoz* income distribution must be one of equalisation. There are reasons to suggest that this, indeed, is the case. There is very limited variation in the average size of personal plots between collective farms. The range of feasible technologies would also be limited in view of the very small size of the plots. The lower the return from collective labour the greater will be the incentive to work the personal plots intensively. All these factors would tend to combine to make the effect of personal plots an equalising one as far as the *inter-Kolkhoz* distribution of income is concerned.

It may seem paradoxical that the distribution of income in the non-socialised part of the *Kolkhoz* is possibly more egalitarian than in the socialised part. The distribution of income in the socialised sector, *in principle*, is proportional to the individual member's capacity to work. Individuals differ in terms of their capacity. In the socialised sector their differences in capacity are expressed fully as individuals work with the relatively large amounts of socially owned capital and resources. Thus the resulting distribution of income can be as unequal as the distribution of the ability of individuals.

In the non-socialised sector there are such severe limitations on the volume of the means of production per person that the differences between the capacity to work of individuals cannot be fully translated into differential results of work. As a consequence, the distribution of income can be less unequal than that of ability to work. Thus, the result surmised by us on the basis of highly inadequate evidence is not unexpected. What must be noted is that the equalising influence of earnings from personal plots is due to the strictly egalitarian limitation on the size of such plots – a limitation that arises from the fact that most means of production in the economy are socially owned.

9.8 Accumulation

We have discussed the fact that Central Asian agriculture in the past had been subjected to a very low rate of surplus extraction compared to the rest of the USSR. Direct taxation of *Kolkhoz* income in Central Asia continues to be low. The weighted average rate of income tax in the five *Kolkhozy* was 4.1 per cent of value added in 1976. The highest rate of 6.6 per cent obtained in the prosperous *Kolkhoz* Kholkabad while *Kolkhoz* XXII Party Congress, because of its low rate of profit, paid no taxes at all.

Whether the terms of trade between collective agriculture and the state is a mechanism that represents a concealed tax or not is very difficult to say. To establish criteria of quantifying such an extraction of surplus is extremely difficult. Even if one could agree about such principles the inadequacy of available information would make it impossible to apply them. The burden of our preceding discussion, however, has been to show that the prices received by the products of collective agriculture in Central Asia probably represent no substantial degree of concealed taxation. Whether such taxation is incorporated in the prices of the products sold to agriculture is unknown.

Whatever the amount the state extracts on the aggregate must be balanced against a substantial state contribution to the collective consumption and investment of the *Kolkhoz.* The contribution to collective consumption takes the form of payments for teachers, doctors and staff at the *Kolkhoz* schools and hospitals and a major share of the costs of pensions and family support allowances. The contribution to investment consists of the financing of large-scale irrigation and land improvement projects.

As far as we could determine, it is not the usual practice to charge the users for water or other benefits of such state-financed projects. Yet another form of state contribution is cheap long-term credit to finance capital investment. The rate of interest on such credit is only

1 per cent while the repayment period can be as long as 20 years.

On the whole collective agriculture appears to be making a fairly low, and perhaps zero, aggregate transfer to the state. The inadequacy of information on which the conclusion is based must be emphasised yet again, however. But the available evidence does not suggest that the situation is one of 'primitive socialist accumulation' imposed on agriculture.

The information that we have about the savings and investment behaviour of collective agriculture refers only to the five *Kolkhozy*. The weighted averages of the relevant magnitudes for the five *Kolkhozy* were as follows for 1976 (all percentages of value added):

Total investment	21.1
Increase in reserve	3.1
Internal saving	20.4
Gross long term borrowing from credit institutions	3.8

The rate of investment is quite high, especially when one considers that the data do not include any large-scale development projects undertaken by the state. The internal saving rate also is very high, viz. over 20 per cent of value added. While three of the five *Kolkhozy* borrowed from the financial institutions, the weighted average of *net* borrowing for all five *Kolkhozy* was less than one per cent of value added.

Behind the average picture depicted above there was considerable variation among the five collective farms. The saving rate ranged from 10 per cent to 30 per cent of value added and the rate of investment from 14 per cent to 41 per cent (see Appendix Table). Over a single year variations of this magnitude are to be expected for enterprises engaged in such an inherently volatile activity as agriculture. What would be useful to know is whether the weighted average for the five *Kolkhozy* is a reasonable approximation of the behaviour of collective agriculture in Central Asia. This, unfortunately, we do not know.

9.9 Some concluding observations

The Soviet Central Asian republics were a backward and poor region as recently as four decades ago with a living standard not qualitatively different from that of its immediate neighbours such as Afghanistan. Within a relatively short period these republics achieved a remarkable transformation. By 1976 per capita income in Uzbekistan, the largest of the republics, was above current US $1300 at the official rate of exchange. Income per capita for the region as a whole was not much different. Even recognising the problems of the rate of exchange and differences in the method of accounting one cannot escape the conclusion that the standard of living in Central Asia today is much higher than that in most of the neighbouring Asian countries. The comparison is even more favourable to the Soviet Central Asian republics when one examines the provision of collective consumption and the social services.

The transformation in Central Asia has taken place during a period of a high and accelerating rate of growth of population. The high population growth not only raised the required rate of growth in overall income for a given rate of growth in per capita income but also raised the dependency ratio and made it necessary to devote a greater proportion of resources to education, child care and various related services.

Prosperity in the Central Asian republics has been achieved not through the classical path of industrialisation but through rural development. The industrial progress of the region has been very substantial, but rapid growth in agriculture has been a key element in the progress of the republics. A distinctive and related feature of their experience has been the continued predominance of the rural sector. While the region was not behind (and indeed Uzbekistan, the leading republic, was ahead of) the rest of the USSR in terms of urbanisation immediately before the revolution, the region remains predominantly rural and now is far behind the rest of the USSR in terms of urbanisation. The lower rate of urbanisation has not implied a lower rate of growth: the experience has not been the well-known one of stagnation and failure to industrialise. What the Central Asian republics experienced was rapid agricultural growth leading to a rising standard of living in the rural areas and the consequent absence of a push out of the rural sector. The rising living standards in rural areas also made it possible to keep a low urban-rural income differential – a factor which weakened the pull of the urban areas. Thus Central Asia experienced economic development with rural development, and in this sense it deserves to be studied carefully by those countries which cannot hope to develop through the classical type of industrialisation.

The chief explanation for the material progress of agriculture was the specialisation in cotton. This was characterised both by a very high rate of increase in yield per hectare and by a growing area; starting from a low level, the region soon surpassed all other significant producers of cotton in the world in terms of output per hectare.

The specialisation and prosperity in cotton were brought about through very high price incentives. While collectivised agriculture in much of the rest of the USSR was squeezed by a policy of 'primitive socialist accumulation', cotton was not subjected to adverse terms of trade. For the two decades starting the middle of the 1930s, cotton enjoyed extraordinarily favourable terms of trade in comparison to the other major products of Soviet agriculture. As a result Central Asia became increasingly more specialised in cotton and its agricultural income increased very rapidly.

The truly remarkable nature of the Central Asian agricultural growth is fully revealed only in the context of the vast institutional change that preceded it – the kind of institutional change that is known frequently to have produced at least temporary disruptions in output elsewhere. In less than a decade Central Asia experienced a transition from backward feudal relations to institutions such as collective and state farms with only a very brief interregnum during which private peasant ownership existed. The achievement is even more remarkable in that the rural population did not undergo a long political struggle which would have prepared them ideologically for collectivation. And yet the people were able to adapt to these collective institutions relatively quickly and achieve a rapid growth in output. A proper and complete explanation of the phenomenon is beyond the scope of this paper. However, available evidence suggests that production gains did not immediately follow collectivisation. There was a period of hesitation and decline in yield until policies towards agriculture changed from discrimination to encouragement through incentives. The exceptional experience of Central Asia probably demonstrates that advanced forms of organisation such as the collectives can be introduced successfully only if the overall policies provide appropriate incentives and rewards. That so many experiences of collective agriculture in the last half century have failed to serve as vehicles of agricultural growth is probably due more to inappropriate overall policies towards agriculture than to any inherent difficulties with collective institutions.

The collective farms have been the main form of organisation in Central Asian agriculture. Over the years the state farms have increased their share of total farm households, but to this day the collective farms cover nearly three quarters more of the farm households than do state farms. Compared to the rest of the USSR, agriculture in Central Asia has a greater preponderance of the collective farms. While no conclusive generalisations can be made on the question, available evidence provides no basis for the conclusion that the performance of an average collective farm has been inferior to that of an average state farm. Indeed, under certain plausible assumptions about relative factor scarcities, the opposite may be true.

Collective agriculture has facilitated growth with a relatively egalitarian distribution of income. Poverty as it exists in much of Asia is unknown. The relative egalitarianism in the distribution of income has been reinforced further by the provision of social consumption which is perhaps even more equally distributed.

All this is not to deny the existence of significant inequalities in the distribution of earnings. The spread between real earnings of the top management and the earnings of the field workers would appear to be quite large. Similarly, skill differentials seem to be considerable. Such differentials are officially justified by the need to provide material incentives for work and higher productivity.

While the productivity achievements of the system have been truly remarkable in terms of output per hectare of land (which, contrary to an impression of a vast expanse of territory, is a relatively scarce factor), output per worker continues to be very low compared to the agriculture in advanced countries. There are also some important questions concerning the efficiency of utilisation of scarce capital inputs such as tractors. For further material progress the collective agriculture of Central Asia must overcome these problems and challenges.

Notes

1. There are 15 republics in the Union of Soviet Socialist Republics (USSR). Administratively, each republic is divided into a number of *oblasts* (provinces) which in turn are divided into *raions* (districts). Sometimes a republic contains one or more Autonomous Soviet Socialist Republics (ASSRs) each of them consisting of a separate nationality that deserves special status. Sometimes *oblasts* are accorded autonomous status.
2. *Kolkhozy* and *Sovkhozy* are plurals of *Kolkhoz* and *Sovkhoz* respectively. Similarly, *Kolkhozniki* is the plural form of *Kolkhoznik* (which means *Kolkhoz* member).

3. These two republics respectively account for 60 and 15 per cent of the population of the entire Central Asian economic region.
4. Tajikistan, Uzbekistan, Turkmenistan and Kirghizia (in that order) have the highest ranks among the 15 Soviet republics in terms of birth rate and population growth. For sources of all these demographic data see A. R. Khan and D. P. Ghai, *Collective Agriculture and Rural Development in Soviet Central Asia* (London: Macmillan, 1979).
5. As we shall see later, the terms of trade for agriculture improved significantly over the decade. Since the sectoral shares are derived from the current price estimates this means that the decline in the share of agricultural value added *at constant price* was greater than is shown by the figures.
6. It is not very satisfactory to speculate about a trend on the basis of only four observations. But the last two rows of Table 9.2 suggest that the per capita income and consumption disparities between Uzbekistan and the USSR have widened a little between 1965 and 1976. We hope that the discussion in the following pages will make it clear that in our view the disparity in per capita income between Central Asia and the USSR
   (a) became reduced gradually over the two decades leading to the mid-1950s; and
   (b) may have increased slightly thereafter.
7. *Novy Mir,* 10/59 quoted in Alec Nove and J. A. Newth, *The Soviet Middle East* (London: George Allen and Unwin, 1967) p. 103.
8. Thus, let us compare Uzbekistan with the USSR. In 1970 *Kolkhoz* payment per day in Uzbekistan was 4.24 rubles and that in the USSR was 3.90 rubles. *Sovkhoz* payment per day was 4.43 rubles in the USSR and 4.30 rubles in Uzbekistan (assuming that the income ratio of 0.97 in state agriculture applies to *Sovkhoz* income ratios of Uzbekistan to the USSR). The share of *Sovkhoz* was 0.35 for the USSR and 0.28 for Uzbekistan. Thus the weighted average agricultural earning was 4.25 rubles per day in Uzbekistan – only 4 per cent above the 4.09 rubles in the USSR. In these calculations it has been assumed that *Kolkhoz* and *Sovkhoz* make up all the agriculture. Even if allowance were made for the small amount of the other forms of agricultural organisations, the above calculations would remain pretty much unchanged.
9. Thus in 1970 the average number of family members per earning member in a *Kolkhoz* was 3.93 in Uzbekistan and 2.80 in the USSR as a whole. Therefore, although the earning per day per *Kolkhoz* worker was 9 per cent higher there, income per capita from *Kolkhoz* in Uzbekistan was 22 per cent lower than in all USSR.
10. Per capita earnings from personal plots are probably lower in Central Asia than in the rest of the USSR. Average size of the personal plot per household is 2.36 times higher in USSR than in Central Asia where the household size is considerably larger than in the rest of the USSR. See the table on the profile of five *Kolkhoz* in the Appendix for comparative data on the size of personal plots.
11. Perhaps the most cogent statement of the doctrine that was to be implemented later is contained in the writings of E. Preobrazhensky (*The New Economics,* Clarendon, 1965). He forcefully argued, in this book first published in 1924, that to develop a socialist economy in the Soviet

Union, resources must be extracted from the then non-socialist parts of the economy – the petty bourgeois and, above all, the peasantry. He describes the methods of transferring resources from the then private peasantry to the state. Besides considering taxation, profits from foreign trade and borrowing, he proposed the method of 'non-equivalent exchange' as the most powerful instrument to effect such transfer. Such non-equivalent exchange between the town and the country would constitute the manipulation of the prices offered by the state to the peasantry and the prices charged by it on the sale of industrial goods to the peasantry. For an account of this theory and how it became a reality in the Soviet Union a few years after the publication of Preobrazhensky's book see (in addition to Preobrazhensky's own work): M. Lewin, *Russian Peasants and Soviet Power* (London: George Allen and Unwin, 1967) and A. Ehrlich, *The Soviet Industrialisation Debate* (Harvard Universty Press, 1967). Stalin's own account of the justification for the implementation of such a policy is to be found in: J. V. Stalin, 'The Right Deviation in the C.P.S.U.(B)' in *Problems of Leninism* (Moscow: Foreign Languages Publishing House, 1953).

12. This is the all USSR ratio for 1976. For Central Asia the ratio was lower.
13. Meat, milk, grains and cotton (in that order) are the four biggest items of state procurement in terms of value in recent years and together they account for nearly three-quarters of the total state payments for procurement. In the earlier years the ordering in value terms was different for obvious reasons.
14. It should be mentioned that whether cotton was actually subsidised depends on the appropriateness of the exchange rate. If in 1952 the appropriate exchange rate was such that a ruble should have exchanged for 24 per cent fewer dollars than it officially did, then the procurement price on cotton would have been the same as the international price at the appropriate exchange rate. Whatever the appropriate exchange rate the rate of concealed taxation (or subsidy) on cotton was way below (above) that on grains and meat. It should be reiterated that what we call 'concealed tax' is the difference between actual income and the income that would accrue if the producers could sell internationally at the going price. One can easily think of reasons why such a situation would be undesirable for the nation to adopt: this would imply acceptance of the existing international specialisation as the optimal solution. There is also no stipulation that the average prices (themselves subject to many errors mentioned above) would remain the same if a major producer/consumer such as the USSR were to buy or sell more internationally. In spite of all these qualifications it is quite clear that, however correct or incorrect the interventions were, the effect was asymmetrical on the incomes of the producers of cotton as compared to other agricultural producers.
15. To make this argument there is no need to go so far as to argue that in the absence of the price incentives the state would have failed to achieve an expansion in cotton acreage by other means. The evidence in the preceding section shows, however, that such other means in the 1930s failed to secure an increase in cotton acreage with increasing yield.
16. The trend in labour requirements for sugar beet, potato and vegetables

is similar to that for grains. The relative labour intensity of cotton, especially in harvesting, becomes obvious to even a casual observer of the Central Asian rural scene in the autumn. The present authors witnessed a great deal of cotton picking by hand. All the officials emphasised that the target of increasing the share of mechanical picking was a very important one. In some places the present authors were told that there were technical problems of designing harvesters that would pick high quality cotton without waste.

17. The cost ratios of cotton to grain in the Central Asian republics are as follows:

Uzbekistan	3.53
Tajikistan	3.62
Kirghizia	5.69
Turkmenistan	4.03

But the grain prices are higher in these republics than in the USSR on the average. We have no information on the grain procurement prices in the republics in 1962, but it is quite certain that the price/cost ratio in these places had also become more favourable for grain by 1962.

18. As already hinted at, procurement prices differ from one region to another. The main reason for such variation is to cover differences in cost due to differential land rent arising out of differences in land quality. Since the payment of land rent does not exist in the USSR this device provides a useful instrument to prevent inequalities of income due to differential rent. The point will be discussed in greater detail later.

19. *NK SU 60*, p. 355. It is not clear how costs are calculated. Presumably labour costs are included. Since labour payment is partly dependent on prices received, the argument that prices are set to cover costs involves an element of circularity.

20. At the Central Asian Institute of Scientific Research on Agricultural Economy the authors were told that this system was instituted in March 1965 to replace a previous system of incentive payments. At present cotton, wheat, rice, meat, wool and milk are covered by such a system of incentive payments.

21. These facts emerged through the discussion the authors had with the staff members of the Central Asian Institute of Scientific Research on Agricultural Economy, Tashkent.

22. We recognise that a comparison of efficiency must be based on a more general measurement and not on the output per unit of a single factor. But to arrive at more general measures we need to quantify relative social scarcities of different factors. On this we have few guidelines either for Central Asia or for the countries with which comparisons are to be made. Later in this chapter we shall try to guess which factors in Central Asia are relatively more scarce and, on that basis, arrive at some tentative judgments on the question of efficiency.

23. The average yield in the collective farms of the republics was about 3.27 tons and that in the five farms 3.42 tons, i.e. 5 per cent higher. The data for the five farms are summarised in the Appendix Table at the end of the paper.

24. The US figures have been calculated from CIA, *USSR Agriculture Atlas* (Washington, DC) 1974.

25. In terms of horsepower an average Soviet tractor is bigger than an American one.
26. In some of the collective farms visited by the authors manual harvesting of cotton was widespread. To give an example, *Kholkhoz* Kholkabad in Samarkhand Oblast, Uzbekistan has more than 18 cotton harvesters per thousand hectares of cotton land. This is nearly 40 per cent above the average for the republic. Still only 63 per cent of the cotton at this *Kolkhoz* was harvested mechanically. The proportion must be lower for an average *Kolkhoz*. The authors were actually able to observe large numbers of *Kolkhozniki*, frequently women, picking cotton.
27. All these USSR data do not provide convincing evidence of the decline in personal plots. Compare, for example, the data shown in Morosov, *Soviet Agriculture* (Moscow, 1977). The contribution of personal plots to marketed output of all crops declined from 13 per cent in 1940 to 11 per cent in 1960 and 10 per cent in 1975. In view of the steady decline in the labour intensity of agriculture over the period, such a minor reduction is probably quite consistent with a fairly rapid increase in the output of personal plots per household. In meat, milk and poultry the share declined more sharply. But these are the sectors in which collective agriculture experienced a rapid acceleration in growth in recent decades and newly available technology is frequently more suitable to large-scale production.
28. At the huge Tashkent *Kolkhoz market* the authors saw tomatoes selling at 12 kopecs per kg. at the state shop (located inside the market) and at 60–70 kopecs at the stalls set up by the *Kolkhozniki*. The quality difference was striking. For cucumber the price difference was 50 per cent and the difference in quality was less striking.
29. At the Dushanbe *Kolkhoz market* the authors saw the *Kolkhozniki* selling a wide variety of home-made goods, e.g., honey, jam and jelly.
30. There is a considerable theoretical literature concerning the allocation of labour between collective and personal land in a Soviet type *Kolkhoz* under alternative assumptions about the behaviour of the entities involved. The interested reader is referred to E. D. Domar, 'The Soviet Collective Farm as a Producer's Co-operative', *American Economic Review* (September 1966), Walter Oi and Elizabeth Clayton, 'A Peasant's View of a Soviet Collective Farm', *American Economic Review* (March 1968) and Michael Bradley, 'Incentives and Labour Supply on Soviet Collective Farms' in *Canadian Journal of Economics* (August 1971).
31. These figures were supplied to the authors by the *Kolkhoz* department of the Ministry of Agriculture, Uzbekistan.
32. Note, however, that tying taxes to profits may easily produce a perverse result by inducing the *Kolkhoz* to pay high incomes to workers and report low profit. *Kolkhoz* XXII Party Congress paid more than 85 per cent of value added to the *Kolkhozniki*. This is way above the average of 66 per cent obtaining in the other four collective farms.
33. As the authors found, it is usual for the *Kolkhoz* management to organise such work in slack seasons. They provide transport to take 'surplus' workers to construction sites and generally try to take advantage of lucrative outside work by making internal adjustments.

Appendix Table A Profile of Five Soviet Central Asian Collective Farms with Some Comparative Data for the Republics and all USSR

	Karl Marx (Tajikistan)	*Rossiya (Tajikistan)*	*XXII Party Congress (Tajikistan)*	*Leninism (Uzbekistan)*	*Kholkabad (Uzbekistan)*	*Average of the Five* Kolkhoz	*Average for all* Kolkhoz *in Uzbekistan 1976*	*Average for all* Kolkhoz *in Tajikistan 1973*	*Average for all* Kolkhoz *in USSR 1976*
No. of households	2338	1420	1891	854	1045	1510	866	861	486
Average size of a household	6.84	7.75	6.35	5.90	6.77	6.77	—	—	—
Total no. of workers:									
working in *Kolkhoz*	4833	1104	2838	1515	1911	2440	1110	1027	534
working outside *Kolkhoz*	1513	2260	1210	265	587	1167	—	—	—
No. of earners per household:									
working in *Kolkhoz*	2.07	0.78	1.50	1.77	1.83	1.61	—	—	—
working outside	0.65	1.59	0.64	0.31	0.56	0.77	—	—	—
Total	2.72	2.37	2.14	2.08	2.39	2.38	—	—	—
No. of earners per household in standard employment intensity (250 man-days per year):									
working in *Kolkhoz*	1.18	0.53	1.06	1.30	1.41	1.07	—	—	—
Female workers as per cent of total workers in *Kolkhoz*	48.8	44.4	43.1	56.2	51.8	48.0	—	—	—

Appendix Table (continued)

	Karl Marx (Tajiki-stan)	*Rossiya (Tajiki-stan)*	*XII Party Congress (Uzbeki-stan)*	*Leninism (Uzbeki-stan)*	*Kholkabad (Uzbeki-stan)*	*Average of the Five* Kolkhoz	*Average for all* Kholkhoz *in Uzbeki-stan 1976*	*Average for all* Kholkhoz *in Tajiki-stan 1973*	*Average for all* Kolkhoz *in USSR 1976*
Average no. of man-days worked by a *Kholkhoz* worker:	142.6	169.4	177.2	183.8	193.0	166.3	252.6	251.8	247.9
(Male)	—	—	—	(215.9)	(202.2)	—	—	—	—
(Female)	—	—	—	(158.9)	(184.5)	—	—	—	—
Area of collective sown land[1] (hectares)	3736	1400	3639	2617	3152	2909	1777	1950	3539
Area under personal plots (ha)[2]	309	240	245	146	128	214	97500	27100	43000
Collective land per *Kolkhoz* worker (in ha)	0.77	1.27	1.28	1.73	1.65	1.19	1.60	1.90	6.63
Average personal plot per household (in ha)	0.13	0.17	0.13	0.17	0.12	0.14	0.12	0.13	0.33
Hectares of *total* sown land:									
per household	1.73	1.15	2.05	3.24	3.14	2.06	2.17	2.39	7.61
per person	0.25	0.15	0.32	0.55	0.46	0.30	—	—	—
Per cent of collective land under cotton	60.4	43.0	41.8	45.9	76.2	54.9	63.3	40.0	—

continued overleaf

Appendix Table (continued)

	Karl Marx (Tajikistan)	*Rossiya (Tajikistan)*	*XXII Party Congress (Tajikistan)*	*Leninism (Uzbekistan)*	*Kholkabad (Uzbekistan)*	*Average of the Five* Kolkhoz	*Average for all* Kolkhoz *in* Uzbeki*stan 1976*	*Average for all* Kolkhoz *in* Tajiki*stan 1973*	*Average for all* Kolkhoz *in* USSR *1976*
Cotton yield (raw) cotton, tons hectare)	3.44	3.25	3.24	4.30	3.14	3.42	3.30	3.08	—
Average basic procurement price per ton of raw cotton (rubles)	660.08	430.69	482.72	492.1	502.92	564.09	—	—	—
Realised 'premia' per ton as per cent of basic procurement price (unweighed average for all qualities)	68.6	77.4	52.7	58.8	101.6	62.7	—	—	—
Per cent of collective land under grain	—	30.0	26.9	19.0	23.8	18.2	17.4	39.6	57.2
Grain yield (tons hectare)	—	1.28	1.59	1.42	1.84	1.58	1.69	1.14	1.75
Hectares of collective sown land per tractor	23	30	21	29	18	22.3	21	29	92
No. of cattle:	2000	1300	3567	909	2360	2027	835	1288	1722
including milk cows	604	300	930	300	545	536	226	351	568
No. of sheep and goats	—	5900	—	1491	—	1478	1307	4500	1842

Appendix Table (continued)

	Karl Marx (Tajikistan)	*Rossiya (Tajikistan)*	*XXII Party Congress (Tajikistan)*	*Leninism (Uzbekistan)*	*Kholkabad (Uzbekistan)*	*Average of the Five* Kolkhoz	*Average for all* Kolkhoz *in Uzbekistan 1976*	*Average for all* Kolkhoz *in Tajikistan 1973*	*Average for all* Kolkhoz *in USSR 1976*
Average payment per man-day (rubles)	5.49	4.02	4.07	7.00	6.15	5.33	4.96	4.37	4.77
Average annual earning per worker from collective *Kolkhoz* labour (at actual intensity of employment)	783	681	721	1287	1187	885	1253	1100	1183
(male)	—	—	—	1511	1244	—	—	—	—
(female)[3]	—	—	—	1112	1135	—	—	—	—
Average annual family income from collective *Kolkhoz* labour	1621	531	1082	2277	2172	1426	—	—	—
Average annual income from personal plot per family[4]	936	1200	878	600	700	900	—	—	—
Average family earning from outside employment[5]	611	1299	554	479	798	737	—	—	—
Total household income from above three sources	3167.3	3030.5	2513.7	3355.9	3669.9	3063	—	—	—

continued overleaf

Appendix Table (continued)

	Karl Marx (Tajikistan)	*Rossiya (Tajikistan)*	*XXII Party Congress (Tajikistan)*	*Leninism (Uzbekistan)*	*Kholkabad (Uzbekistan)*	*Average of the Five* Kolkhoz	*Average for all* Kolkhoz *in Uzbekistan 1976*	*Average for all* Kolkhoz *in Tajikistan 1973*	*Average for all* Kolkhoz *in USSR 1976*
Per capita income from above three sources	463.1	391.0	395.9	568.8	542.1	452.4	—	—	—
Gross money income (thousand rubles)	6528	1527	4247	3998	5508	4362	—	—	—
Current inputs, amortisation etc. (thousand rubles)	1331	423	1852	1178	1363	1229	—	—	—
Value added in collective production (thousand rubles)	5197	1104	2395	2820	4145	3133	2058[6]	1714[6]	846[6]
Per cent of value added allocated to:									
Labour payment	73	68	85	69	55	69	64	66	76
Investment (financed by internal saving)	16.4	23.3	9.9	21.8	18.0	17.3	—	—	—
Education, health, welfare	6.8	7.2	4.8	4.1	8.9	6.6	—	—	—
Tax	4.0	1.4	—	5.2	6.6	4.1	—	—	—
Reserve fund etc.	—	—	—	—	11.7	3.1	—	—	—

Appendix Table (continued)

	Karl Marx (Tajikistan)	*Rossiya (Tajikistan)*	*XXII Party Congress (Tajikistan)*	*Leninism (Uzbekistan)*	*Kholkabad (Uzbekistan)*	*Average of the Five* Kolkhoz	*Average for all* Kolkhoz *in* Uzbekistan *1976*	*Average for all* Kolkhoz *in* Tajikistan *1973*	*Average for all* Kolkhoz in USSR *1976*
Investment financed by long-term borrowing as percentage of value added	—	18.0	3.9	10.6	—	3.8	—	—	—
Total investment as percentage of value-added	16.4	41.3	13.8	32.4	18.0	21.1	—	—	—
Total internal saving as per cent of value added	16.4	23.3	9.9	21.8	29.7	20.4	—	—	—
Gross value of output (rubles) per sown hectare of collective agriculture	1747	1091	1167	1528	1747	1499	—	—	—
Gross value of output (rubles) per hectare of personal plot[7]	7200	7059	6754	3529	3529	5833	6364	—	—
Output per hectare of personal plot as the ratio of that for collective land	4.1	6.5	5.8	2.3	3.3	4.25	—	—	—

continued overleaf

Appendix Table (continued)

	Karl Marx (Tajikistan)	*Rossiya (Tajikistan)*	*XXII Party Congress (Tajikistan)*	*Leninism (Uzbekistan)*	*Kholkabad (Uzbekistan)*	*Average of the Five* Kolkhoz	*Average for all* Kolkhoz *in Uzbekistan 1976*	*Average for all* Kolkhoz *in Tajikistan 1973*	*Average for all* Kolkhoz *in USSR 1976*
No. of brigades (including firma)	47	21	37	18	27	30	—	—	—
Average no. of workers per brigade	103	53	77	84	71	81	—	—	—
Monthly average earnings (rubles) of:									
Chairman	500	300	412	500	480	438	—	—	—
Chief book-keeper/ Chief economist	—	—	—	315	380	348 (U)	—	—	—
Brigadier	300	175	170	280	173	220 (U)	—	—	—
Tractor driver	350	245	180	260	175	242 (U)	—	—	—
Average worker (for average no. of days worked)	65	57	60	107	99	78 (U)	—	—	—
Minimum guaranteed income (at the rate of working of an average worker)[8]	29	40	36	43	45	39 (U)	—	—	—

Appendix Table (continued)

	Karl Marx (Tajikistan)	*Rossiya (Tajikistan)*	*XXII Party Congress (Tajikistan)*	*Leninism (Uzbekistan)*	*Kholkabad (Uzbekistan)*	*Average of the Five* Kolkhoz	*Average for all* Kolkhoz *in Uzbekistan 1976*	*Average for all* Kolkhoz *in Tajikistan 1973*	*Average for all* Kolkhoz *in USSR 1976*
Average worker for 25 days' work in a month	137	101	102	175	154	134 (U)	—	—	—
Minimum guaranteed income for 25 days' work in a month	60	70	60	70	70	—	—	—	—

continued overleaf

Notes

1. Sown area refers to land under crops, orchards and vineyards.
2. Personal plot includes total area under such plots. Sown area under personal plots is 85 per cent of total in Uzbekistan, 89 per cent in Tajikistan and 84 per cent for all USSR.
3. Female earnings are calculated on the assumption that the distributions of labour force among categories earning different rates of income are the same for male and female. This assumption is almost certainly inaccurate – the differential between sexes is, therefore, greater than is estimated here.
4. Personal plot earnings are estimated at market prices. Where estimates were given by the responding *Kolkhoz* at state prices a 30 per cent mark up was added.
5. It is assumed that daily earning per worker outside the *Kolkhoz* is 20 per cent higher than that in the *Kolkhoz*.
6. From the statistical reports it is not quite clear if these gross values of output or values added. We think they refer to the latter.
7. Value of output per hectare of personal plot refers to total, not just sown land. To convert these into values per hectare of sown land a multiplier of 1.18 for Uzbekistan and 1.12 for Tajikistan should be used.
8. The lowest minimum guaranteed monthly income (frequently 70 rubles, but sometimes 60 rubles) is paid for 25 days of work per month. Our figures are proportionate to average number of days worked.
9. The entries in the column 'Average of the Five *Kolkhoz*' usually show weighted averages. Those which are not so have been marked (U), meaning unweighted.
10. — means not available or not shown and . . . means zero.
11. Sources for the Uzbek, Tajik and all USSR figures are usually: *Narodnoe Khozyaistvo Uzbekskoi SSR Za 60 Let. Sovietski Tajikistan Za 50 Let and Narodnoe Khozyaistvo SSSR Za 60 Let.*

10 Cuban Agriculture and Development: Contradictions and Progress

Arthur MacEwan

10.1 Introduction

When the Cuban revolutionary government came to power at the beginning of 1959, it quickly proceeded to implement a land reform and other social programmes which would radically alter the country's social structure. These programmes and the extensive redistribution of income associated with them were major accomplishments in and of themselves. In addition, they had important consequences for the course of Cuban agriculture and the Cuban economy generally.

On the one hand, the land reform, social programmes and income redistribution upset the existing system of economic organisation. Yet as these changes were being implemented, they were not accompanied by any satisfactory moves to establish a new system of organisation. An important contradiction was set in motion: the successful implementation of social reforms tended to disrupt the operation of the economy. And the operation of this contradiction – along with the severe problems resulting from the conflict between Cuba and the United States – contributed to serious production difficulties in the early 1960s and general economic stagnation throughout the decade.

On the other hand, the land reform, social programmes and income redistribution created the possibility of establishing a basis for economic growth. Through these changes, the government gained both the effective administrative control, and the political foundations it needed to implement a marked acceleration of capital accumulation. Moreover, the political transformation set in motion by these changes helped create the possibility of mass participation in economic decision making.

During the 1970s, it appears that the successful capital accumulation of the preceding decade is beginning to bear fruit and that a new system of economic organisation – with mass participation as an important component – is being established. Significant production problems persist, most notably in sugar, but a major economic expansion has taken place in the 1970s. And if the new system of organisation is secured and extended, there is good reason to anticipate continuing expansion. If so, the Cuban experience could, in the view of the present author, take on special significance as an example of economic growth combined with major social reform and popular participation.

This essay is going to present a broad overview of Cuban economic strategy and development through the 1960s and early 1970s with the purpose of establishing the argument outlined in the preceding paragraphs. The remainder of the introduction will briefly describe the land reform, income redistribution and other social programmes of the early years of the revolution. These changes, as already suggested, set the context for later events. Section 2 will describe the major policies and course of the economy in the 1960s and begin to develop an explanation of the economic weaknesses of that period in terms of the interrelation between economic organisation and popular participation. Section 3 will deal with the progress and reorganisation of the 1970s; by examining planning and political changes, it will be possible to develop the analysis of organisation and popular participation. The final section will offer some observations on the historic origins of Cuba's economic problems and progress and on Cuba's prospects for future development.

Throughout, the focus will be on agriculture. In 1958, about half of Cuba's 6.5 million people lived in the rural areas. Agriculture employed 40 per cent of the workforce and directly accounted for 25 per cent of national income. Moreover, the major industries – most particularly sugar – were closely tied to agriculture. Agriculture had played a central role in Cuba's economic history, and it has continued to play an important role during the socialist period.

But Cuban rural life and agriculture are not things apart from the rest of the nation's activities. The policies and problems of the countryside are policies and problems of the whole economy. While concentrating on the rural economy, it is thus necessary to examine – and learn a good deal about – the entire development process in Cuba.

(a) The land reform: a dual process[1]

In May of 1959, less than five months after its seizure of power, the revolutionary government promulgated a far-reaching agrarian reform law. Cuban agriculture, like agriculture in much of Latin America, had been dominated by large estates; more than 70 per cent of Cuba's 10 million hectares of farm land was held in 10 per cent of the farms. The May 1959 law proscribed estates of more than 400 hectares; these lands had constituted 63 per cent of all agricultural holdings. Four years later, at the end of 1963, a second agrarian reform reduced the upper limit on land holdings to 67 hectares, thus eliminating middle level capitalist farms as well as the large estates.[2]

In changing the structure of land holdings in Cuba, the agrarian reform involved a dual process. It created a public agricultural sector out of many of the large estates. It also created – or greatly enlarged – the class of small peasants who owned and tilled their own land.

From the outset of the reform in 1959, the government had attempted to avoid breaking up large agricultural production units. The government's position was partly determined by a desire to take advantage of economics of scale (often only potential) and by a desire to use the social organisation of large farm units as the basis for providing services (schools, health clinics, etc.) to the rural population. In addition, the estates had been worked in large part by an agricultural proletariat, and there seems to have been no move among this group for dividing the land of the estates into separate, small units.

Indeed, while the land reform was very popular, it was not actively implemented by the rural workers themselves. There were few, if any, spontaneous land take-overs; the government led in implementing the reform. While for a short period it attempted to establish co-operatives on the large estates, by mid-1961 the government was operating state farms on most of the large agricultural production units. After the second reform in 1963, 76 per cent of the nation's total lands and 63 per cent of cultivated lands were in the public sector; almost all of this was operated as state farms.

The private agricultural sector was not only reduced in size during this reform process, but it was also changed to an exclusively small-farm agriculture. At the beginning of 1959, there had been 40–45,000 peasants owning farms of less than 67 hectares. The 1959 reform added about 110,000 peasants to this category by giving title on the lands they worked to renters, sharecroppers and squatters. By the end of 1963, a few thousand more farmers had entered this category by having the size of their holdings reduced in the second reform. Thus when the reforms were completed there were roughly 160,000 private peasant farms in Cuba.

The second reform was accompanied by the government's assurance to the small farmers that the expropriation of land would go no further. Farms under 67 hectares would be safe. At the same time the government made it clear that it expected private agriculture to diminish in importance over time, and that Cuba's agriculture of the future would be entirely public. Further transition, however, was to be a voluntary process.

Through the 1960s and 1970s, Cuban agriculture would build on the dual basis of large scale state farms and small peasant operations. The rapid and drastic reorganisation of Cuban agriculture had created the opportunity for implementing a socialist growth strategy. The government, through the state farms, directly affected production decisions and controlled a large part of the agricultural surplus; thus, it could initiate planning. The government's influence over the decisions of small farmers, though indirect, was strong because of its control over credit, purchases of products, sale of inputs, and the supply of labour. Moreover, in 1961 the government had created the National Association of Small Farmers (ANAP) which worked to organise the peasants politically and to involve them in the development plans.

But changing the structure of land holdings would prove only the beginning of the creation of socialist agriculture. The rapid changes had destroyed not only the old system of ownership, but also the systems of allocation, decision-making, work motivation – in short, the system of organisation in Cuban agriculture. Thus, land reform embodied a serious contradiction: its success engendered considerable disruption which would threaten expansion of production.

(b) Redistribution and social programmes

The contradiction in land reform was actually one aspect of a general contradiction manifested during the years immediately following the establishment of the revolutionary government. Along with the land reform, an entire set of social programmes was being implemented in Cuba which, taken all together, resulted in a major redistribution of income and alteration of social relations.

A large part of the redistribution of income can be accounted for by the combined effect of the land reform and the government's employment programme. The former eliminated rent payments for the tens of thousands of renters and sharecroppers, producing an immediate rise in their net incomes; also, many small farmers were given new lands. The creation of the state farms was a major factor allowing the government to implement its employment policy of guaranteeing jobs to those who would work. Employment on the areas covered by the state farms grew from 50,000 prior to 1959 to 150,000 by 1962. On an economy wide basis, unemployment, by conservative estimates, had run at annual rates of 12 to 16 per cent of the labour force. It had dropped to no more than 8 per cent by 1963 and continued downward thereafter. Also, the sharp seasonal fluctuations of employment had been substantially reduced by 1961.[3]

The cumulative impact of these and other redistributive measures was great. According to various estimates the income of the working population rose by at least 25 per cent between 1958 and 1962, while national income was roughly the same in those two years.[4] Some estimates place the increase in income of the poorest groups as high as 100 per cent in this period.[5] It seems reasonable to conclude that in less than five years, the distribution of income in Cuba became substantially more equal.

For our purposes, it should be emphasised that poverty in Cuba had been concentrated in the countryside; the rural-urban income inequality had been typical of many poor countries. Consequently, it was in the countryside where the income redistribution had its greatest positive impact. Likewise, it was in the countryside where other social programmes – those not immediately and directly affecting the distribution of income – had their greatest impact.

Among these other social programmes, the literacy campaign of 1961 stands out as the most dramatic. At the beginning of 1961, there were one million illiterates in Cuba – roughly a quarter of the adult population; more than three-quarters of these illiterates lived in the rural areas. During 1961, a mass mobilisation was undertaken to eliminate illiteracy; this involved some 250,000 volunteers as teachers. The campaign resulted in the reduction of illiteracy to about 5 per cent by early 1962. In addition, the literacy campaign provided a catalyst for the continuing expansion of education, integrated the masses into solving the socio-economic problems of the nation, and trained large numbers of people, not only in the tasks of education, but in the tasks of organisation and leadership as well.[6] The expansion of medical services and their redirection to the countryside, although somewhat less dramatic than the literacy drive, was similarly important.[7]

All of the early reform programmes had the combined function of altering both infrastructure and ideology. They served to create a land tenure structure and a workforce which could be organised into an effective development strategy. They also seemed to establish a new ideology among the Cuban people in at least two important ways. First, the Cuban people developed a sense that the accomplishments of the revolution belonged to them. They viewed themselves as 'owners' and beneficiaries of the socialised enterprises. The reforms and the equality became entrenched; they became viewed as 'rights' by the Cuban people.[8] Second, the Cuban people began – but only began – to develop the idea that social change was their own responsibility.[9] Some programmes, most particularly the agrarian reform, had been initiated from above. And the revolutionary struggle itself had directly involved only a handful of people. But in the implementation of change, as illustrated most effectively by the literacy campaign, success depended on the active involvement of the population. Consequently, the early reforms created the motivation for and potential embryo of organisation based on participation and planning by working people themselves.

10.2 Production and accumulation in the 1960s

Socialist organisation in Cuba, however, would not arise immediately from destruction of the old system of organisation or from the foundations laid by the early reforms. Into the early 1960s there was, in fact, no coherent strategy for economic development. Economic goals were articulated – especially the need to diversify agriculture – but no new system of organisation was formed. Without organisation the goals would become counter-productive and combine with the disruptive aspects of the social reforms in contributing to the severe production crisis in 1962 and 1963.

The effort to diversify agriculture illustrates the problem. The Cuban leadership and people generally identified the problems of the country with the domination of their economic life by sugar. And, indeed, sugar had been the mechanism by which Cuba had been tied into a subordinate role in the international division of labour since the end of the eighteenth century. Thus, it is not surprising that Cuban economic policy in the 1959–61 period was 'anti-sugar'. The resulting programme of diversification led to taking 200,000 hectares – roughly 15 per cent of the area cultivated with cane – out of sugar with the intent of devoting it to other crops. Yet, because no new system of economic organisation had been established, there was no mechanism to increase cane yields, and for some of the 'other crops' sufficient experience and technical skills were lacking. In addition, paying little attention to sugar and stressing programmes in construction and industry, the government contributed to a severe agricultural labour shortage.[10]

The labour shortage had more fundamental roots, however. It was the direct consequence of providing employment and income security – the social reforms – and thus undermining traditional mechanisms of labour allocation and incentives without providing a new system of organisation. While data are inexact, the agricultural labour shortage seems to have been severe. In 1963, a census taken by local cane commissions placed the need for cane cutters at 350,000 equal to the 1959 supply; yet in 1963, there were only 260,000 cane cutters available.[11] In addition, the labour shortage took the form of a decline in the intensity and hours of work on state farms. An investigation conducted on 136 state farms in 1963 showed, for example, that employees worked only 4.5 to 5 hours per day. And a later government report indicates such problems were typical, not exceptional.[12]

Similar difficulties arose in private agriculture. The supply of food crops, coffee and tobacco coming out of the private sector suffered a severe decline after 1960.[13] Although data are incomplete, there is little doubt that a significant part of the supply problem resulted from rising consumption within the private sector. Relieved of rent (or sharecropping payments), many peasants appear to have increased their own consumption of agricultural goods. Furthermore, as industrial production began to falter in the early 1960s and a shortage of manufactured goods developed, the problem of obtaining goods from the peasants intensified. The old mechanisms of exchange had broken down, and no new system of organisation had been created to integrate peasant production with the rest of the economy.

The crisis that developed in 1962 and 1963 was probably all the more severe because its development had been masked by a favourable performance of the economy in the 1959–61 period. Redistribution of income and expansion of government spending had provided a major stimulus to demand. Considerable underutilisation of capacity and unemployment had allowed a marked response to this rising demand, and both agriculture and industry grew by 4–5 per cent a year. Sugar output rose to 6.8 million tons in 1961, and the 1959–61 sugar harvests exceeded those of any other three year period in Cuba's history.[14]

But in 1962 and 1963, a number of factors reversed the situation. The negative impact of diversification began to take its toll. The contradiction embodied in the social reforms began to mature in the form of labour and commodity shortages. Weather was unfavourable. The acute problems arising out of the conflict between Cuba and the United States, including the US trade embargo, seriously disrupted the economy.[15]

(a) Development strategy of the 1960s

Out of the context of crisis, the Cuban government formulated its first comprehensive development strategy. The strategy embodied a response to the problem of organisation that had become so serious, and it was shaped in large part by the equality and social change already attained. Three factors defined the Cuban approach to development in the 1960s: a heavy emphasis on agriculture generally and sugar in particular; a combined reliance on highly centralised economic control and 'moral' incentives; and an extremely high rate of investment. In addition, the high rate of foreign assistance and trade stability provided by the Soviet Union was a distinctive feature in Cuba's strategy.

(b) Substance of strategy[16]

The relation with the Soviet Union combined with a reaction against the early error of over-diversification to yield an agriculture-based development strategy for the 1960s. In particular the return to an emphasis on sugar was viewed by the government as the only short-run means by which to obtain the foreign exchange necessary for general modernisation and long-run industrialisation. The failure to develop alternatives to sugar in the 1959–63 period was taken as evidence that Cuba's historic and geographic–climatic circumstances necessitated a continued reliance on sugar. Moreover, the Soviet guarantee of a stable market for Cuban sugar eliminated much of the uncertainty traditionally associated with reliance on a primary product export.

Thus, the cornerstone of the Cuban strategy became the goal of producing 10 million tons of sugar in 1970. The goal was almost 40 per cent above the previous peak of 7.2 million tons produced in 1952. The 6.8 million tons produced in 1961 was Cuba's second largest sugar output, and in 1964, when the target was set, sugar output was only 4.5 million tons. The 10 million tons target was an ambitious one.

The agricultural emphasis of Cuba's programme in the 1960s was not, however, confined to the sugar sector. Other objectives included: a near doubling of coffee production; a 200 per cent increase in the area devoted to rice cultivation; meeting all internal needs for fruits and vegetables and beginning to export some of these products; providing the raw material needs of the textile industry; and establishing an export base in the cattle industry.

In addition to these output goals, the plan called for the rationalisation and modernisation of agriculture. In particular, in sugar cane, mechanisation was to be given high priority as the principal means of relieving the labour shortage. Mechanisation was also called for in other branches of agriculture; and new seed varieties, genetic improvement of livestock, the use of modern inputs and expansion of technical education were all given emphasis.

While agriculture was the foundation and key to Cuban development efforts, it would not stand alone. Extensive infrastructure expansion was included in the plan – in the areas of water storage, irrigation, transport facilities, for example. And linkages between agriculture and industry (e.g., fertilisers and food processing) as well as within agriculture (sugar by-products as cattle feed) were important components of the plan. The general expansion of industry was to wait, however, until some point in the 1970s when agriculture would provide a firm foreign exchange source on which to draw.

This need for foreign exchange in industrialisation provided a strong rationale for the direction of the Cuban plan. There was, in addition, an ideological factor pushing in the direction of an agricultural emphasis. Cuba's leaders saw themselves as having firm roots and strong political support in the countryside. In spite of their early statements about the need to industrialise and the negative historical role of sugar, the major reforms of the 1959–63 period had had a decidedly rural emphasis. Thus, the strategy for the 1960s did fit with the well established rural bias of the revolution.

(c) Planning and incentives

Ideological factors also played a role on the matter of how the plan for the 1960s would be organised and considerable ideological controversy surrounded the government's decision to rely on highly centralised planning and moral incentives. Yet, while much of the debate of the early 1960s was highly ideological, the issue at hand was a very practical one: how was it possible to do away with inequality and insecurity, as the Cuban revolution had begun to do, and still get people to work hard? Once direct personal gain was reduced as a motivation for work and as a guide to allocation decisions, how would people work effectively for the collective welfare? These were the questions which had been forced to the forefront by contradictions embodied in the early reforms, and the Cuban answer was the official adoption of a programme emphasising 'moral' incentives in the context of highly centralised planning.

For the Cuban leaders, 'moral' incentives meant that, as far as possible, people should be motivated to work by their commitment to the welfare of society at large, rather than by that to their own private interests. This incentive system tied work and politics together, since it was only through political development – identification of the individual with society – that such commitment could become effective. The practical foundations for such incentives were the relative equality and collective pay-offs which had been achieved by the Cuban people through the revolution. More than any political exhortation, these tangible factors provided the basis for collective identity and the workability of moral incentives.[17]

While the Cuban position on incentives is viewed in ideological terms and is sometimes criticised as being 'idealistic', it was in large part the consequence of prior steps taken by the revolution. The social reforms of the early period had, as we have noted, destroyed the old system of economic organisation. The work process – incentives and decision making – had already been drastically altered by the early 1960s. Within the framework of the already accomplished changes, alternatives were surely possible, but, barring a reversal of the basic programmes of the revolution, all alternatives would have accepted a large role for moral incentives.

Likewise, the centralisation of economic decisions was, to a large extent, a consequence of the particular circumstances which existed in the 1960s. In part, the Cuban leaders viewed centralised planning as tied to the system of incentives. Using devices such as prices and enterprise profits as keys to a system of decentralised decision-making was necessarily bound-up with individual-gain motivation and the dominance of market relations. And individual-gain motivation and market relations, however controlled, were seen as necessarily connected to inequality.

Moreover, decentralisation, in the absence of an effectively operating system of workers' participation at the local level, would have relied heavily on enterprises' managers and technicians. But in the early 1960s, workers' participation had not been established. Reliance on managers and technicians would have exalted their positions exacerbating inequality (of status and authority as well as income). Centralised planning was seen as a safeguard against the emergence of local elites.[18]

Finally, the combined impact of the various reforms, the nationalisation of industry, the US trade embargo, and the introduction of rationing had thoroughly upset the structure of prices and greatly limited the usefulness of the market as a meaningful co-ordinator of economic activity. In the early 1960s, had the market been allowed to operate, prices might have fluctuated sharply, and many enterprises would have been unable to operate. The general instability of the period would have been aggravated, and the drive to eliminate unemployment could have been hampered.

Thus, the Cubans turned to centralised planning in the 1960s. Individual enterprises would be guided by quantitative production targets established under the authority of central planners. This general procedure would be followed in both industry and agriculture.

1. The Big Push Highly centralised planning, whatever its other merits or shortcomings, would give the Cuban government far reaching control over the nation's economic surplus and would allow the direction of that surplus toward capital accumulation. Investment as a percentage gross material product rose from 16.4 per cent in 1962 to 20 per cent in 1965 to 31 per cent in 1968. This rise of investment was primarily the result of increased domestic savings. Foreign finance did cover a substantial share of the investment – 36 per cent in 1968; but its relative importance declined substantially after 1962 and 1963 and showed no trend after 1964; see Table 10.1

Along with the rise in investment, there was a substantial shift in the direction of state investment. The share of investment going to agriculture rose substantially after 1963, while the share (and absolute volume) going to industry fell; see Table 10.2. This emphasis on agriculture in the 'big push' of the 1960s is seen in several other statistical series; tractor imports rose from 1950 in 1962 to 7275 in 1970, and during the decade five times more

Table 10.1 Investment and Related Aggregates, 1962–8 (million *pesos*)

Year	*Gross Material Product*	*Gross Investment*[1]	*'Aid'*[2]	*Saving*[3]	*I/GMP*	*S/GMP*	*A/I*
1962	3689.2	607.6	277.0	330.6	.164	.089	.455
1963	3736.7	716.8	322.2	394.6	.192	.106	.449
1964	4074.6	794.9	304.5	490.4	.195	.120	.383
1965	4136.5	827.1	175.6	651.5	.200	.158	.212
1966	3985.5	909.8	327.7	582.1	.228	.146	.360
1967	3612.5	979.0	294.1	684.9	.271	.189	.300
1968	4000.0	1240.0	450.9	789.1	.310	.197	.364

Source 'Aid' figures are from Junta Central de Planificación *Anuario Estadístico de Cuba*, 1973, p. 186; other data are from A. Ritter, *The Economic Development of Revolutionary Cuba*, (New York: 1974), Praeger Publishers, p. 140.

Notes 1. State sector investment.
2. Aid equals the trade deficit.
3. Savings equals investment less aid.

tractors were imported than had existed in the nation in 1960; fertiliser supply rose from 444,100 metric tons in 1963 to 1487,800 in 1968; water storage capacity was increased from 28 million cubic metres in 1958 to 324.7 million in 1966 to 1762 million in 1970; and so on.[19]

Table 10.2 Distribution of State Investment, 1962–6 (per cent breakdown)

	1962	*1963*	*1964*	*1965*	*1966*
Agriculture	29.4	24.3	30.5	40.5	40.5
Industry	23.1	31.6	29.1	18.1	16.7
Transport and communications	9.5	9.6	9.1	11.7	14.3
Housing and social services	13.5	11.5	11.4	9.4	9.6
Education and culture	8.1	7.0	5.3	5.0	4.4
Construction	4.3	5.8	4.6	3.8	2.2
Other	12.1	12.2	10.0	11.5	12.4

Source C. Auroi, *La nouvelle agriculture cubaine*, Editions Anthropos, Paris, 1975, p. 90.

As far as the particular emphasis on sugar was concerned, Fidel Castro later described the actions taken as follows:

> The first plan for the development of the sugar industry was put into practice in the 5 year period, from 1966–1970. The objectives were those of increasing installed capacity, replacing the industry's obsolete equipment, massively introducing advanced techniques in cane planting and cultivation, and solving the problem of mechanising the harvesting. In those years, 334 million went into reconstruction and 235 million into enlargement. The lands earmarked for cane growing increased by 35 per cent, new varieties of cane were introduced, and the extension of irrigation and the use of herbicides was begun. The use of fertiliser was considerably increased. In this period, new machines were designed to mechanise the harvesting. In 1970, Cuba grew it biggest sugar harvest. The process made in agricultural yield and the use of technical means was considerable. Cultivation was mechanised and the mechanisation of harvesting was started. The bulk-sugar shipping system was enlarged.[20]

In addition to heavy investment programmes, the 'big push' in agriculture involved a huge labour mobilisation. Bringing labour from other sectors of the economy was – for the period prior to successful mechanisation – the way the Cubans tried to solve the problem of seasonality of sugar production and the continuing labour shortage in the countryside. Also, the labour mobilisations were tied to the general programme of developing worker motivation based on moral incentives; students, women outside the paid work force, and workers regularly employed elsewhere were involved in a national effort which depended upon and contributed to a commitment to the collective welfare.

These mobilisations of volunteer workers reached their high point in 1970 when 1.2 million people were mobilised during the course of the harvest. At the peak of cane cutting in 1970, the daily average of volunteer workers was 200,000. Prior years had seen smaller mobilisations, but as early as 1968 over a quarter of a million people had been involved in the sugar harvest in this manner.[21] (These mobilisations, of course, were a supplement to the workers regularly employed in sugar cane.) As with the rapid acceleration of capital accumulation, the Cuban government had been able to use its authority to deploy Cuba's labour towards the objects of the national development strategy.

2. Weakness of the 1960s strategy Cuba's development strategy during the 1960s, however, suffered from serious weaknesses, and production goals went by the wayside early on. Except for the first year of the new strategy's implementation, 1965, targets for sugar output were not attained; see Table 10.3. Indeed, 1966 and 1969 were poor years by historical standards, let alone the standards of the plan. The 1970 harvest was a record of 8.5 million tons, but it was far short of the targeted 10 million.

Table 10.3 Plan Targets and Sugar Production, 1965–70 (thousand metric tons)

	Plan Target	*Actual Production*	*Deficit*	*% Deficit*
1965	6.0	6.2	+0.2	+3
1966	6.5	4.5	−2.0	−31
1967	7.5	6.2	−1.3	−17
1968	8.0	5.2	−2.8	−35
1969	9.0	4.5	−4.5	−50
1970	10.0	8.5	−1.5	−15

Source Planned output figures are from a speech by Fidel Castro printed in *Verde Olivo*, 20 June, 1965, and cited by Auroi, op cit, p. 66. Actual output figures are from Junta Central de Planificación, *Anuario Estadístico de Cuba*, 1973, p. 124.

In other parts of agriculture, the plan also fell short. In rice, for example, the plan called for 200,000 hectares in cultivation by 1970, but only 187,000 were under rice in that year; and in the preceding five years, the figure had been far lower. In citrus fruits only 46,000 hectares were cultivated, as compared to the plan target of 100,000.[22] The generally poor performance of agriculture in the 1960s is revealed in Table 10.4. Throughout the 1962 to 1969 period, per capita agriculture production was well below the level of the mid-1950s.

The Cuban economy as a whole did somewhat better than agriculture. Industry appears to have grown by a bit more than 2 per cent a year after the 1962/63 crisis. But nowhere could the performance be classified as favourable.

Any analysis of the Cuban situation in the 1960s made from the perspective of the late 1970s benefits from the knowledge of what followed the production deficiencies of the former decade. Through the early 1970s, the economy grew rather rapidly. Whatever weaknesses existed in the 1960s did not prevent some subsequent

Table 10.4 Agricultural Production in the 1960s

	Total Agricultural Production	*Agricultural Production Per Capita* (1952–6 = 100)	*Sugar Harvest* (million metric tons)
1960	114	101	5.9
1961	122	106	6.8
1962	100	85	4.8
1963	86	72	3.8
1964	93	75	4.4
1965	112	89	6.1
1966	94	72	4.5
1967	116	88	6.1
1968	106	79	5.3
1969	99	73	4.4
1970	151	107	8.5

Source Ritter, op cit, p. 113, based on various UN sources. Ritter notes that the data on which the indices are based are 'fragmentary and incomplete'. Many important products are excluded from the calculations, and very crude estimates for other products are included.

successes, and the explanation of the 1960s which follows must be seen in this light.

3. The organisation problem A central component in the advancement of economic activity is the creation of patterns of social relations in which people interact with one another in an effective and disciplined manner. These patterns of social relations are the organisational foundation of economic development. Traditionally, in capitalist development, the market, in combination with the authority of owners of capital, has served as a principle mechanism of organisation. Between, and within, enterprises, the market coordinates and controls economic activity; though within the firm direct, top-down exercise of authority by managers also plays a major role.

Socialists, in Cuba and elsewhere, reject the role of the capitalist market as the organisational mechanism of development. In their view, such mechanisms run contrary to socialist principles of planning for social welfare and democratisation of economic activity. In Cuba, moreover, regardless of such matters of principle, the early reforms had wiped out the traditional mechanisms of economic organisation. It would, therefore, have been necessary for the

Cubans to find some mechanism of organising economic development which would be compatible with the changes in social relations that had already been brought about.

Experience in Cuba suggests that a socialist solution to the problem of organising development rests on the effective exercise of power and participation by working people. In the 1960s, working class participation had not been well established, and in what follows particular weaknesses of the period will be traced to this factor. Conversely, in the 1970s, there were major steps taken towards the development of working class power and, as we shall see in section 3 below, the economic success of the latter is, to some degree, related to political changes.

4. Ineffective Investment We have noted above the success which Cuban planners attained during the 1960s in directing the nation's surplus toward capital accumulation. But one of the most notable features of the period was the failure to establish the connection between accumulation and production. An outstanding example is provided by the under-utilisation of tractors. In 1970, following the huge build-up of the supply of tractors on a national level, only one-quarter of tractor capacity was actually used. 'Breakdowns, repairs and maintenance' along with 'administrative' problems accounted for most of the under-utilisation.[23] Another important example is Cuba's inability to mechanise the sugar harvest. The combines were there, but they were too heavy, complicated and fragile for the Cuban conditions. And in the sugar mills, the investment technically necessary for the 10 million tons harvest was achieved, but its use and productiveness were another matter.[24]

These problems can be traced to organisational deficiencies. Investing is a process of change, and changing the way things are done is a complex matter, the success of which depends upon effective co-ordination and administration. More tractors and combines cannot bring in the crop unless they are at the right place at the right time. A mass mobilisation of labour is ineffective unless those involved acquire skills, work together and are transported to the places they are needed. And cutting sugar cane does not meet targets unless the mills are operating with sufficient capacity.

The deficiencies in organisation which made investment ineffective appeared so sharply in the 1960s because the 'big push' of the period was so concentrated in a single sector of the economy. The experience and skills – relating to both technical operations and group activity – which constitute organisational capacity are not readily transferable from one activity to another. Thus, the major drive undertaken in sugar ran up against the organisational limits of the sugar industry.

(d) Weak design of strategy

The basic design of the Cuban strategy, the heavy imbalance in favour of sugar, is called into question by the ineffectiveness of investment. Perhaps the poor design of the strategy comes out most clearly with the failure to mechanise the sugar cane harvest. When the 10 million tons goal was established as the keystone of the 1960s' strategy in 1964, Cuban leaders believed that the sugar harvest could be mostly mechanised by 1970. Yet, in actuality only 2 per cent of the 1970 harvest was accomplished by mechanised means; as noted above, the investments were made, but their use was poorly organised. This failure to mechanise meant that a much greater labour mobilisation was required than had originally been anticipated. Consequently, the harvest placed a severe strain on the organisational capacity of the nation; and this strain had its impact throughout the economy, but especially in the sugar mills.

A recognition of the organisational limits on expansion would have led to a rather different balance in the sectoral emphasis of Cuba's development strategy. The adoption of an extremely unbalanced development strategy seems particularly inappropriate when organisational capacity is an active constraint on development. Administrative officials in the Ministry of Sugar or on state farms had no experience in directing and co-ordinating a massive labour mobilisation. Dock workers, students, and office workers have little familiarity with the work organisation of the sugar harvest. People can learn to handle these tasks, but learning takes time. The 'big push' in sugar did not allow much time.

Another weakness in the plan for the 1960s arose out of the high degree of centralisation. The planning authorities sometimes articulated plans for the agricultural sector which were so detached from the possible as to become inoperative. Local plans, in some cases, ended up operating independently of any overall framework.[25]

One manifestation of over-centralisation seems to have been an insufficient emphasis on secondary crops, particularly those crops in which the private sector played an important role. Table 10.5 includes data for the period 1962–70 on the state collection of

Table 10.5 Collection of Selected Categories of Agricultural Commodities, 1962–70 (thousand metric tons)

	1962	*1963*	*1964*	*1965*	*1966*	*1967*	*1968*	*1969*	*1970*
Grains – total	340.2	319.7	174.6	82.3	96.6	117.1	111.5	189.3	299.4
– private	140.6	116.5	52.2	39.0	32.6	31.4	20.0	16.2	11.0
Tubers – total	345.2	429.7	440.6	402.2	545.9	393.5	410.5	326.2	196.7
– private	249.8	288.5	261.0	225.5	252.8	171.8	160.3	130.8	75.3
Vegetables – total	—	—	195.0	220.4	230.9	282.7	216.7	126.7	136.2
– private	—	—	142.4	161.2	162.4	188.4	138.7	82.1	62.7
Fruits – total	—	365.6	291.7	325.4	321.3	318.1	385.0	297.4	289.7
– private	—	268.9	184.0	224.9	217.7	213.1	240.1	171.0	125.7

Source Junta Central de Planificación, *Boletín Estadístico*, 1970, pp. 70–8

secondary crops in four major categories. In each case, the end of the period shows a marked deterioration of the total collected, and the decline is most pronounced in the private sector. It would seem that the central authorities did not develop a pricing system that could elicit the supply of these crops. In addition, a money pricing system alone could not have dealt with the problem; the existence of rationing and the non-market allocation of various goods made money prices have a limited relation to the real value of payments received by the private sector.

(e) Poor labour mobilisation

In terms of organisation, the problems of over-centralisation can be seen as the inability of central planners to give accurate, useful direction to activities at the local level. Within the context of the Cuban incentive system, there was little connection between particular problems and responsibility for solving them. When tractors lay idle, no individual or specific group paid a price. When transport facilities were inadequate, the costs were not felt by those who had created the problem. The only incentive to solve these problems of disorganisation was a moral commitment to society at large. Yet, when organisational structures are weak, the task is all the greater and the commitment must be more powerful. With insufficient direction coming from above, and without individual incentives, there was no way for people at the local level to know how to make their commitment effective. Collective pay-offs of the revolution and political development even if they can motivate activity, cannot direct it.

The clearest manifestation of incentive deficiencies in the late 1960s was the rising rate of worker absenteeism. The problem was economy-wide and is illustrated by the situation in agriculture. In the May–June period of 1969, absentee rates for permanent farm labourers in Camagüey province did not fall below 35 per cent. And at the height of the 1970 harvest, the rate of absenteeism – justified and unjustified – was found to be almost 29 per cent.[26]

The matter of motivation should not be separated from the problem of organisation. In a capitalist society, pre-revolutionary Cuba for example, the market simultaneously provides the basis for organisation and motivation. If an alternative system of incentives were to be developed in Cuba, it would require some institutional structure to substitute for the organisational–incentive structures of the market. Economic planning *per se* does not constitute a solution to the problem of organisation. Indeed, for planning to be effective, it is necessary to have accomplished a solution to the organisation problem. Without mechanisms to co-ordinate the activities of various enterprises and without a system for organising work (specifically one for motivating workers and directing their activity), a plan remains little more than an expression of official aspirations.

As economic problems became more and more serious during the 1960s, the Cuban government attempted to control the situation and push the plan by increasing central control, using the military to direct activity, and resorting to political exhortation. In themselves, such methods provided no solution to the organisation problem. As far as they could be accompanied by an effective organisation of the economy, that organisation would be based on top-down directives, a classic 'command system'. Had the Cuban government continued to move in this direction, it is likely that such organisation would have led toward rising inequality, as the distinction between those who directed and those who worked became increasingly pronounced.

If an alternative to both 'commandism' and reliance on the market were possible in Cuba, it would have to be sought in terms of what was missing in the Cuban economy during the 1960s. From the point of view of the goals of socialist development the most outstanding omission in the Cuban strategy was any programme for the involvement of workers themselves in the direction of economic activity. It is at least possible that were workers more involved in exercising economic authority, many failures of the 1960s would not have occurred; the separation of decisions from local conditions and the resulting errors might have been avoided. People might have been motivated in carrying out decisions in which they themselves had a hand. And, most important, through participation, they might have more quickly gained the skills to organise their own activity. In this sense working class power might, in the view of the present author, have provided a key to overcoming the economic weaknesses of the 1960s.

10.3 Reorganisation and progress in the 1970s

The Cuban experience in the 1970s lends some support to this hypothesis that working class power participation may provide a solution to the organisation problem and thus engender economic expansion. Economic expansion certainly took place in the early 1970s. Table 10.6 gives annual growth rates by sector for the 1969 to 1975 period. The large sugar harvest of 1970 seems to have initiated a period of growth. The economy as a whole grew by 88 per cent between 1969 and 1975 (an average annual rate of 11 per cent). Industrial production roughly doubled, and construction activity grew by more than threefold. Agriculture remained a lagging sector, but even there output expanded by over 30 per cent between 1969 and 1975.[27]

For the years since 1975, data are rather limited, and thus the period up to 1975 will be the focus for the analysis here. It is known that the rate of economic expansion slowed down substantially in 1976 and 1977, overall growth of the economy being only 4.0 per cent and 3.8 per cent, respectively. In 1978, however, a very large sugar harvest moved the economy ahead more rapidly (though no overall growth rate figures are yet available). All in all, what is known about the Cuban economy in these more recent years does not seriously conflict with an analysis based on the period up through 1975.

The rapid expansion which took place in the early 1970s can be partially attributed to the massive investments of the 1960s. The stock of plant and equipment was probably increased by 50 per cent to 75 per cent in that earlier decade. In addition there was a tremendous rise in the educational attainment level of the work force; primary schools, for example, were graduating four times as many students in 1970 than at the time of the revolution, eleven years earlier. The capital accumulation of the 1960s may have had its impact delayed by the various weaknesses in organisation discussed above. Nonetheless, it appears that the consequence of disorganisation was that it postponed, rather than eliminated, the productive impact of investment.

Furthermore, during the early 1970s some major changes began to take place which would alter the organisational foundation of the economy. These changes were of two sorts. First, the major advances which had been accomplished in education and the continuing 'special programmes' of the revolution began to have their impact on the quality of work. Second, a dual set of reforms was instituted which restructured planning and incentives.

(a) The role of education and special programmes

In Cuba, the form and content of education are still evolving. But there have been major efforts to develop a type of curriculum and a

Table 10.6 Annual Growth Rates, 1969–75

	Gross Social Product[1]	*Gross Material Product*	*Agriculture*	*Industry*	*Construction*	*Transportation*[1]
1969	−1.3	−1.7	−3.8	1.5	−14.8	−1.0
1970	15.5	21.4	18.9	25.9	−2.6	30.4
1971	7.3	4.2	−8.4	4.4	31.5	0.5
1972	16.2	9.7	5.3	6.7	40.1	3.7
1973	14.4	12.5	4.3	11.9	27.8	2.2
1974	10.3	8.3	4.0	8.2	14.3	2.8
1975[2]	9.0[3]	8.1	4.0	8.0	15.0	12.0

Source A. MacEwan, *Agriculture and Development in Cuba*, unpublished manuscript, Appendix Table A.1.

Notes 1. Sectors not included in GMP are in current prices.
2. Based on rates for first half of the year.
3. Does not include commerce.

form of education which prepare students for the reality of Cuba's egalitarian economy. Two programmes developed in the 1970s demonstrate this mode of education. One is the work-study programme in the universities which involves all students in half-time productive labour while they are in school. The other is the schools in the countryside movement at the secondary level in which students regularly work in agricultural production as part of their education. Both programmes are designed to emphasise the importance of productive work for all members of society and combat the development of an elitist ideology, which so often accompanies educational attainment. These programmes simply highlight tendencies which have been present in Cuban education since 1959. When they are combined with the quantitative achievements of the system, it would appear that education in Cuba has made an important contribution to the socialist organisation of the economy in the 1970s.[28]

A variety of special programmes has played a role similar to that of education. The literacy campaign of 1961 was the prime example of a special programme. As noted above, in addition to its formal accomplishments, the literacy campaign developed leaders, organisational skills among the participants, and new attitudes toward social problems. Throughout the 1960s, other special programmes received considerable attention. Most notable were special youth brigades involved in agriculture and construction and the mass mobilisations for the sugar harvests. Their function as training grounds for the development of organisational skills was of considerable importance. (The mass mobilisations for the sugar harvests were, however, undoubtedly reduced in their social effectiveness by the economic problems of the period; new leaders and positive social attitudes, for example, do not emerge as well from failures as from successes.) At no time did the special programmes constitute more than a small share of total production activity. But they received considerable attention and set a tone for normal production activity.

Special programmes have been continued in the 1970s. The most important special programme in the 1970s has been the use of small brigades in construction. In an effort to attack Cuba's continuing housing shortage, these brigades – selected from a workplace to build housing for the members of that workplace – have, as the special programmes before them, involved a major exercise in creating new forms of work processes. Workers have been brought together in a situation where they must develop new leaders, make decisions about their hours and methods of work, and bear direct responsibility for their actions.

One example of work reorganisation in these construction brigades has been the integration of women into types of labour traditionally considered men's preserve. By approaching the work process in new ways, the brigades have created the opportunity for old ideas about work to be set aside. The point here is not simply that there has been progress in the status of women, but that such progress is an example of the way special programmes have been playing a role in relation to organisational development since the early years, and their impact seems to be becoming visible in the 1970s.

(b) The dual reform of politics and planning[29]

Yet, in the 1970s the impacts of education and special programmes, as well as the impact of the capital accumulation of the 1960s, might have been considerably different were it not for the political changes and planning changes which have been introduced. It is useful to see these two sets of changes as the two parts of a dual process. If the political changes are seen in isolation, they may seem purely formalistic. If the planning (or economic) changes are examined alone, they could imply a rejection of the central features of what had previously constituted the Cuban approach to development.

One aspect of the political change in the 1970s has been an increasing emphasis on the activities of the mass organisations. Cuba's mass organisations include the Association of Small Farmers (ANAP), the labour unions (which include state farm workers), the neighbourhood club-like Committees for Defence of the Revolution (CDRs), the Women's Federation and the various student organisations.[30] All of these groups played roles in implementing the programmes of the revolution in the 1960s. But during the 1970s, they have widened their activities and increased their decision-making authority.

Labour unions, in particular, have developed to take on a more significant function in plants, offices and farms. Their formal realm of authority remains as it was in the 1960s; they continue to focus on solving problems in the workplace and on mobilising workers for greater productivity – for example, by handling the political offensive against absenteeism – yet, in carrying out these tasks, the role of the unions has become primary; they are more heavily relied upon and more active than in the 1960s. With increased activity and responsibility comes increased authority.

Moreover, the authority of unions has been enhanced by the re-establishment of local (contested) elections. Also, in preparation for a national union congress in 1973 and in connection with changes in planning and incentive structures (to be discussed below), there have been extensive discussions and debates in workers' assemblies. The elections and assemblies have combined to give workers a much greater role in decision-making.[31]

In some cases, as in that of the CDRs which include upwards of 70 per cent of the population, the extension of decision-making procedures to include the populace has been accomplished simply by widening the scope of action of mass organisations.[32] In the case of the small farmers, there does not appear to have been major change in the role of ANAP during the early 1970s. However, in light of the government's goal of phasing out private farming and the greater use of prices in the 1970s (see below), it is noteworthy that ANAP's importance has not diminished. In the early 1970s, the organisation still included 232,000 members (162,000 farm owners, the rest family members) and continued its tasks of encouraging production and procurement and organising political education among its members.[33]

The general rise in the role of the mass organisations is one component of political change in the 1970s. The other is the institution of a formal political electoral system called *Poder Popular* (Popular Power). The two components are interdependent. The activity of the mass organisations provides a grass roots foundation for *Poder Popular*, and *Poder Popular* provides an extension of grass roots activity beyond the local level.

Poder Popular is a pyramidal political structure. Local assemblies are popularly elected; they elect district assemblies; and so on. After many years without formal mechanisms for popular participation in political affairs, the institution of *Poder Popular* indicates a significant shift in Cuban policy.

But the institution of *Poder Popular* goes beyond political affairs *per se*, because the realm of authority of the elected assemblies includes matters of production and allocation. Local assemblies have authority over local enterprises – for example, local retail outlets, consumer services, factories producing for local consumption (perhaps a bakery). District assemblies have authority over district enterprises; and so on up to the National Assembly which, as the highest legislative authority, exercises control over national economic matters. Thus, *Poder Popular* is the mechanism of popular participation in economic planning.[34]

But political reforms were only one component of the programme for economic recovery in the 1970s. In fact, the political reforms were all the more significant because they were combined with certain changes in incentives and planning structures. As the weaknesses of the 1960s system of moral incentives had become apparent, the government's initial response had been to heighten the military campaign atmosphere in the economy and to resort to coercion via 'anti-loafing' laws. While these measures may have raised labour force participation rates, they seem to have done little for productivity. Since 1970, the Cubans have moved in another direction: towards a greater reliance on more traditional material incentives. The new position was articulated at the National Labour Congress in late 1973, where a series of resolutions called for tying pay more closely to work and reducing the free distribution of certain goods and services. It seems, however, that the consequent steps have not meant any marked increase in inequality of money incomes. Instead, adjustments have been made within the framework of the existing wage spread to 'norm' wages, or tie them more directly to output.

To some extent, money wage differences in the 1960s had been tied to productivity differences, but the scarcity of consumer goods had meant that money wage differences were not equivalent to real income differences. In the 1970s, the expansion of the economy has increased the availability of consumer goods, so money wage differences have become real differences. The greater output of the Cuban economy has thus made the new incentive system operative without the re-creation of any serious poverty; more goods means the lower limit on incomes can be maintained and that greater productivity can be materially rewarded.

The shift away from moral incentives has limits. There does not appear to have been any major widening of the wage structure. Health care and education are still free; housing is kept free or cheap; employment is guaranteed; and so on. Moreover, on an ideological level moral incentives have not been abandoned as a goal. But the task is now seen by the Cubans as one of developing, rather than imposing, moral incentives.

Nonetheless, the incentive changes have been significant, and their significance is all the greater because they have been accompanied by a decentralisation of planning procedures. A primary purpose of the decentralisation has been to increase local accountability in order to raise productivity. By introducing more thorough receipt-and-payments relations at all levels, errors and inefficiencies should be more readily identified. In addition, with achievements more clearly measured, rewards can be tied to those achievements. A system of limited bonuses has been introduced in which a share of the profits obtained by a unit – a state farm, for example – remains in the hands of the collective of workers to be used for solving the social problems (e.g. housing) of the group and rewarding outstanding workers.

Using profits as an indicator of success means that a more important role has been given to prices in co-ordinating economic activity. Prices, however, instead of being determined on the market, are the instruments of control used by the central planners. Thus, prices are used to bring local decisions into line with the overall social goals of planning. (Prices and local decisions based on prices are not the basis for investments, except for limited expansion and modernisation programmes.)[35]

These incentive and planning changes would seem to be a radical departure from the Cuban practice of the 1960s. And, in part, they do represent a recognition of errors of the earlier era. However, the planning and incentive changes are also the consequence of material changes, and, in part, represent but an adjustment to new circumstances.

To begin with, the successful expansion of the educational system has created a work force with the capabilities – both technical and organisational – for handling a decentralised planning system. In addition, the success of education and the general ideological development of the revolution reduce the likelihood that decentralisation will create a bureaucratic, elite strata of managers and technicians. Better economic performance of the early 1970s has also been the basis for the planning and incentive changes, by providing stability, by assuring employment, and, as noted above, by allowing material incentives to operate.

But, of greatest importance, the political reforms of the 1970s have radically altered the context and therefore the implications of the entire planning and incentive process. Had decentralisation with material incentives been implemented in the absence of popular participation, it could have led toward greater managerial authority, increasing worker alienation, and worsening inequality. In the 1970s, the increased role of the Cuban populace in all types of decision-making tends to pre-empt such consequences and makes it possible for the new planning system to become an instrument for developing socialist social relations as well as raising productivity.

(c) Re-balancing the Cuban economy

The planning and political reforms of the 1970s apply to all sectors of the Cuban economy, agricultural as well as non-agricultural. State farms are treated in the same general manner as other state enterprises, and even in the private sector production has been affected by the restructuring of the system. In addition, there has been a redirection of the Cuban economy in the 1970s, away from the heavy reliance on sugar and agriculture and towards a greater role for industry.

Among the clearest differences in the Cuban economy of the early 1970s as compared to earlier years is that the expansion of the rest of the economy appears to have been firmly established independently of agriculture. The rapid growth rates of non-agricultural sectors have been seen in Table 10.6 above. The differential growth rates of agricultural and non-agricultural sectors noticeably altered the economy's structure. In the 1962–9 period, agriculture accounted for between 20.8 and 22.5 per cent of Gross Material Product; there was only a slight downward trend in this figure. After 1969, however, the decline in agriculture's share of output was large and steady. By early 1975, agriculture accounted for only 14.8 per cent of GMP; see Table 10.7.

The changing structure of the Cuban economy reflects an alteration in planning priorities. More emphasis on industrial diversification, housing, and consumption account for much of the structural change. Agriculture has by no means been abandoned, but the programme for the 1970s is much more of a balanced expansion than it was in the 1960s.[36]

Within agriculture, the performance has been mixed, with serious difficulties persisting in the sugar sector. Average sugar output in the 1971–5 period was only 5.6 million metric tons. The favourable 1977 output of 6.3 million metric tons and the extremely high 7.3 million metric tons attained in 1978 have raised the average for the period since 1971 to 5.9 million metric tons; see Table 10.8. But these two good years hardly establish a trend. In the 1970s, there has been no decisive break with the long run stagnation of sugar

Table 10.7 Agriculture's Share in Total Output, 1962–75

	Agriculture as a per cent of Gross Material Product	*Agriculture as a per cent of Gross Social Product*
1962	22.1	14.7
1963	21.7	14.2
1964	21.7	14.0
1965	22.5	14.9
1966	22.0	14.3
1967	20.8	13.9
1968	21.4	13.9
1969	21.0	13.5
1970	20.6	13.9
1971	18.1	11.9
1972	17.3	10.8
1973	16.1	9.8
1974	15.4	9.3
1975	14.8	n.a.

Source MacEwan, op cit, Appendix Table A.1.

production. Weather has been unfavourable, but it would seem that problems of effective organisation persist. While there has been progress in mechanising the harvest – 39 per cent of the cane was cut by machines in 1978 – efficiency has not been high. And expansion elsewhere continues to pose a labour shortage problem for sugar.

Difficulties also persist in cattle production, Cuba's other traditionally important agricultural activity. Considerable effort has been devoted to genetic improvement of the herd. But between 1967 and 1970 the herd declined from 7 to 5 million, and during the early 1970s a restrictive slaughter policy was followed. Consequently, beef production dropped by 16 per cent between 1970 and 1974.[37]

By contrast, other branches of livestock did not do so badly in the early 1970s. Poultry production rose by 28 per cent between 1970 and 1974, and thus accounted for a greater value of output than did beef production. Egg production, which had already done well in the 1960s, rose by 20 per cent in this period. Pork production, in spite of a counter-revolutionary sabotage effort, increased by almost 50 per cent (though pork production was still only 20 per cent of beef or poultry in value terms). And while beef production did badly, milk production rose by 16 per cent in the early 1970s.[38]

Table 10.8 Sugar Output, 1949–76 (million metric tons)

1949	5.1	1959	6.0	1969	4.5
1950	5.4	1960	5.9	1970	8.5
1951	5.8	1961	6.9	1971	5.9
1952	7.3	1962	4.9	1972	4.3
1953	5.2	1963	3.9	1973	5.3
1954	5.0	1964	4.5	1974	5.9
1955	4.6	1965	6.2	1975	5.7[1]
1956	4.8	1966	4.5	1976	5.7[1]
1957	5.7	1967	6.2	1977	6.3[1]
1958	5.9	1968	5.2	1978	7.3

Source For 1949, International Bank for Reconstruction and Development, *Report on Cuba,* p. 799; for 1950, Ritter, op cit, p. 113; for 1951–73, Junta Central de Planificación, *Anuario Estadístico de Cuba, 1973,* p. 124; for 1974, National Bank of Cuba, *Development and Prospects of the Cuban Economy,* Havana, 1975; for 1975 and 1976, Economic Intelligence Unit, *Cuba, Dominican Republic, Haiti, Puerto Rico, Annual Supplement for 1977,* p. 9, London, 1978; for 1977, estimated on the basis of information in Fidel Castro's 24 December 1977 speech in *Granma Weekly Review,* 1 January 1978 and Economic Intelligence Unit, *Quarterly Report on Cuba, Dominican Republic, Haiti, Puerto Rico, First Quarter 1978,* p. 9, London, 1978; and for 1978, *Granma Weekly Review,* 2 July 1978.

Note 1. These data are not from official Cuban sources; their validity is questionable.

In the production of food crops, Cuba also did relatively well in the beginning of the 1970s. Table 10.9 presents data on the state collection of various categories of crops. It is clear from the data that the fall in the 1960s in the collection of these crops (see Table 10.5 above) has been reversed. With many of these crops, as the data in the table indicate, the private farmers play an important role, and that role has remained strong in the recent period. Also, coffee and tobacco production did well in the early 1970s, each rising by about 35 per cent between 1970 and 1974.[39]

It is useful to speculate on the reasons for the relatively favourable performance of secondary crop production generally and in the private sector particularly. Several factors would seem to be involved. To begin with, these crops (with the exception of rice) are in general cultivated in relatively small units and require intensive cultivation. Such processes are not effectively directed by central authorities who are out of touch with local conditions. The move toward decentralisation may have been particularly important for this category of production.

Table 10.9 Collection of Selected Categories of Agricultural Commodities, 1969–74 (thousand metric tons)

	1969	*1970*	*1971*	*1972*	*1973*	*1974*
Grains – total	189.3	299.3	297.2	251.0	244.0	347.7
– private	16.2(9%)	11.0(4%)	15.1(5%)	19.2(8%)	13.7(6%)	n.a.
Tubers – total	326.2	196.8	239.3	383.5	397.5	289.5
– private	130.8(40%)	75.3(38%)	82.6(35%)	168.1(44%)	181.9(46%)	n.a.
Vegetables – total	126.7	136.2	172.2	150.0	239.8	265.9
– private	82.1(65%)	62.7(46%)	74.0(43%)	63.3(42%)	123.6(52%)	n.a.
Fruits – total	297.4	289.6	296.9	372.4	434.0	463.4
– private	171.0(57%)	125.7(43%)	130.1(44%)	161.7(43%)	207.9(48%)	n.a.

Source For 1969–73, Junta Central de Planificación, *Anuario Estadístico de Cuba, 1973*, pp. 91–3; for 1974, crop values in constant prices are reported in National Bank of Cuba, op cit, p. 40. The per cent increases were used to obtain these 1974 volume figures. The 1974 data may not be consistent with the figures for earlier years.

There has been a move towards the development of a variety of locally organised development plans involving both state farms and private producers. These plans serve, on the one hand, as mechanisms of decentralisation. On the other hand, within the local context, they would appear to tie the private producers more closely to the state so that the production practices of the farmers can be improved and the state can more readily influence production and marketing decisions.

The control attained through the local plans is part of a general effort by the government to limit sales outside the state procurement system. The new function of prices allows such a policy to be implemented more readily. In addition, the increased availability of consumer goods has provided the inducement to peasants to trade within the official framework.

Similar considerations probably explain the favourable performance in non-cattle livestock branches of the economy. Decentralisation and the greater use and effectiveness of price incentives have given impetus to the expansion of these relatively new activities.

Sugar and cattle production, by their nature in Cuba, are not so responsive to the types of planning changes that have been introduced in the 1970s. These are large scale activities and (especially sugar) are more directly tied to the overall national planning system. Also, sugar and cattle are Cuba's older, more thoroughly established agricultural activities. Changes in production techniques and modes of organisation may be expected to be more rapid and successful in new activities; old sectors are unlikely to lead in the process of modernisation. Still, the sugar and cattle production activity is not immune to change. The beginnings of substantial progress in mechanising the sugar harvest is a positive sign; even if total output does not grow substantially, productivity should increase a great deal.

Cuban plans for the late 1970s and early 1980s call for a continuation of the diversification and industrial emphasis established in the first part of the decade. Sugar is still important – it continued to account for over 75 per cent of export earnings in the early 1970s – and some mechanism will be needed to spread progress to that basic sector.[40] Yet, the favourable developments of the early 1970s give some basis for a positive appraisal of the Cuban development potential. The political and planning changes may provide a system of economic organisation through which the steady expansion of the Cuban economy can take place.

Notes

1. This brief description of land reform draws on José Acosta, 'Las leyes de reforma agraria en Cuba y el sector privado campesino', *Economía y Desarrollo,* No. 12, 1970; 'La revolución agraria en Cuba y el desarrollo económico', *Economía y Desarrollo,* No. 17, May–June 1973; and 'Cuba: De la Neocolonia a la Construcción del Socialismo (II)', *Economía y Desarrollo,* No. 20, 1974; on James O'Connor, *The Origins of Socialism in Cuba* (Ithaca and London: Cornell University Press, 1970), Chapter 5; on Andrés Bianchi, 'Agriculture', in D. Seers et al, *Cuba: The Economic and Social Revolution* (Chapel Hill: University of North Carolina Press, 1964); and on Carlos Rafael Rodríguez, 'El nuevo camino de la agriculture Cubana', *Cuba Socialista,* No. 27, November 1963; and 'The Cuban Revolution and the Peasantry', *World Marxist Review,* October 1965, Vol. 8, No. 10.
2. For readers unfamiliar with the Cuban situation, an upper limit of 67 hectares would seem rather high. In Europe or Asia, for example, 67 hectares would constitute a large, if not gigantic estate. Yet in Cuba, owing the the extensive methods of farming and the general backwardness of agriculture, few farmers who held less than 67 hectares could reasonably be classified as 'rich peasants'. It should be emphasised that the 67 hectare limit applied to total land, arable and non-arable. On the private lands taken altogether, roughly 40 per cent of the land could be classified as arable, but on the larger peasant farms this percentage is probably smaller. For the private sector as a whole, the amount of arable land per worker was a bit higher than in the public sector (not counting pasture and forest land) – about 3.5 hectares per worker as compared to roughly three in the private sector. With the upper limit at 67 hectares, the average private farm is only 15 hectares, and there would seem to be considerable inequality in the size of private holdings. However, the inequality is accounted for in large part by regional differences. And, of special importance, the correlation between farm income and farm size is not high; tobacco farmers, who operate on small plots, are probably Cuba's richest farmers. While the large private farmers in Cuba – those at or near the 67 hectare limit – are probably relatively well off, the above would imply that they do not constitute a class of wealthy farmers. Also, by historical standards, in Cuba anything less than 67 hectares – with exceptions in some areas – was a small or, at most, middle level peasant operation.
3. Regarding the expansion of employment on state farms, see O'Connor, op cit, p. 221. O'Connor in Appendix B, and Carmelo Mesa-Lago, *The Labour Force, Employment, Unemployment and Underemployment in Cuba, 1899–1970* (Beverly Hills and London: Sage Publications, 1972) provide details on unemployment changes.
4. See Felipe Pazos, 'Desarrollo insufficiente y depauperacion económica', *Cuadernos,* Supplemento del No. 47, March–April 1961, p. 48, and 'Comentarios a dos Artículos Sobre la Revolución Cubana', *El Trimestre Económico,* Vol. XXXIX, No. 113, January–March 1962, p. 9, and Seers, op cit, pp. 32–3.

5. See Nelson Amaro and Carmelo Mesa-Lago, 'Inequality and Classes', in C. Mesa-Lago (ed.), *Revolutionary Change in Cuba* (Pittsburgh: University of Pittsburgh Press, 1971), p. 361.
6. Richard Fagen, *The Transformation of Political Culture in Cuba* (Stanford: Stanford University Press, 1969) places the literacy campaign in the context of general political and ideological change in Cuba, and he provides much information on the campaign.
7. See David Barkin, 'The Redistribution of Consumption in Cuba', *Review of Radical Political Economics*, Vol. 4, No. 5, 1972 for details on the health programmes. The data on rural hospitals (to be found in Junta Central de Planificación, *Anuario Estadístico de Cuba*, 1973, pp. 256–7) show that the number of general hospitals in rural areas increased from three in 1958 to 38 in 1963.
8. See the argument of Maurice Zeitlin, *Revolutionary Politics and the Cuban Working Class* (New York: Harper and Row, 1970) for elaboration and substantiation.
9. See Fagen, op cit, for more discussion.
10. See 'informe de la delegación de Cuba al XIII Período de Se siones de la C.E.P.A.L.', Peru, 1969, in *Economía y Desarrollo*, No. 1, January–March 1970, pp. 51–7, and La Delegación de Cuba al Seminario Latinoamerica no sobre Reforma Agraria, 'Una evaluación de la reforma agraria en Cuba', in *Economía y Desarrollo*, No. 11, May–June 1972, pp. 176–88. The 200,000 hectares figure is from the former of these sources, p. 54; Bianchi, op cit, p. 85, reports total 1958 cane cultivation of roughly 1.3 million hectares.
11. See O'Connor, op cit, Chapter 8 for a more thorough discussion of the labour shortage. The data are his, p. 220.
12. See Mesa-Lago, *The Labour Force* . . ., op cit, pp. 49–50.
13. See Archibald Ritter, *The Economic Development of Revolutionary Cuba* (New York: Praeger, 1974) pp. 188–90, and explanatory notes in Junta Central de Planificación, *Anuario Estadístico de Cuba*, 1973.
14. See Ritter, op cit, pp. 111–16.
15. It is impossible to establish the relative importance of diversification, the contradictions of reform, weather and the conflict with the US in bringing on the 1962/63 crisis. The fact that I have focused on internal matters should in no way be taken to imply a judgement that external matters were less important.
16. On the formulation and substance of the 1960s strategy, see 'Informe de la delegación . . .', op cit, La Delegación de Cuba . . ., op cit, and Fidel Castro's speech of 29 May 1970, as printed in R. E. Bonachea and N. Valdes (eds.), *Cuba in Revolution* (Doubleday, Garden City, 1972).
17. See, for example, Fidel Castro's speech of 26 July 1968, as printed in M. Kenner and J. Petres, *Fidel Castro Speaks* (New York: Grove Press, 1969).
18. This point was made by Ernesto Guevara in an article in *Nuestra Industria Revista Económica*, February 1964.
19. For tractor and water storage data, see La Delegación de Cuba al Seminario Latinoamerica no sobre Reforma Agraria, 'Una Evaluación de la Reforma Agraria en Cuba' in *Economía y Desarrollo*, No. 11,

May–June 1972, pp. 184–5. For fertiliser data, see 'Informe de la delegación . . .', op cit, p. 59.

20. Fidel Castro, 'Report of the Central Committee of the Communist Party of Cuba to the First Congress' in *First Congress of the Communist Party of Cuba* (Moscow: Progress Publishers, 1976) p. 74.
21. Informe de la Delegación . . . , op cit, p. 136 and Informe de la Delegación de Cuba a la XI Conferencia regional de la F.A.O., 'Dos años de desarrollo agropecuario cubano', in *Economía y Desarrollo,* No. 4. October–December 1970, pp. 27–9.
22. Junta Central de Planificación, 1973, op cit, various tables.
23. *Pensamiento Crítico,* October 1970, p. 183 as cited by Ritter, op cit.
24. Carmelo Mesa-Lago, 'Economic Policies and Growth' in C. Mesa-Lago (ed.), op cit, p. 305; and Fidel Castro's speech of 26 July 1970, as printed in Bonachea and Valdes (eds.), op cit.
25. See I. Joshua, *Organisation et Rapports de Production dans une Economie de Transition (Cuba)* (Paris: Centre d'Etudes de Planification Socialiste, Sorbonne, 1968).
26. Ritter, op cit, p. 282
27. It should be noted that 1969 was a rather poor year. Even with an adjustment of the base, however, the growth of the early 1970s would still be impressively rapid.
28. For the details of the data on educational achievements see Junta Central de Planificación, *Boletin Estadístico,* 1970, pp. 216–17.
29. The most useful source on these matters is Castro, 1976, op cit.
30. The concept of a 'mass' organisation in Cuba means an organisation with open membership. It is to be distinguished from the concept of a select organisation – the Communist Party of Cuba. To enter the Communist Party a person must be selected by the other members of his or her work place (or, in some cases, the other members of a mass organisation) and approved by the Party.
31. See Mesa-Lago, *Cuba in the 1970s* (University of New Mexico Press, 1974), pp. 72–89.
32. See, for example, *Bohemia,* 12 October 1972, for an account of the activities of the CDRs.
33. Castro, 1976, op cit, pp. 193–7.
34. While these political developments represent an opening up of Cuban politics, all decision-making in Cuba still takes place under the guidance of the Communist Party, also, while elections in *Poder Popular* are not uncontested, no organised opposition is allowed.
35. See A. López Coll and A. Santiago, 'Notas sobre el processo de planificación en Cuba', *Economía y Desarrollo,* No. 29, May–June 1975.
36. On priorities in the late 1970s, see Castro, 1976, op cit; National Bank of Cuba, *Development and Prospects of the Cuban Economy,* 1975; and 'Las tareas principales del quinquenio 1976–80', in *Economía y Desarrollo,* No. 36, July–August 1976.
37. National Bank, op cit, pp. 47–50.
38. Ibid.
39. Ibid., pp. 40–1.
40. 'Las tareas . . .', op cit, and Castro, 1976, op cit.

Index